Spectroscopy

VOLUME THREE

SPECTROSCOPY

VOLUME ONE

Atomic spectra; Nuclear magnetic resonance spectroscopy; Nuclear quadrupole resonance spectroscopy; Electron spin resonance spectroscopy; Mössbauer spectroscopy

VOLUME TWO

Molecular spectra; Symmetry and group theory; Microwave spectroscopy; Infrared and raman spectroscopy; Far-infrared spectroscopy; Force constants; Thermodynamic functions.

VOLUME THREE

Molecular quantum numbers; Electronic spectra of diatomic molecules; Dissociation energies of diatomic molecules; Electronic spectra of polyatomic molecules; Fluorescence and phosphorescence spectroscopy; Astrochemistry; Photoelectron spectroscopy.

Spectroscopy
VOLUME THREE

EDITED BY

B.P. STRAUGHAN Ph.D., F.R.I.C.

Department of Chemistry,
University of Newcastle upon Tyne,
England

AND

S. WALKER, M.A., D.PHIL., D.Sc.

Chairman, Department of Chemistry,
Lakehead University,
Ontario, Canada

LONDON
CHAPMAN AND HALL
A HALSTED PRESS BOOK
JOHN WILEY & SONS INC., NEW YORK

First published 1976
by Chapman and Hall Ltd
11 *New Fetter Lane, London EC4P 4EE*

Set by EWC Wilkins Ltd, London and Northampton,
and printed in Great Britain by
Butler & Tanner Ltd, Frome and London

ISBN 0 412 13380 6 (*cased edition*)
ISBN 0 412 13390 3 (*Science Paperback*)

Distributed in the U.S.A. by Halsted Press,
a Division of John Wiley & Sons, Inc., New York

Library of Congress Cataloging in Publication Data
Main entry under title:

Spectroscopy.

Previous editions by S. Walker and H. Straw published in 1962 and 1967, entered under: Walker, Stanley.
Includes bibliographies.
1. Spectrum analysis. I. Straughan, B.P.
II. Walker, Stanley. Spectroscopy.
QC451.W33 1976 535'.84 75–45328
ISBN 0–470–15031–9 (v. 1)
0–470–15032–7 (v. 2)
0–470–15033–5 (v. 3)

Preface

It is fifteen years since Walker and Straw wrote the first edition of 'Spectroscopy' and considerable developments have taken place during that time in all fields of this expanding subject. In atomic spectroscopy, for example, where the principles required in a student text have been laid down for many years, there have been advances in optical pumping and double resonance which cannot be neglected at undergraduate level. In addition, nuclear quadrupole resonance (n.q.r.) and far infrared spectroscopy now merit separate chapters while addtional chapters dealing with Mössbauer spectroscopy, photoelectron spectroscopy and group theory are an essential requisite for any modern spectroscopy textbook.

When the idea for a new edition of Spectroscopy was first discussed it quickly became clear that the task of revision would be an impossible one for two authors working alone. Consequently it was decided that the new edition be planned and co-ordinated by two editors who were to invite specialists, each of whom had experience of presenting their subject at an undergraduate level, to contribute a new chapter or to revise extensively an existing chapter. In this manner a proper perspective of each topic has been provided without any sacrifice of the essential character and unity of the first edition.

The expansion of subject matter has necessitated the division of the complete work into three self contained volumes.

Volume 1 includes atomic, n.m.r., n.q.r., e.s.r. and Mössbauer spectroscopy.

Volume 2 contains chapters on molecular symmetry and group theory, microwave, infrared and Raman, far-infrared spectroscopy, force constants, evaluation of thermodynamic functions.

Volume 3 centres on the information which results when a valence electron(s) is excited or removed from the parent molecule. It includes electronic spectroscopy, quantum numbers, dissociation energies, fluorescence and phosphorescence spectroscopy, astrochemistry, photoelectron spectroscopy.

The complete work now provides a single source of reference for all the spectroscopy that a student of chemistry will normally encounter as an undergraduate. Furthermore the depth of coverage should ensure the books' use on graduate courses and for those starting research work in one of the main branches of spectroscopy.

A continued source of confusion in the spectroscopic literature is the duplication of symbols and the use of the same symbol by different authors to represent different factors. The literature use of both SI and non SI units further complicates the picture. In this book we have tried to use SI units throughout or units such as the electron volt which are recognised for continued use in conjunction with SI units. The symbols and recognised values of physical constants are those published by the Symbols Committee of the Royal Society 1975.

B.P. Straughan
S. Walker

October, 1975

Acknowledgements

Although not involved in the production of this second edition, we would like to express our sincere thanks to Mr. H. Straw whose vital contribution to the first edition of Spectroscopy helped to ensure its widespread success and hence the demand for a new edition. One of us (S.W.) wishes to thank his wife, Kathleen, without whose help at many stages part of this work could not have gone forward.

Contributors to Volume Three

CHAPTERS ONE AND THREE

Mr. H. Straw, Leicester Polytechnic, and
Professor S. Walker, Lakehead University, Ontario, Canada

CHAPTER TWO

Professor S. Walker, Lakehead University, Ontario, Canada

CHAPTER FOUR AND APPENDIX

Dr. J.K. Burdett, University of Newcastle upon Tyne

CHAPTER FIVE

Dr. D. Phillips and Dr. K. Salisbury, University of Southampton

CHAPTER SIX

Professor S. Walker and Dr. H. Walker, Lakehead University, Ontario, Canada

CHAPTER SEVEN

Dr. P.M.A. Sherwood, University of Newcastle upon Tyne

Contents

1 Molecular quantum numbers of diatomic molecules

1.1 FORMATION OF MOLECULAR QUANTUM NUMBERS

In the study of atomic spectra it was necessary to introduce the principal quantum number (n), the azimuthal quantum number (l), the electron and nuclear spin quantum numbers (s and I, respectively) or some combination of these. In addition, to account for the behaviour of the spectral lines of atoms in the presence of various electric or magnetic fields further quantum numbers were required.

When two atoms combine to give a diatomic molecule, these quantum numbers may be related to a new set of quantum numbers which characterize the electronic energy states of the molecule. In some respects diatomic molecules behave like atoms (see Vol. 1), and their energy distribution seems to follow a reasonably similar type of pattern. This is strikingly illustrated by comparing the electron states of the molecules with those of the atoms containing the same number of electrons. This comparison may be observed in Fig. 1.1. The main way in which they differ results from the presence of two nuclei; these produce a cylindrically symmetrical force field about the internuclear axis.

In an atom $l_1, l_2, l_3, \ldots$ couple together to give L, while $s_1, s_2, s_3, \ldots$ give a resultant electron spin quantum number S. The resultant orbit angular momentum is then $\sqrt{[L(L+1)]}\, h/2\pi$, and the resultant electron spin angular momentum $\sqrt{[S(S+1)]}\, h/2\pi$. When an atom characterized by $\sqrt{[L_1(L_1+1)]}\, h/2\pi$ and $\sqrt{[S_1(S_1+1)]}\, h/2\pi$ combines with another characterized by $\sqrt{[L_2(L_2+1)]}\, h/2\pi$ and $\sqrt{[S_2(S_2+1)]}\, h/2\pi$ the possible values of L and S for the molecule are given by:

$$L = L_1 + L_2, L_1 + L_2 - 1, \ldots, |L_1 - L_2| \tag{1.1}$$

$$S = S_1 + S_2, S_1 + S_2 - 1, \ldots, |S_1 - S_2| \tag{1.2}$$

An atom, however, unlike a diatomic molecule, has a spherically symmetrical force field. In a diatomic molecule, there is a strong electric field along the internuclear axis due to the electrostatic field of the two nuclei. The result is that the orbit and spin orbit angular momenta are uncoupled in this cylindrically symmetrical force field. The effect is that the corresponding $L^*h/2\pi$ and $S^*h/2\pi$ vectors for the resultant orbit and total spin angular momentum, respectively,†

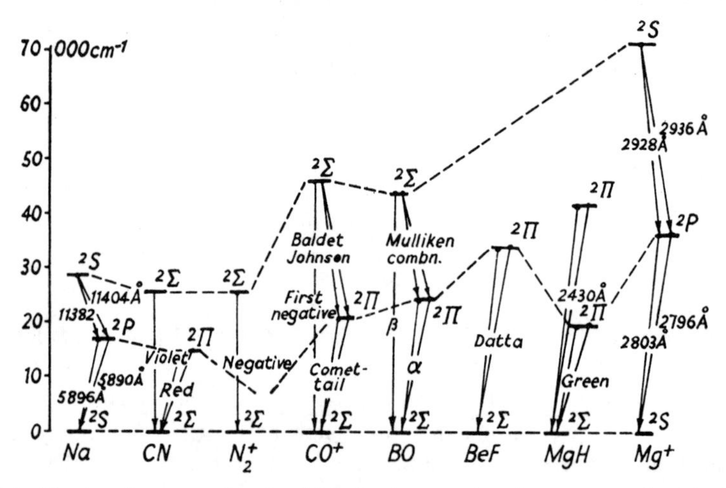

Fig. 1.1 Electronic energy levels of 13-electron molecules compared with those of Na and Mg^+ [1.2]. Corresponding levels are connected by broken lines, and observed lines and band systems are indicated by arrows. The names used to characterize some of the electronic transitions are also given. (Courtesy of Dr. W. Jevons, and the Council of the Physical Society, London)

precess independently about the internuclear axis,‡ and $L^*h/2\pi$ has a constant component in that direction of $M_L h/2\pi$ where $M_L = L, L-1, \ldots, -L$. In an electric field, if the direction of motion of all the electrons is reversed, the energy of the molecule remains unchanged, although the positive M_L value is changed into a negative one. It follows, therefore, that only states with different M_L values possess different energies. These different energy states are characterized by a quantum number Λ which defines the constant component of the resultant orbital angular momentum ($\sqrt{[L(L+1)]}\,h/2\pi$) for each energy state

† The shorthand representation of the form $X^*h/2\pi$ will be used for $\sqrt{[X(X+1)]}\,h/2\pi$, where X may be L, S, J or K.

‡ Whether $L^*h/2\pi$ and $S^*h/2\pi$ are uncoupled depends on the strength of the internal electric field. The latter is dependent on the internuclear distance, and when this is relatively large, then the electric field becomes too weak to uncouple $L^*h/2\pi$ and $S^*h/2\pi$, and they form a common resultant vector. This is dealt with later under Hund's case (c). The treatment given here for the uncoupling of $L^*h/2\pi$ and $S^*h/2\pi$ corresponds to Hund's case (a).

along the internuclear axis where Λ may take the values $0, 1, 2, 3, \ldots, L$.† The precession of the resultant orbit angular momentum about the internuclear axis is illustrated in Fig. 1.2(a).

The actual value of Λ is a measure of the number of units of angular momentum (in $h/2\pi$ units) resulting from projecting the electron orbit angular momentum along the internuclear axis. This component, $\Lambda h/2\pi$, along the axis always remains defined, whereas $L^* h/2\pi$ itself frequently is not.

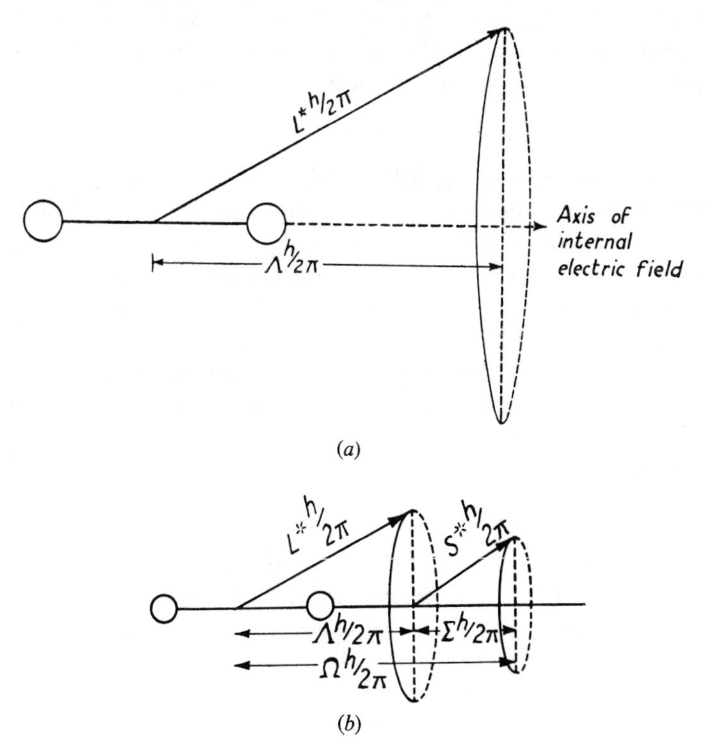

Fig. 1.2 (a) Precession of the resultant orbital angular momentum about the internuclear axis of a diatomic molecule. (b) Vector model for the precession of $L^* h/2\pi$ and $S^* h/2\pi$ in a diatomic molecule.

The electronic states of diatomic molecules are characterized by:

$$\Lambda = 0, 1, 2,$$

and are called a Σ-state, Π-state, and Δ-state, respectively. This is a parallel

† This type of classification applies not only to diatomic molecules but to linear polyatomic molecules as well. In both cases the electric field is cylindrically symmetrical about the axis through the nuclei. For more complex molecules the electronic states are classified on the basis of symmetry properties in terms of group theory.

classification to that employed for atoms, where when:

$$L = 0, 1, 2$$

these are termed S-, P-, and D-states, respectively.

The effect of electron spin cannot be neglected when attempting to define a molecular electronic state by means of quantum numbers; the total electron spin quantum number S must be taken into account. Each electron has a spin of $\pm\frac{1}{2}$ from which the resultant spin of the individual atoms may be evaluated. If S_1 and S_2 are the resultant spins in the separate atoms, then the possible values for the resultant electronic spin, S, of the molecule are given by:

$$S = S_1 + S_2, S_1 + S_2 - 1, \ldots, |S_1 - S_2| \tag{1.3}$$

The value of S may be integral or half-integral, depending on whether the total number of electrons is even or odd. If the component of the electron orbit angular momentum ($L^* h/2\pi$) along the internuclear axis is not zero, that is, it is not a Σ-state, then there exists an internal magnetic field† (acting along the internuclear axis). The field is produced by the orbital motion of the electrons and may be identified with the precession of the orbit angular momentum. The magnetic moment associated with the electron spin can interact with this internal magnetic field causing a precession of the spin angular momentum about the internuclear axis. The component values of this spin angular momentum along this axis are governed by another quantum number Σ ‡ which takes the values:

$$\Sigma = S, S - 1, S - 2, \ldots, -S \tag{1.4}$$

Thus, Σ is analogous to the symbol M_S employed in the case of atoms, and $\Sigma h/2\pi$ is the value of the spin orbit angular momentum about the internuclear axis, there being $(2S + 1)$ possible orientations. The vector model for the precession of $L^* h/2\pi$ and $S^* h/2\pi$ is given in Fig. 1.2(b).

The component of the total electron angular momentum along the internuclear axis may be obtained by coupling Λ and Σ:

$$|\Lambda + \Sigma| = \Omega \tag{1.5}$$

where $\Omega h/2\pi$ is the total electron angular momentum about the internuclear

† The magnetic moment of the spinning electron cannot interact directly with the internuclear electric field as does the induced electric dipole moment associated with the resultant orbit angular momentum vector. The latter is governed by the torque exerted by the electric field which causes it to precess about the internuclear axis and to have quantized components along this axis.

‡ This is the sloping Greek capital Σ, and it is necessary not to mistake this for the upright symbol Σ which is used to represent a molecular state $\Lambda = 0$.

axis.† The combination of Λ and Σ in this way to give Ω is to be compared with the corresponding case of atoms in a strong electric field, where $|M_L| + M_S = M_J$. For each value of Λ there are $(2S + 1)$ sublevels determined by the $(2S + 1)$ values of Σ, and the multiplicity is, thus, $(2S + 1)$; this value is added as a left superscript to the electronic state symbol. For example, a state for which Λ is equal to 1 having a spin of 1 would be the $^3\Pi$-state. The value of Ω is indicated as a right subscript to the main symbol; in this case $\Sigma = 1, 0$, and -1, and $\Omega = 1 + \Sigma$ and has $2 \times 1 + 1$ values which are 2, 1, and 0. The three different electronic energy states for the molecule in this quantum condition would thus be $^3\Pi_2$, $^3\Pi_1$, and $^3\Pi_0$, and would be represented by three different potential energy curves.

The splitting of the energy levels in a multiplet state may be related to the values of the quantum numbers Λ and Σ. For example, the electronic energy of term value T_e of a multiplet term is given approximately by:

$$T_e = T_0 + A\Lambda\Sigma$$

where T_0 is the term value when $\Sigma = 0$. A is known as the *coupling constant* and fixes the magnitude of the multiplet splitting; its value increases rapidly with the size of the atoms as does the multiplet splitting $(T_e - T_0)$.

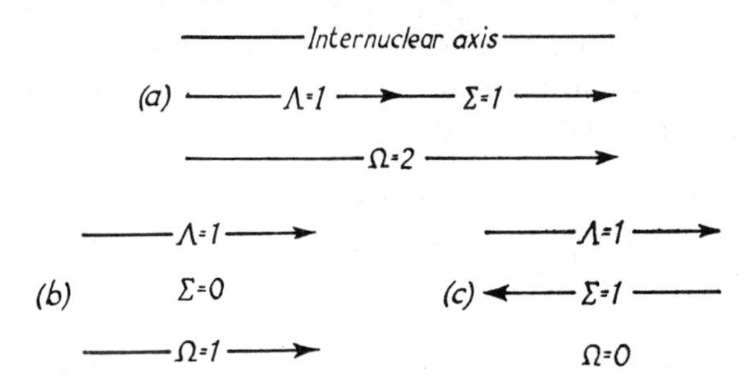

Fig. 1.3 Vector diagram of (a) $^3\Pi_2$, (b) $^3\Pi_1$, and (c) $^3\Pi_0$ states.

The relative orientations of the vectors Λ and Σ in the $^3\Pi_2$, $^3\Pi_1$, and $^3\Pi_0$ states are illustrated in Fig. 1.3, and neglecting all considerations of rotational and vibrational levels the corresponding three electronic states may be represented as in Fig. 1.4.

The coupling constant A is constant for a given multiplet term. A coupling constant may have a positive or negative value, and in the case in Fig. 1.4, if A

† This is really Hund's case (*a*) (see later). In case (*b*) the angular momentum along the internuclear axis is $\Lambda h/2\pi (= \Omega h/2\pi)$ while in case (*c*) $L^* h/2\pi$ and $S^* h/2\pi$ remain coupled, and it is the presence of their resultant vector which gives the value of $\Omega h/2\pi$. Thus, for cases (*a*), (*b*), and (*c*) of Hund's coupling the angular momentum along the internuclear axis in each case may be regarded as being $\Omega h/2\pi$.

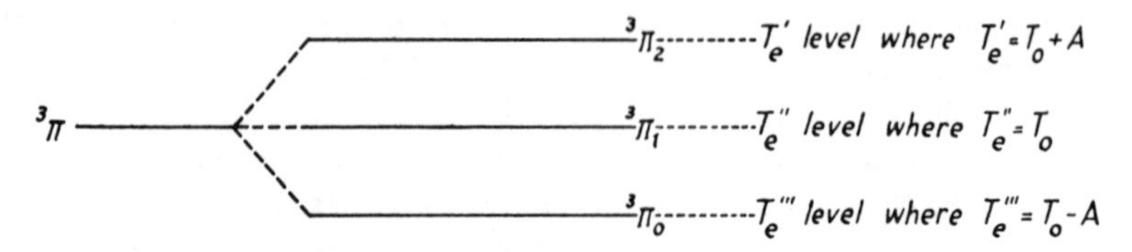

Fig. 1.4 Energy level diagram of a $^3\Pi$-state.

had been negative, then the $^3\Pi_2$ level would interchange places with the $^3\Pi_0$ level.

When $\Lambda = 0$, this means that the component of the resultant electron orbit angular momentum along the internuclear axis is zero; hence, the magnetic field along this line will also be zero and no splitting of the energy levels will occur. For example, if $\Lambda = 0$ and $S = 2$ this would be represented by a $^5\Sigma$-state, but there would be no splitting of this electronic energy state. This is in harmony with putting $\Lambda = 0$ in the formula $T_e = T_0 + A\Lambda\Sigma$, whence $T_e = T_0$. Thus, superficially the superscript symbol may seem somewhat misleading. However, even for Σ-states a small amount of splitting of the electron states may occur when $(2S + 1) > 1$ provided that the molecule is rotating. This type of splitting results from interaction of the magnetic moment due to electron spin with the magnetic field due to molecular rotation.† As regards molecules in other than Σ-states (e.g. Π and Δ) the full splitting of the electronic energy levels required by $(2S + 1)$ generally seems to occur.

1.2 'SCRIPTS' GIVING INFORMATION ON THE WAVEFUNCTION SYMMETRY OF DIATOMIC MOLECULES

Another superscript is used in addition to the one denoting multiplicity and this is a right-hand one with + or − sign. These signs give information on the symmetry characteristics of the wavefunction of a molecular electronic state. If the wavefunction of the electron eigenfunction does not change sign on reflection of the coordinates to the other side of a plane between the two nuclei, then the + superscript is given, and if the sign is changed, the electronic state is characterized by a − superscript.

A right-hand subscript of g or u attached to the electronic energy state symbol is also employed. This is used when the atoms joined by the bond are identical or are isotopes of the same element. Such a system has a centre of symmetry.

If the coordinates of all the electrons x_i, y_i, z_i are replaced by $-x_i$, $-y_i$, and $-z_i$, then provided that the electron eigenfunction remains unaltered in sign by

† This magnetic field is produced by the rotation of the nuclei as a pair about a common centre of gravity, and consequently a perpendicular magnetic field is produced in the direction of the vector representing the nuclear angular momentum.

this reflection through the centre of symmetry, the subscript g is attached. If the sign of the wavefunction changes, a subscript u is employed. The g symbol is derived from the German *gerade* denoting even and u from *ungerade* meaning odd. Thus, a molecule characterized by the symbol g is regarded as having an even state indicating that on reflection of the electron eigenfunction at the centre of symmetry its sign is unchanged and conversely so for the u symbol.

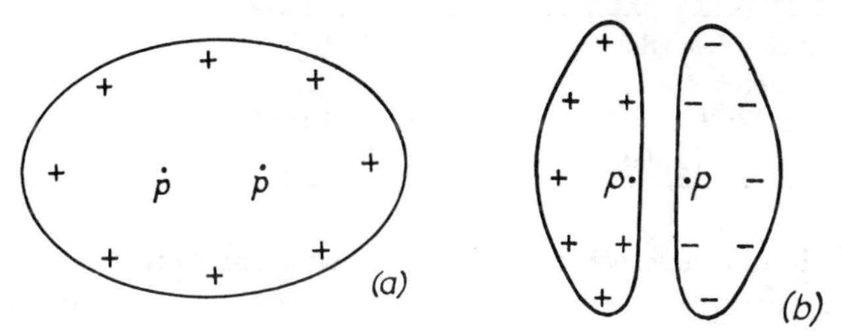

Fig. 1.5 (a) A bonding orbital, e.g. a $^2\Sigma_g^+$-state. (b) An antibonding orbital, e.g. a $^2\Sigma_u^+$-state.

As an example of the use of + and − symmetry and g and u, the molecular orbit of the $\mathbf{H}_2^+$ will now be considered. The bonding orbit is shown in Fig. 1.5(a) and the non-bonding orbit in Fig. 1.5(b). The positive and negative signs inside the boundary indicate that the wavefunction ψ is positive and negative, respectively, in that region, while the p represents a proton. If a plane is placed through the two protons, then on either side of the plane the wavefunction does not change sign on reflection through the plane; hence, a + sign is attached as a superscript to the term symbol. If a straight line is drawn between the centres of the two protons, then the mid-point would be the centre of symmetry, and if this is made the origin of the coordinates (x, y, z) it follows from Fig. 1.5(a) that on replacement of these coordinates by $(-x, -y, -z)$ the wavefunction will still be positive; hence, the symbol g is used for this case. In Fig. 1.5(b), however, by a similar procedure the wavefunction would be negative; hence, for this case, the symbol u would be employed. As there is only one electron, the multiplicity must be $2 \times \frac{1}{2} + 1 = 2$, and this is added as the left-hand superscript; hence, the bonding orbital is described by $^2\Sigma_g^+$ but the antibonding orbital by $^2\Sigma_u^+$.

1.3 CORRELATION BETWEEN ATOMIC AND MOLECULAR STATES

A further link between atomic and molecular spectra is that various rules have been evolved as to which possible molecular electronic states may result from the combination of two atoms in known electronic states to form a diatomic molecule. These rules emerge from quantum mechanics where only certain multiplicities and molecular states are permissible from the combination of any two atomic states. For instance a few of the examples on how atomic and molecular multiplicities may be related are given in Table 1.1.

Table 1.1 Feasible molecular multiplicities resulting from the combination of two atoms of known multiplicity

Atomic multiplicity	*Resulting molecular multiplicity*
singlet + singlet	singlet
singlet + doublet	doublet
doublet + doublet	singlet or triplet
triplet + triplet	singlet or triplet or quintet

In Table 1.2 the molecular electronic states which may result from combining two atomic states of known multiplicity to form a symmetrical diatomic molecule are given. Once, however, P- and D-states are considered, the number of feasible resulting molecular states considerably increases.

Table 1.2 Feasible molecular states of a symmetrical diatomic molecule resulting from the combination of known atomic states

Atomic states	*Resulting molecular state*
$^1S + {}^1S$	$^1\Sigma_g^+$
$^2S + {}^2S$	$^1\Sigma_g^+$ or $^3\Sigma_u^+$
$^3S + {}^3S$	$^1\Sigma_g^+$ or $^3\Sigma_u^+$, or $^5\Sigma_g^+$

For example, the combination of two atoms in a ^{1}P-state may result in $^1\Sigma_g^+$, $^1\Sigma_u^-$, $^1\Pi_g$, $^1\Pi_u$, $^1\Delta_g$ states, while $^1D + {}^1D$ may give nine different molecular states.

1.4 COUPLING OF ANGULAR MOMENTA

1.4.1 Introduction

So far we have considered the rotation of a diatomic molecule with respect to the rotation of the nuclei only. However, this rotation and the rotational energy equation may be influenced by the motion of the electrons. We shall now consider the effect of one on the other and the quantum numbers which characterize the rotational levels in the various types of electronic states.

A molecule has four different sources of angular momenta which are due to (i) the motion of the electron in the orbit, (ii) electron spin, (iii) one or both of the nuclei spinning, and (iv) the rotation of the nuclei as a unit (nuclear rotation). The nuclear-spin angular momentum is often neglected, and the total angular momentum of the molecule may then be regarded as the resultant angular momentum obtained by combining the other three types of angular momenta.

In nearly all diatomic molecules in their ground state the electron spins are paired,† and the electron spin angular momentum $S^*h/2\pi$ is zero. Furthermore, the orbit angular momentum $L^*h/2\pi$ of the electrons is also usually zero. Thus, such molecules have a $^1\Sigma$ ground state. If the nuclear spin is neglected, then the only other source of angular momentum of the molecule is that due to nuclear rotation, and the rotational energy equation for such a case was considered in Vol. 2, p. 84.

For free radicals, and certain electronically excited molecules, however, there would be more than one source of angular momentum. Hund was the first to show how these different sources of angular momenta could be coupled together to give a resultant angular momentum. He considered five different cases, known as Hund's cases (*a*), (*b*), (*c*), (*d*), and (*e*), in which the resultant angular momentum could be formed.

To appreciate Hund's coupling cases it is necessary to realize that both internal electric and magnetic fields can act along the internuclear axis. The electric field results from the electrostatic field due to the two nuclei. This electric field causes an induced electric dipole to be set up in the molecule which can interact with the electric field and make the orbit angular momentum precess about its axis. Furthermore, as a result of the precession of this orbit angular momentum a magnetic field is set up along the internuclear axis. In addition to these fields a further magnetic field may arise because of the rotation of the nuclei as a pair about their common centre of gravity. This magnetic field acts in the direction of the nuclear angular momentum vector, that is perpendicular to the internuclear axis. For cases where $\Lambda > 0$ and $S > 0$ the interaction between the nuclear angular momentum and the orbit and spin angular momenta of the electrons occurs magnetically.

For each of Hund's cases an equation may be obtained for the rotational energy of the molecule in terms of quantum numbers. All the five cases of Hund are either limiting or extreme cases, and many intermediate cases are found in practice. In general, cases (*a*) and (*b*) may be looked upon as the normal coupling cases; these together with case (*c*) will now be considered.

1.4.2 Hund's case (*a*)

This applies to diatomic molecules where the internuclear distance is small enough for there to be a sufficiently strong electric field (resulting from the electrostatic field of the two nuclei) along the internuclear axis to prevent $L^*h/2\pi$ and $S^*h/2\pi$ from directly coupling. The electric dipole moment (associated with $L^*h/2\pi$) induced by this electric field then interacts with the electric field and $L^*h/2\pi$ is caused to precess about the internuclear axis and has an angular momentum $\Lambda h/2\pi$ in the direction of the internuclear axis. The precession of $L^*h/2\pi$ sets up a strong magnetic field along the internuclear axis,

† The O_2 and NO molecules are exceptions.

which enables the magnetic dipole associated with the electron spin vector to interact, and $S^*h/2\pi$ is also caused to precess about the internuclear axis. The magnetic interaction of the $S^*h/2\pi$ and $\Lambda h/2\pi$ vectors is strong. This results in the $S^*h/2\pi$ vector being strongly coupled to the magnetic field along the internuclear axis so that its axial component $\Sigma\, h/2\pi$ is quantized. It follows, therefore, that the total angular momentum quantum number Ω along the internuclear axis is given by:

$$\Lambda + \Sigma = \Omega$$

where $\Omega h/2\pi$ is the total electron angular momentum along the internuclear axis.

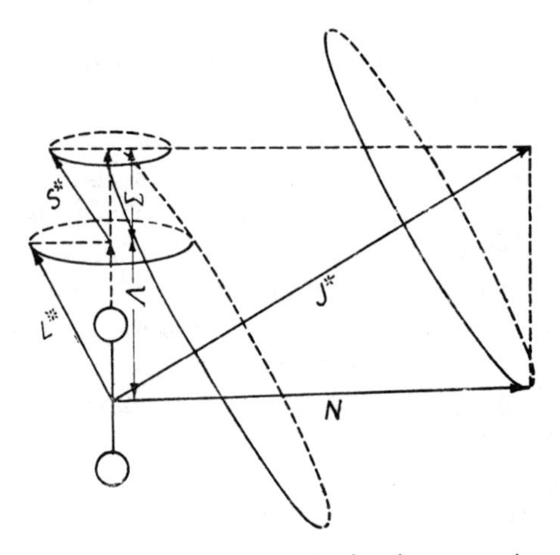

Fig. 1.6 Vector diagram for Hund's case (*a*) L^*, S^*, Λ, Σ, J^*, and N are each in $h/2\pi$ units.

The precession of $L^*h/2\pi$ and $S^*h/2\pi$ is shown in Fig. 1.6, and their components $\Lambda h/2\pi$ and $\Sigma\, h/2\pi$, respectively, along the internuclear axis are indicated. The total electron angular momentum about the internuclear axis is well defined. The nuclei in the rigid diatomic molecule may be regarded as rotating as a whole about an axis perpendicular to the internuclear axis and having an angular momentum of rotation $Nh/2\pi$ which will act perpendicular to the centre of gravity of the molecule. If the interaction between $Nh/2\pi$ and $\Omega h/2\pi$ is very weak, then these two angular momenta may be combined vectorially to form a resultant angular momentum:

$$\sqrt{[J(J+1)]}\, h/2\pi$$

where $J = \Omega, \Omega + 1, \Omega + 2, \ldots$.

The precession of various vectors is given in Fig. 1.6, where the vector parallelogram formed from $\Omega h/2\pi$ and $Nh/2\pi$ has been completed with the broken lines, and the resultant total angular momentum is represented by the diagonal. The coupling vectors $\Omega h/2\pi$ and $Nh/2\pi$ precess about the axis of the resultant total angular momentum.

N is related to Ω and J by the equation:

$$N = \sqrt{[J(J+1) - \Omega^2]} \tag{1.6}$$

From classical theory the rotational energy of a rigid diatomic molecule may be expressed in terms of its angular momentum (P) and its moment of inertia (I) by the equation:

$$E_r = \frac{1}{2}\frac{P^2}{I} \tag{1.7}$$

For the rotation of the molecule (nuclei) as a whole:

$$P = N\frac{h}{2\pi} \tag{1.8}$$

Hence, the rotational energy is given by:

$$E_r = \frac{h^2}{8\pi^2 I}N^2 \tag{1.9}$$

If the value of N^2 is substituted from Equation (1.6) into (1.9), then:

$$E_r = [J(J+1) - \Omega^2]\frac{h^2}{8\pi^2 I} = [J(J+1) - \Omega^2]Bhc \tag{1.10}$$

where $B = h/8\pi^2 cI$ and $J = \Omega, \Omega + 1, \Omega + 2$, etc. For a given electronic state Ω would have a fixed value.

It follows that case (a) demands both $\Lambda > 0$ and $S > 0$, and therefore it cannot apply to any Σ-states, or to singlet states (i.e. $S = 0$) of any kind. The point on Σ-states is to be contrasted with case (b) which will now be considered and which is applicable to Σ, and also to some other states where $\Lambda > 0$.

1.4.3 Hund's case (b)

Again as with case (a) the internuclear distance is small enough for there to be a sufficiently strong electric field along the internuclear axis to prevent $L^*h/2\pi$ and $S^*h/2\pi$ from directly coupling. $L^*h/2\pi$ again precesses about this electric field acting along the internuclear axis. However, case (b) differs from case (a) in that the magnetic field associated with this precession is so weak that the interaction between $\Lambda h/2\pi$ and $S^*h/2\pi$ is negligible compared with that due to the interaction of the molecular rotation angular momentum ($Nh/2\pi$) and $\Lambda h/2\pi$. $\Lambda h/2\pi$ then combines directly with the nuclear angular momentum ($Nh/2\pi$) to give a resultant angular momentum, $K^*h/2\pi$ where the rotational quantum number K takes the values $\Lambda, \Lambda + 1, \Lambda + 2$, etc. $K^*h/2\pi$ is known as the total angular momentum apart from spin. The coupling occurs through the magnetic fields produced by the rotating nuclei and that of the orbital angular momentum, and $\Lambda h/2\pi$ and $Nh/2\pi$ precess around $K^*h/2\pi$. The electron spin vector $S^*h/2\pi$ couples with $K^*h/2\pi$ to give a resultant $J^*h/2\pi$ which is the total angular momentum including electron spin.† The values which J may take for a given value of K are obtained by the rules of vector addition:

$$J = (K+S), (K+S-1), \ldots, |K-S| \tag{1.11}$$

Apart from when $K < S$ each rotational level with a particular K-value has $2S+1$ components. Since

$$K = \Lambda, \Lambda+1, \Lambda+2, \ldots$$

and Λ may have only the values $0, 1, 2, \ldots$ it follows that K may take only integral values. In addition, since S may have integer or half-integer values according to whether there is an even or odd number, respectively, of electrons in the molecule, then it may be seen from Equation (1.11) that J is half-integral when the number of electrons is odd and integral for an even number. In Fig. 1.7 the rotational levels of a ${}^2\Sigma$-state may be observed where for such a state $\Lambda = 0$ and $S = \frac{1}{2}$, and the relation between the J- and K-values indicated in the figure.

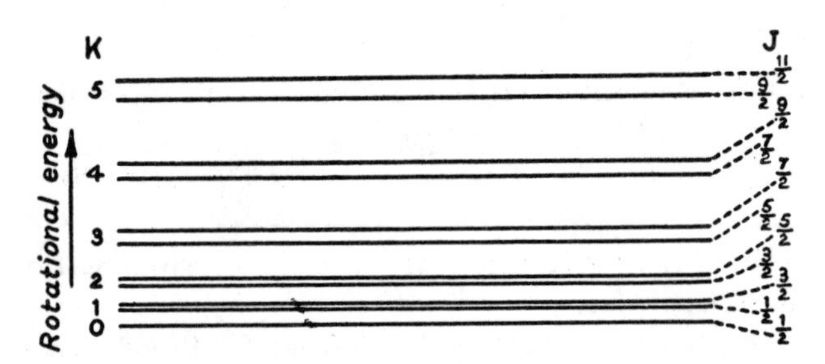

Fig. 1.7 Rotational energy levels of a vibrational level in a ${}^2\Sigma$-state.

The precession of the various vectors is represented in Fig. 1.8(i), and the formation of the total angular momentum vector $J^*h/2\pi$ from $K^*h/2\pi$ and $S^*h/2\pi$ is indicated. Both $K^*h/2\pi$ and $S^*h/2\pi$ precess around $J^*h/2\pi$. The value of N is:

$$N = \sqrt{[K(K+1) - \Lambda^2]} \tag{1.12}$$

and the rotational energy equation is given approximately by:

$$E_r = [K(K+1) - \Lambda^2]Bhc \tag{1.13}$$

where $K = \Lambda, \Lambda+1, \Lambda+2, \ldots$. Hund's case ($b$) applies to molecules in Σ-states, since when $\Lambda = 0$ there is no internuclear magnetic field to make $S^*h/2\pi$ precess around it. Nearly all diatomic molecules have a Σ ground state. In addition, case (b) may also apply to states where $\Lambda = 1, 2, 3, \ldots$, if the internuclear magnetic field is sufficiently weak.

† For this case, since $S^*h/2\pi$ is not coupled to the internuclear axis it follows that the quantum number Σ does not exist.

1.4.4 Hund's case (*c*)

When the internuclear distance is sufficiently large then the electric field along the internuclear axis may be inadequate to destroy the coupling between the orbit and spin angular momenta. This corresponds to Hund's case (*c*) where the orbit and spin angular momenta couple to give a resultant angular momentum $\sqrt{[J_a(J_a + 1)]}\, h/2\pi$ which precesses about the internuclear electric field, while the $L^* h/2\pi$ and $S^* h/2\pi$ vectors precess about the axis of $J_a^* h/2\pi$. The symbol J_a is employed to indicate its relationship to the J employed in atomic spectra which is also formed from coupling L and S.

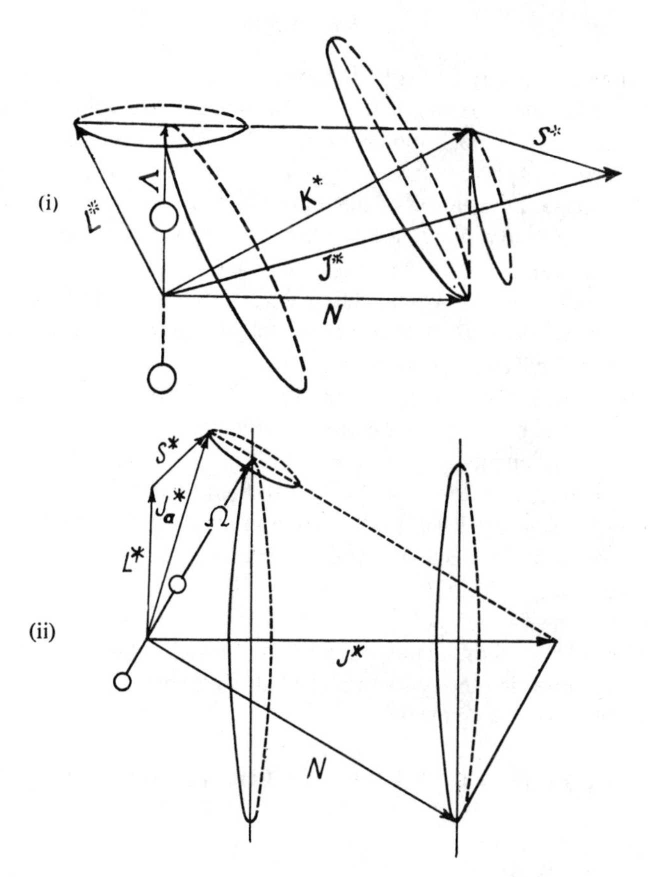

Fig. 1.8 (i) Coupling and precession of the vectors in Hund's case (*b*). (ii) Coupling and precession of the vectors in Hund's case (*c*).

The component of $\sqrt{[J_a(J_a + 1)]}\, h/2\pi$ along the internuclear axis is $\Omega h/2\pi$ where Ω takes the values:

$$J_a, J_a - 1, \ldots, \tfrac{1}{2} \text{ or } 0$$

The $\Omega h/2\pi$ and $Nh/2\pi$ vectors then combine to give the resultant $\sqrt{[J(J+1)]}\,h/2\pi$ and precess about it. These points may be observed in Fig. 1.8(ii). It should be noted that for case (*c*) Λ has no meaning, and hence the symbols Σ, Π, and Δ cannot be employed to represent such states.

On comparison of Fig. 1.6 and 1.8(ii) it follows that in Hund's case (*c*) N must be related to Ω and J by the same equation as that given for case (*a*), that is Equation (1.6). In addition, the same rotational energy equation, that is Equation (1.10), must apply to both these cases.

It is necessary to reconcile Equations (1.10) and (1.13) with the one already quoted for the rotational energy of a rigid molecule, that is with:

$$E_r = J(J+1)Bhc \tag{1.14}$$

As already indicated nearly all diatomic molecules are in a $^1\Sigma$ ground state, that is a singlet, and for such a state $J \equiv K$. The resultant electron spin (S) and the resultant orbit angular momentum component $\Lambda h/2\pi$ of the electrons are both zero, and, hence, so is Ω; therefore, Equations (1.10) and (1.13) reduce to (1.14), and the total angular momentum of the molecule may be attributed solely to the rotation of the rigid molecule. Since $\Lambda = 0$ and $\Omega = 0$ for a $^1\Sigma$ state, then $J = 0, 1, 2, 3$, which again were the ones previously quoted.

In an electronic transition from a detailed analysis of the rotational structure of the bands the values of Ω, Λ, and the multiplicity are allotted, and the two electronic states involved in the transition are identified.

One of the most helpful factors in this type of analysis is to note which lines are missing from the bands in the electronic spectrum and to relate them to the particular rotational energy levels involved. In Hund's coupling case (*a*) J takes the values Ω, $\Omega+1$, $\Omega+2$, etc., and if $\Omega > 0$, then there will be at least one 'missing' rotational energy level. In the case considered on p. 5, where $\Lambda = 1$ and $S = 1$, there were three sub-electronic states $^3\Pi_0$, $^3\Pi_1$, and $^3\Pi_2$ where $\Omega = 0$, 1, and 2, respectively. For the sub-state $^3\Pi_1$ there will be one missing rotational level, the $J = 0$ level, while for the sub-state $^3\Pi_2$ the $J = 0$ and the $J = 1$ levels would be absent. Thus, knowledge of which rotational levels are absent assists in inferring the types of molecular states between which the electronic transitions are taking place.

1.5 SELECTION RULES OF DIATOMIC MOLECULES

1.5.1 General

As is typical of spectra, selection rules limit the possible transitions–for both absorption and emission–between the various electronic states, and for a diatomic molecule some of these are:

$$\Delta\Lambda = 0, \pm 1 \tag{1.15}$$

$$\Delta S = 0\dagger \tag{1.16}$$

$$\Delta \Sigma = 0 \tag{1.17}$$

$$\Delta \Omega = 0, \pm 1\ddagger \tag{1.18}$$

Thus, in theory transitions such as:

$$^2\Delta_{5/2} \longleftrightarrow {}^2\Pi_{3/2}, {}^2\Delta_{3/2} \longleftrightarrow {}^2\Pi_{1/2}$$

would be permissible whereas the transitions:

$$^3\Delta_3 \longleftrightarrow {}^2\Pi_{1/2}, {}^2\Delta_{5/2} \longleftrightarrow {}^2\Pi_{1/2}$$

would be forbidden.

For transitions between Σ^+- and Σ^--states the selection rules are:

$$\Sigma^+ \longleftrightarrow \Sigma^+, \Sigma^- \longleftrightarrow \Sigma^- \text{ but } \Sigma^+ \not\longleftrightarrow \Sigma^-$$

where the sign $\not\longleftrightarrow$ indicates that the transition is forbidden. For a symmetrical diatomic molecule further restrictions are g → u, u → g, but g $\not\rightarrow$ g and u $\not\rightarrow$ u. In addition to this there are further selection rules which govern the fine structure of the individual bands. The selection rules for the rotational quantum number are governed by whichever of Hund's coupling cases is concerned:

(i) For case (*a*): $$\Delta J = 0, \pm 1 \tag{1.19}$$

although the $\Delta J = 0$ is forbidden when $\Omega = 0$ in both electronic states.

(ii) For case (*b*): $$\Delta K = 0, \pm 1 \tag{1.20}$$

where $\Delta K = 0$ is forbidden for a $\Sigma - \Sigma$ transition.
Additional selection rules which determine the fine structure are considered on p. 18.

In order to gain some indication of the complexity of the electronic-energy states of diatomic molecules and the possible electronic transitions involved, together with application of the selection rules, one example will now be considered. Figure 1.9 is the energy-level diagram for the N_2 molecule where the heavy (full) horizontal lines represent some of the several detected electronic states. The short thin lines indicate the vibrational levels relating to a particular electronic state.

The $X^1\Sigma_g^+$ is the ground electronic state and transitions may occur to this from some of the excited states. In addition, transitions occur between the excited electronic states. Two particularly well-known emission transitions to

† Although transitions between states of different multiplicity are forbidden (i.e. $\Delta S = 0$), in practice the rule is sometimes violated as in the case of the Cameron bands of CO which are attributed to a $^3\Pi - {}^1\Sigma^+$ transition. Usually, however, whenever the ΔS selection rule is violated $\Delta S = \pm 1$.

‡ Equations (1.15) and (1.16) apply only to Hund's case (*a*) and (*b*); (1.17) requires that both electronic states belong to case (*a*), and (1.18) to case (*c*).

spectroscopists are the first and second positive nitrogen bands which correspond to the $B^3\Pi_g \rightarrow A^3\Sigma_u^+$, $C^3\Pi_u \rightarrow B^3\Pi_g$ transitions. These satisfy the selection rules:

$$\Delta\Lambda = 0, +1, \Delta S = 0, \text{ and } u \longleftrightarrow g$$

and are, therefore, permitted transitions. The Vegard–Kaplan bands, however, which are also experimentally observed, correspond to the forbidden transition $A^3\Sigma_u^+ \rightarrow X^1\Sigma_g^+$. A number of electronic transitions are indicated in the figure, and the names allotted to the main ones are given. It will be observed that the lowest-lying triplet state is designated by the symbol A placed in front of its energy state symbol and the two succeeding triplet levels in order of increasing energy by B and C.

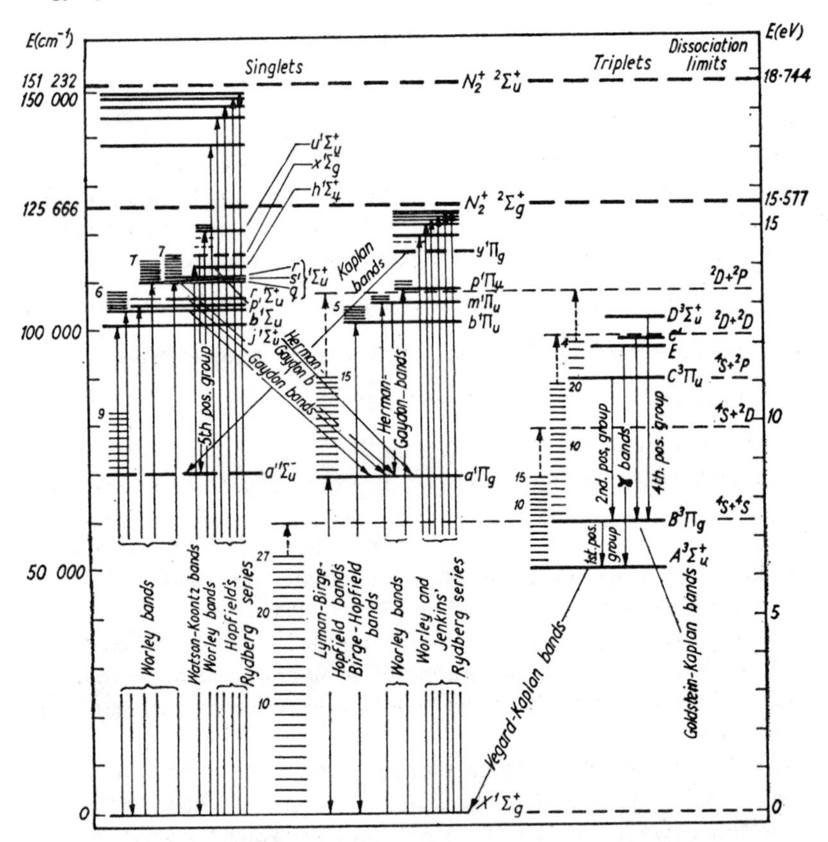

Fig. 1.9 Energy level diagram and electronic transitions of the N_2 molecule. The heavy (full) horizontal lines give the electronic states; the shorter thinner lines give the vibrational levels in each of them. The broken horizontal lines indicate dissociation limits. (After Herzberg [1.1])

On the extreme right of the diagram the atomic states of the dissociation products of some of the molecular states are given of the N_2^+ ion and N_2

molecule, and these are indicated by broken horizontal lines. It will be seen that the atomic states of the dissociation products are ^4S, ^2D, and ^2P; these correspond to the three lowest energy states of the nitrogen atom. The ^2D lies 2.383 eV above the ^4S state which is the ground state, while the ^2P lies 3.574 eV above it. In Fig. 1.9 two ionization limits are indicated by means of the heavy broken horizontal lines across the whole of the figure, and these correspond to the two states $^2\Sigma_g^+$ and $^2\Sigma_u^+$ of the ionized nitrogen molecule (i.e. N_2^+).

1.5.2 Symmetry properties of rotational levels and their selection rules

The total eigenfunction ψ of a diatomic molecule is given to a first approximation by:

$$\psi = \psi_e \frac{1}{r} \psi_v \psi_r$$

where r is the internuclear distance and ψ_e, ψ_v, and ψ_r are, respectively, the electronic, vibrational, and rotational eigenfunctions. The rotational levels of a diatomic molecule may be classified as being either positive (+) or negative (−) according to the behaviour of ψ on reflection at the origin of the coordinates. The reflection applies to all the particles within the molecule including the nuclei, although the rotational and vibrational eigenfunctions depend solely on the coordinates of the nuclei. Reflection at the origin is achieved by replacing the coordinates (x, y, z) by $(-x, -y, -z)$, the potential energy of the system remaining unchanged as does also the magnitude of ψ. When this operation is considered for a particular rotational level, then if the sign of the total eigenfunction remains unchanged, this level is characterized by being allotted a positive sign. If, however, the sign of ψ alters, a negative sign would be employed.

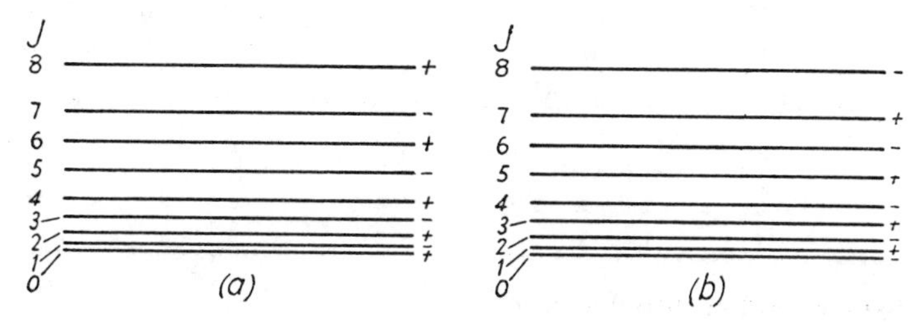

Fig. 1.10 The symmetry properties of rotational levels: (a) where the sign of ψ_e is unchanged on reflection at the origin; (b) where the sign of ψ_e is changed on reflection.

The vibrational part, $(1/r)\psi_v$ of the total eigenfunction is unaffected by reflection, and if ψ_e remains unchanged, then the sign of the rotational level is determined solely by ψ_r. A detailed investigation of ψ_r shows that for even values of J the rotational eigenfunction is unaltered while for odd J-values ψ_r becomes $-\psi_r$. Thus, for the case where ψ_e remains unchanged the rotational levels are characterized by + and − according as J is even or odd, respectively, and this is the case given in Fig. 1.10(a). If the sign of ψ_e also changes on reflection, then even J-values are characterized by a negative sign and odd J-values by a positive one; this is illustrated in Fig. 1.10(b). Before the sign of ψ_e is considered further it is necessary to appreciate what is meant by inversion. Inversion may be achieved by the following operations: (a) rotation of the molecule through 180° about an axis which is perpendicular to the internuclear axis, then (b) reflection of the particles across the plane passing through the internuclear axis and perpendicular to the rotational axis. Only (b) could affect the sign of ψ_e. If these ideas are applied to Σ^+- and Σ^--states, it follows from previous considerations that the sign of ψ_e would be unaltered for a Σ^+-state but would change for a Σ^--state. Thus, for Σ^+-states the $J = 0, 2, 4, \ldots$ rotational levels are characterized by a positive sign. The $J = 1, 3, 5, \ldots$ levels, however, would have a negative sign.

Rotation–vibration and electronic changes are governed by the selection rule:

$$+ \longleftrightarrow -$$

that is, positive levels combine only with negative. Transitions of the type $+ \longleftrightarrow +$ and $- \longleftrightarrow -$ are forbidden for these types of changes but allowed for Raman spectra in which the $+ \longleftrightarrow -$ transitions are forbidden. These selection rules are given in Table 1.3 together with those for J and a and s (see below).

Table 1.3 Selection rules for infrared, Raman, and electronic transitions

Infrared rotation–vibration		*Raman rotation–vibration*		*Electronic*	
Rotator with $\Lambda = 0$	*Rotator with* $\Lambda \neq 0$	*Rotator with* $\Lambda = 0$	*Rotator with* $\Lambda \neq 0$	*Case where* $\Lambda' = \Lambda'' = 0$, *e.g.* $^1\Sigma - {}^1\Sigma$	*Case where* either Λ' *or* $\Lambda'' \neq 0$, *e.g.* $^1\Pi - {}^1\Sigma$
$\Delta J = \pm 1$	$\Delta J = 0, \pm 1$	$\Delta J = 0, \pm 2$	$\Delta J = 0, \pm 1, \pm 2$	$\Delta J = \pm 1$	$\Delta J = 0, \pm 1$
$+ \longleftrightarrow -$	$+ \longleftrightarrow -$	$+ \longleftrightarrow +$ $- \longleftrightarrow -$	$+ \longleftrightarrow +$ $- \longleftrightarrow -$	$+ \longleftrightarrow -$	$+ \longleftrightarrow -$
a ⟷ a s ⟷ s	a ⟷ a s ⟷ s	a ⟷ a s ⟷ s	a ⟷ a s ⟷ s	a ⟷ a s ⟷ s	a ⟷ a s ⟷ s

1.5.3 Symmetric and antisymmetric selection rules for the rotational levels of homonuclear molecules

For a homonuclear diatomic molecule, such as $^1H_2$, $^{16}O_2$, $^{14}N_2$, etc., the rotational levels are characterized in addition to + and − by s (symmetric) and a (antisymmetric). When an exchange of the positions of the nuclei in the molecule leaves the total eigenfunction unchanged the rotational levels are said to be symmetric, while if ψ changes sign the rotational levels are classified as antisymmetric. For a given electronic state of the molecule either the positive rotational levels are all symmetric and the negative all antisymmetric or conversely these relations are reversed. Both of these possibilities are shown in Fig. 1.11(a) and (b). Whether (a) or (b) applies in a particular example depends on whether the nuclei in question obey the Bose–Einstein or the Fermi–Dirac statistics [1.1], but this will not be considered further.

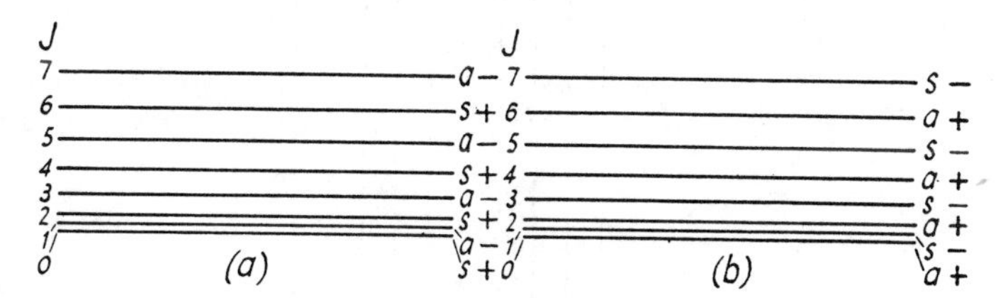

Fig. 1.11 Symmetry properties of rotational levels of homonuclear molecules: (a) where the positive levels and negative levels are symmetric and antisymmetric, respectively; (b) where the negative and positive levels are symmetric and antisymmetric, respectively.

If the possibility of nuclear spin is neglected, a rigid selection rule operates for absorption or emission of radiation between symmetric and antisymmetric levels and is:

$$\mathrm{s} \leftrightarrow \mathrm{s},\ \mathrm{a} \leftrightarrow \mathrm{a},\ \mathrm{a} \not\leftrightarrow \mathrm{s}$$

This with other selection rules is given in Table 1.3.

If all the molecules of a given type were in symmetric states at some given time, and providing the nuclei of these molecules did not possess a nuclear spin, then since the selection rule prohibits intercombinations between symmetric and antisymmetric states, the molecules would always remain in a symmetric state. In addition, if this were so, every second line would be missing from the electronic band spectrum of that species. In Fig. 1.12 some of the transitions between the $^1\Sigma_g^+$ and the $^1\Sigma_u^+$ electronic states are given. The rotational levels have been characterized by + and − and also by symmetric (s) and antisymmetric (a). The selection rule for rotational transitions between such electronic states is $\Delta J = \pm 1$; transitions between antisymmetric states have been indicated by broken lines and those between symmetric states by full lines. It will be observed from Fig. 1.12 that if, in fact, all the molecules were in either symmetric or antisymmetric states, then either the transitions indicated by broken lines or those by the full lines would be absent, respectively, from the electronic band spectrum. This absence of every second line in the band spectrum has been observed for several homonuclear diatomic molecules including $^{16}O_2$ and $^{12}C_2$. If, however, the symmetry of such molecules is destroyed by isotopic substitution, then the question of symmetric and antisymmetric rotational levels does not arise, and all the rotational lines are observed.

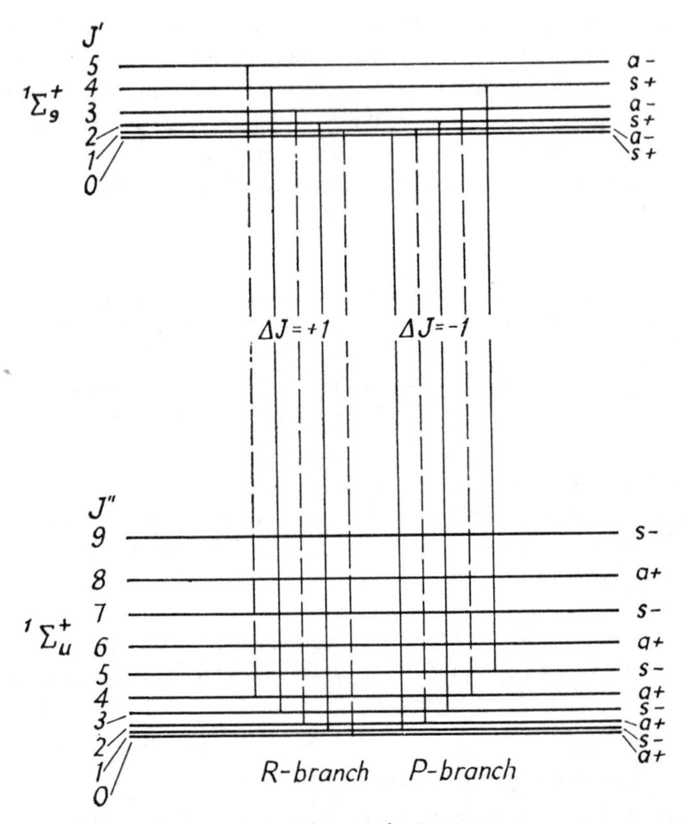

Fig. 1.12 Transitions between the $^1\Sigma_g^+$ and $^1\Sigma_u^+$ electronic states where the full lines indicate s $\longleftrightarrow$ s transitions and the broken lines a $\longleftrightarrow$ a transitions.

For a homonuclear diatomic species in which the nuclei spin the band spectrum is characterized by alternately strong and weak rotational line intensities. The explanation of this intensity alternation was first given by Hund and depends upon the fact that the selection rule prohibiting the combination of symmetric and and antisymmetric states is no longer rigid. The result of this is that both symmetric and antisymmetric term systems can appear but do so with different statistical weights.

The pure rotational Raman spectra of $^{16}O_2$ and $^{14}N_2$ given in Fig. 1.13 illustrate in the case of $^{16}O_2$ the absence of every second rotational line and for $^{14}N_2$ the alternately strong and weak rotational line intensities [1.3].

1.6 Λ DOUBLING

It was assumed in the treatment of Hund's cases (*a*) and (*b*) that the interaction between the angular momentum ($Nh/2\pi$) due to the rotation of the nuclei as a

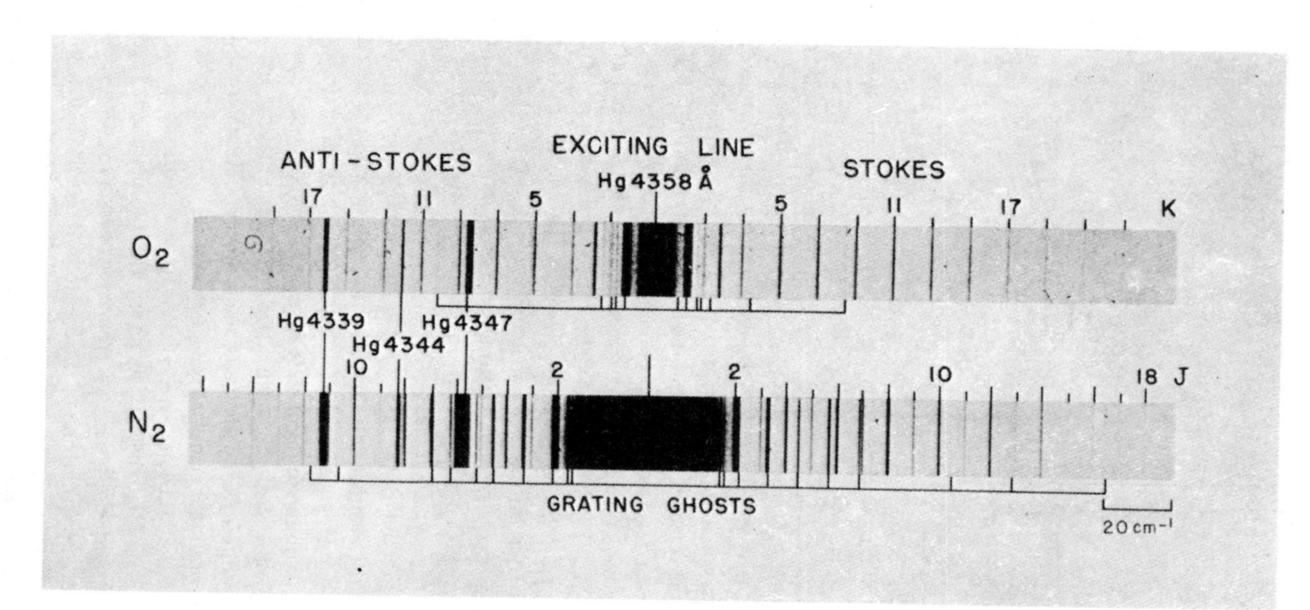

Fig. 1.13 Rotational Raman spectra (see Vol. 2) of the O_2 and N_2 molecules [1.3]. In the case of the O_2 spectrum the rotational quantum number K values (Hund's case b) are given on the upper edge of the spectrogram. It will be noted that only rotational lines with odd K values are observed. The N_2 spectrum consists of widely spaced lines exhibiting the intensity alternation which are strong when J is even and weak when J is odd. The nuclear spin of nitrogen is 1. (Courtesy of Dr. B.P. Stoichett).

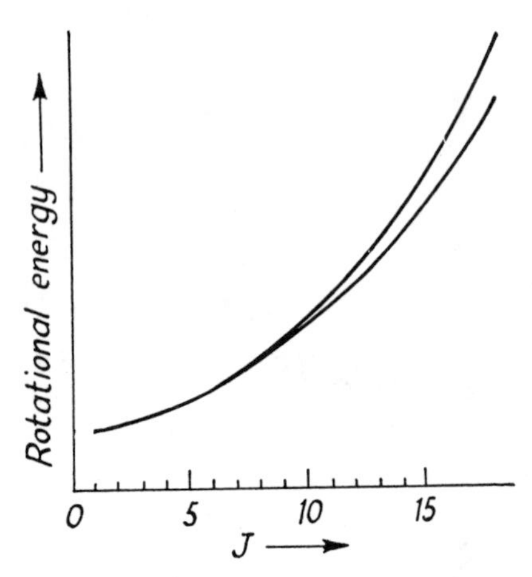

Fig. 1.14 Λ doubling in a $^1\Pi$-state. The doublet splitting is exaggerated.

whole in the rigid diatomic molecule and that of the orbit angular momentum ($\sqrt{[L(L+1)]}\,h/2\pi$) could be neglected. This is only strictly exact when the component of the orbital angular momentum about the internuclear axis is zero, that is when $\Lambda = 0$, and this is the case of a Σ electronic state. For Π, Δ, . . . electronic states where $\Lambda \neq 0$, when the molecule is rotating (i.e. $J \neq 0$), splitting of each rotational level occurs into two components. For very low J-values the splitting is negligible and may not necessarily be detectable, but as the speed of rotation increases (i.e. for higher J-values) the splitting may be detected in the electronic spectrum of the molecule. However, the splitting of the rotational levels is usually only of the order of 1 cm^{-1} or less, and for the simplest cases the splitting increases as a simple function of J. For example, for a $^1\Pi$ electronic state the splitting is approximately proportional to $J(J+1)$ and for a $^1\Delta$-state to $J^2(J+1)^2$.

This small splitting of the rotational energy levels which is produced by the interaction of the electron orbit angular momentum with the rotation of the molecule as a whole is termed Λ doubling. It applies to all molecules where $\Lambda > 0$. The Λ doubling for a $^1\Pi$-state may be observed in Fig. 1.14 where the rotational energy is plotted against J. It will be noted that the splitting is negligible at low J-values. At the higher J-values, however, for each value of J, two values of the rotational energy may be observed.

In Fig. 1.15 the rotational energy levels for the electronic states $^1\Sigma^+$, $^1\Sigma^-$, $^1\Pi$, $^1\Delta$, $^1\Sigma_g^+$, $^1\Sigma_u^-$, $^1\Pi_g$, and $^1\Pi_u$ are given. The J-values are indicated on the figure together with the symmetry properties of each rotational level. It will be noted that there is no splitting of the Σ rotational levels. This is to be contrasted with each of the Π-states. The $^1\Sigma^+$-, $^1\Sigma^-$-, $^1\Pi_g$-, and $^1\Delta_u$-states are those for heteronuclear molecules. For Σ^+-states the $J = 0, 2, 4, \ldots$ rotational levels are characterized by a positive sign whereas the $1, 3, 5, \ldots$ levels have a negative sign. For a Σ^--state, however, the opposite applies. Thus, the signs of the

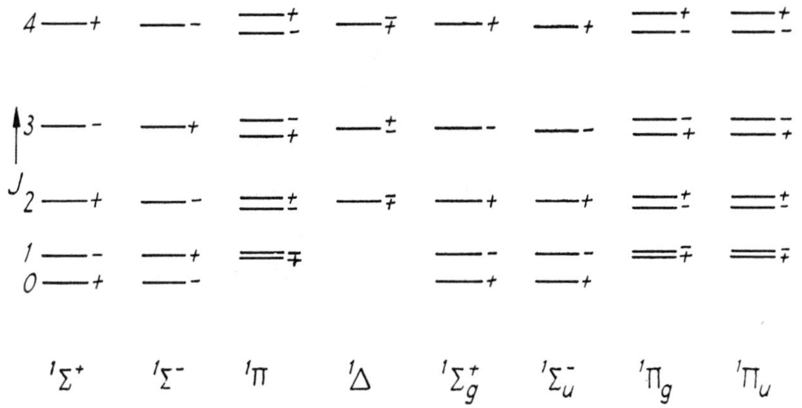

$^1\Sigma^+$ $^1\Sigma^-$ $^1\Pi$ $^1\Delta$ $^1\Sigma_g^+$ $^1\Sigma_u^-$ $^1\Pi_g$ $^1\Pi_u$

Fig. 1.15 Classification of some of the lower rotational energy levels in the electronic states of some homonuclear and heteronuclear molecules [1.2].

rotational levels in Σ^+ and Σ^--states alternate. It may be seen also from Fig. 1.15 that the signs alternate for the Δ- and Π-states as well, although in addition in these cases each of the rotational sublevels is of opposite sign.

As regards $\Sigma - \Sigma$ electronic transitions the position as outlined on p. 20 is unaltered since there is no Λ doubling. However, for other types of electronic transitions as for example a $^1\Pi_u - {}^1\Sigma_g^+$ the Λ doubling has to be taken into account. This is illustrated in Fig. 1.16 for a homonuclear diatomic molecule, where the transitions may be accounted for in terms of the selection rules:

$$\Delta J = 0, \pm 1; + \longleftrightarrow -$$

In addition, the selection rules a $\longleftrightarrow$ a and s $\longleftrightarrow$ s apply although these symbols have not been inserted in the figure. If the transitions were a $^1\Pi - {}^1\Sigma^+$ for a heteronuclear molecule, then the a and s symbols would not apply although the transitions would still be those indicated in the figure.

1.7 c AND d CLASSIFICATION OF ROTATIONAL LEVELS

In addition to the + and − classification, the letters c and d are employed to distinguish the components resulting from Λ type doubling of the rotational levels in electronic states with $\Lambda > 0$, which fall in or between Hund's cases (*a*) and (*b*). For a $^1\Pi$-state c and d characterize, respectively, the same set of rotational sub-states as + and −. For a $^2\Pi_{1/2}$-state, however, the reverse correspondence would be the case. The c and d classification was evolved before the theory was sufficiently developed to give the + and − one, but a definite relationship exists between the two systems. The subscripts c and d are allotted on the basis of transition properties as made from the observed band structure itself. The assignment of c and d to four singlet electronic states for Hund's case

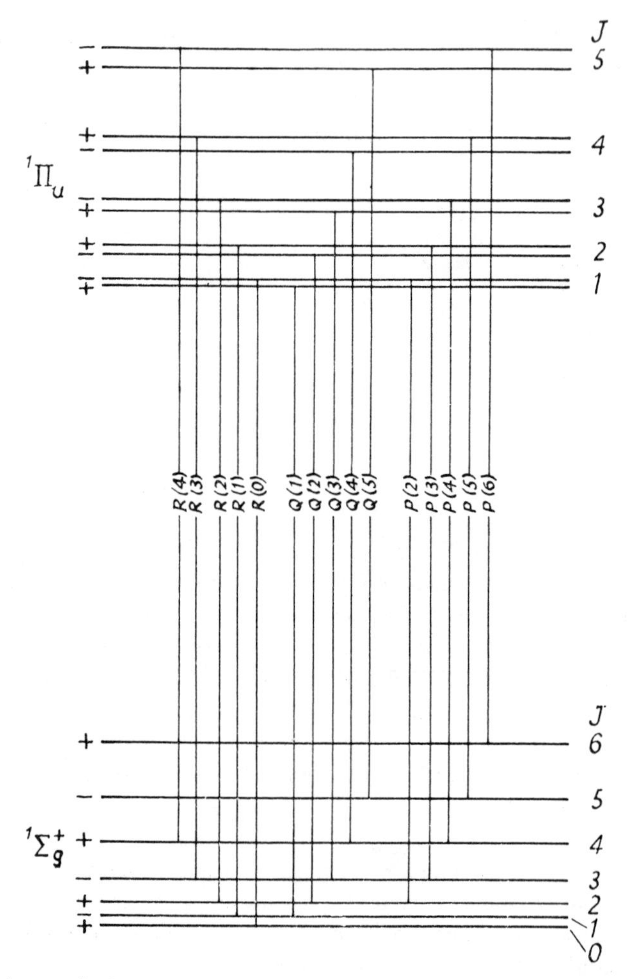

Fig. 1.16 $^1\Pi_u - {}^1\Sigma_g^+$ electronic transition where Λ doubling has been taken in account.

Table 1.4 Comparison of the c and d classification with the + and − one for four electronic states where Hund's case (*b*) applies

Rotational quantum number K	*Electronic states*			
	$^1\Sigma^+$	$^1\Sigma^-$	$^1\Pi$	$^1\Delta$
0, 2, 4, 6, . . .	c, +	d, −	d, +	d, −
			c, −	c, +
1, 3, 5, 7, . . .	c, −	d, +	d, −	d, +
			c, +	c, −

(*b*) is given in Table 1.4 and is compared with the corresponding + and − classification. For examples of the application of the c and d classification Herzberg [1.1] should be consulted.

REFERENCES

1.1 Herzberg, G., *Spectra of Diatomic Molecules,* 2nd Edn., Vol. 1, Van Nostrand, New York (1950).

1.2 Jevons, W., *Report on Band Spectra of Diatomic Molecules*, Physical Society, London (1932).

1.3 Stoicheff, B.P., *Advances in Spectroscopy*, H.W. Thompson (editor), Interscience, London (1959), p. 91.

2 Electronic spectra of gaseous diatomic molecules

2.1 INTRODUCTION

The absorption and emission of energy in the range 10 Å to about 10 000 Å is almost entirely restricted to changes in electronic energy with accompanying vibrational and rotational energy changes. Unlike the microwave, infrared, and Raman changes there is no simple criterion as to whether a molecule may exhibit an electronic spectrum. The criterion is very difficult to evaluate and involves a number of approximations.

If the electronic energy alone changed during a transition, then only one line would result, but since changes in electronic energy are normally accompanied by vibrational and rotational energy changes, a whole set of very closely spaced lines is obtained. The group of lines resulting from transitions between the different rotational levels in two vibrational levels in each of two electronic states is termed a *band*. The sum of all the bands for the transitions between two electronic states is termed a *band system*.

If any change in rotational energy is neglected, representation of a transition between two vibrational levels in different electronic states may be made as in Fig. 2.1 using potential energy diagrams. For a diatomic species each electronic state may be associated with a particular potential energy curve, and in Figs. 2.1(a), (b), and (c) two electronic states are shown in each case, corresponding to the ground and excited states, respectively. In curves (a) the positions of the minima in the upper and lower states lie very nearly one above the other, that is they have almost equal internuclear distance in the two states, while in (b) and (c) the minima of the curves representing the upper electronic levels are displaced to different extents. Transitions between the two electronic states are represented by vertical lines drawn in accordance with the Franck–Condon principle. This requires that the internuclear distance does not alter during the small period of time (about 10^{-16} s) required for the transition. In Fig. 2.1(a), (b), and (c) absorption transitions are indicated from the lowest vibrational level in the ground electronic state of the molecule.

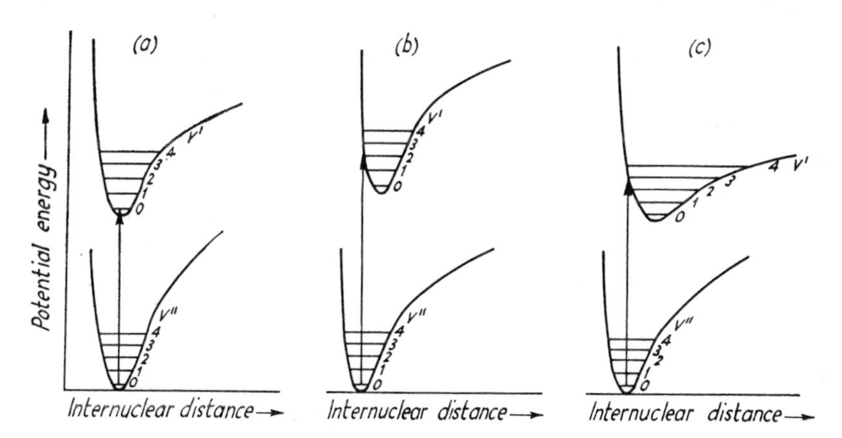

Fig. 2.1 Absorption transitions from the lowest vibrational level in the ground electronic state of the molecule.

2.2 INSTRUMENTATION

2.2.1 Introduction

Before the analysis of the band spectrum of a diatomic molecule is examined we shall consider the theoretical and practical aspects of two kinds of spectrographs termed the prism and grating types which are employed in the study of the electronic spectra of both diatomic and polyatomic molecules. The instruments considered will be mainly those employed in electronic spectra studies ranging from 2000 Å to about 1.8 μ although the principles dealt with will, in the main, apply equally to the vacuum ultraviolet and infrared regions. The spectrometers used in other regions will be considered in the appropriate chapters. References which include details and principles of instrumentation in the various spectroscopic regions are Sawyer [2.1], volumes on molecular spectroscopy [2.2, 2.3], Strouts, Gilfillan, and Wilson [2.4], and Harrison, Lord, and Loofbourow [2.5]. Before the prism and grating spectrographs are described, it is useful to revise some of the basic theory of refraction and dispersion.

2.2.2 Theoretical aspects of prism instruments

Index of refraction

When monochromatic light passes through a transparent isotropic medium at constant temperature the ratio of the sine of the angle of incidence i to the sine of the angle of refraction r (see Fig. 2.2) is constant for that medium. This ratio is known as the *refractive index* n:

$$n = \sin i/\sin r \tag{2.1}$$

In the wavelength region for which the medium is transparent the refractive index increases as the wavelength of the light decreases. A useful empirical relation between the refractive index n and the wavelength λ is given by the *Hartmann dispersion formula*:

$$n = n_0 + c/(\lambda - \lambda_0) \tag{2.2}$$

where n_0, c, and λ_0 are constants over a restricted wavelength range for the material of the medium. This formula is of use in obtaining an expression for (a) the angular dispersion of a prism, and (b) a formula employed in the determination of wavelengths of lines and bands.

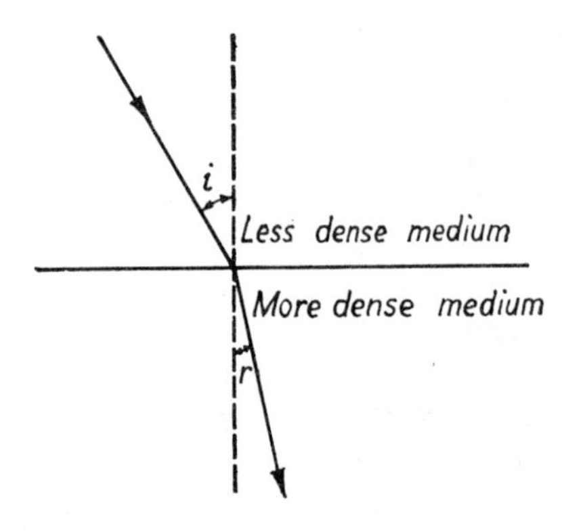

Fig. 2.2 Refraction of a ray on passing from a less to a more dense medium.

Angle of minimum deviation

The best optical image in the case of a prism can be obtained if the prism is traversed by parallel light, and if the light rays pass through the prism parallel to its base, that is, with equal refraction at each surface. These conditions can be realized for only one wavelength and cannot be true simultaneously for a range of wavelengths passing through a prism spectrograph. In the construction of a prism instrument, however, an approximation to these conditions is made.

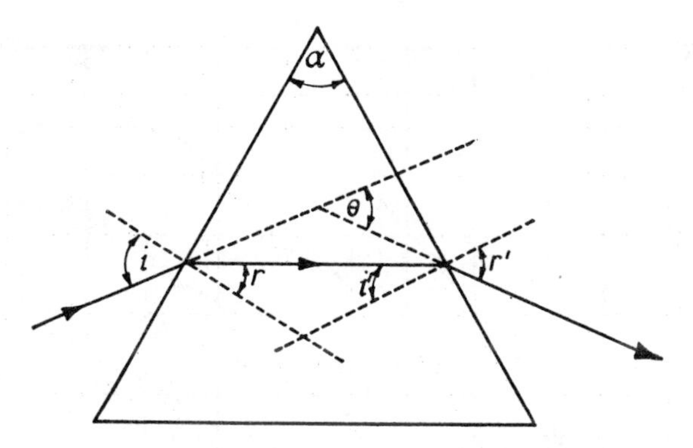

Fig. 2.3 Ray of light passing through a prism at minimum deviation.

When the incident and emergent rays are symmetrically disposed with respect to the prism, that is, when $i = r'$ and $i' = r$ as in Fig. 2.3, the deviation θ is less than for any other angle of incidence i, and the prism is said to be set for minimum deviation.

Dispersion by a prism

The dispersion by a prism is expressed as the deviation $d\theta/d\lambda$ in units of radians per Ångström unit (rad/Å); this measures the angular separation of two light rays which differ in wavelength by an amount $d\lambda$. The value of $d\theta/d\lambda$ is not constant but depends on the wavelength and is greatest in the ultraviolet and decreases with increasing wavelength. For the ray of minimum deviation it may readily be shown that:

$$n = \sin i/\sin r = \sin \tfrac{1}{2}(\alpha + \theta)/\sin \tfrac{1}{2}\alpha \tag{2.3}$$

On differentiation of Equation (2.3) we get:

$$\frac{dn}{d\theta} = \frac{\cos \frac{1}{2}(\alpha + \theta)}{2 \sin \frac{1}{2}\alpha} = \frac{\sqrt{[1 - \{\sin \frac{1}{2}(\alpha + \theta)\}^2]}}{2 \sin \frac{1}{2}\alpha} \tag{2.4}$$

and on substitution from Equation (2.3) it follows that:

$$\frac{dn}{d\theta} = \frac{\sqrt{(1 - n^2 \sin^2 \frac{1}{2}\alpha)}}{2 \sin \frac{1}{2}\alpha} \tag{2.5}$$

If the Hartmann dispersion formula [Equation (2.2)] is differentiated then:

$$\frac{dn}{d\lambda} = -\frac{c}{(\lambda - \lambda_0)^2} \tag{2.6}$$

where $dn/d\lambda$ is the characteristic dispersion of the prism material at the wavelength λ.

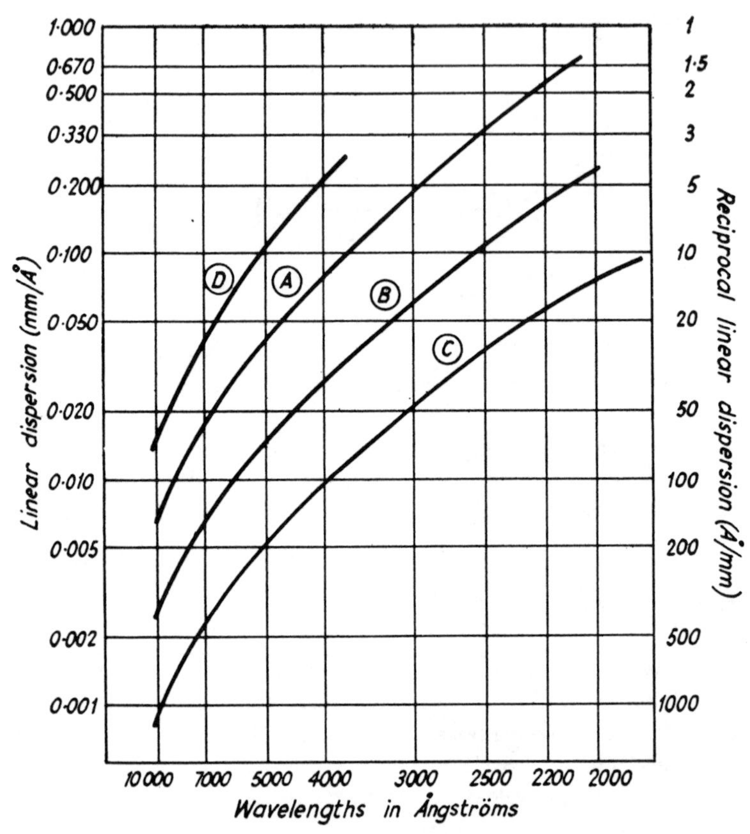

Fig. 2.4 Linear and reciprocal linear dispersion of four Hilger prism spectrographs, where A, B, C, and D are for the large, medium, and small quartz, and large glass spectrographs, respectively. (Courtesy of Hilger and Watts Ltd.).

The angular dispersion $d\theta/d\lambda$ may be written in the form:

$$\frac{d\theta}{d\lambda} = \frac{d\theta}{dn} \cdot \frac{dn}{d\lambda} \tag{2.7}$$

which on substitution for $d\theta/dn$ and $dn/d\lambda$ from Equations (2.5) and (2.6) gives:

$$\frac{d\theta}{d\lambda} = \frac{-2c \sin \frac{1}{2}\alpha}{(\lambda - \lambda_0)^2 \sqrt{(1 - n^2 \sin^2 \frac{1}{2}\alpha)}} \tag{2.8}$$

Thus, the deviation of a ray of wavelength λ depends on the geometry of the prism and the prism material. It should be noted that the size of the prism does not enter into this expression for $d\theta/d\lambda$.

In practice, in the literature it is more usual to find dispersion described in terms of reciprocal linear dispersion in units of Å/mm of photographic plate

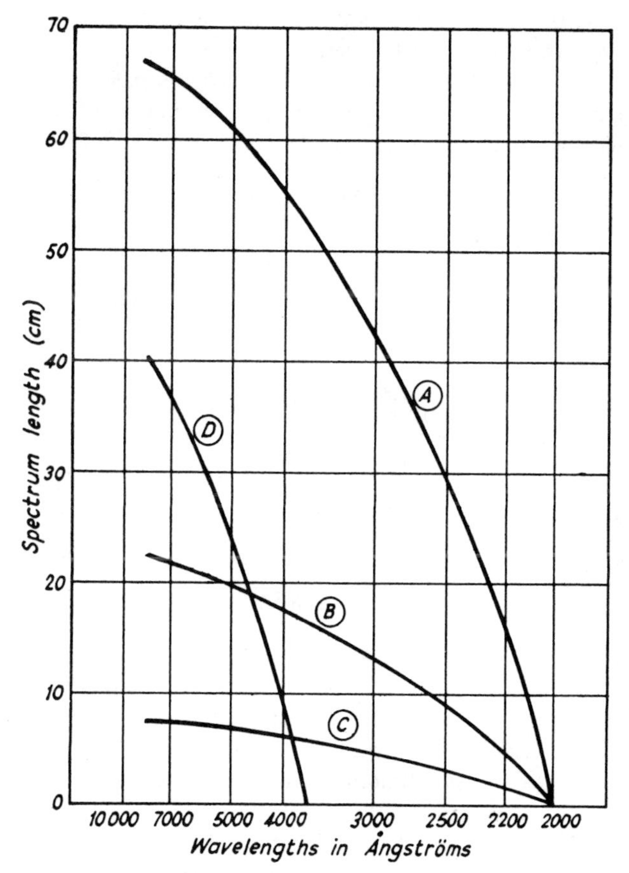

Fig. 2.5 Spectrum lengths of four Hilger prism spectrographs, where A, B, C, and D are for the large, medium, and small quartz, and large glass spectrographs, respectively. (Courtesy of Hilger and Watts Ltd.).

rather than the angular dispersion of the prism in units of rad/Å. The reciprocal linear dispersion is defined as $d\lambda/dx$, where dx is the distance in millimetres on the photographic plate between rays with wavelengths λ and $(\lambda + d\lambda)$. When the plate tilt is zero the angular dispersion $d\theta/d\lambda$ is related to the reciprocal linear dispersion by the equation:

$$d\lambda/dx = \frac{1/f}{d\theta/d\lambda} \tag{2.9}$$

where f is the focal length of the (camera) lens bringing light to focus on the photographic plate of the spectrograph. For convenience the dispersion, both linear $dx/d\lambda$ (mm/Å) and reciprocal linear $d\lambda/dx$ (Å/mm), of a spectrograph may be presented in the form of a graph of dispersion plotted against wavelength. This is illustrated in Fig. 2.4 for the linear dispersion of four Hilger prism instruments. Alternatively, a useful graph is that which gives the plot of spectrum length in centimetres against wavelength in Ångström units, and from this graph the spectrum length between any two wavelengths may be read off. For example, the curves in Fig. 2.5 show the length in centimetres of the spectrograms produced by four Hilger prism spectrographs between 2000 Å to about 8000 Å.

Resolving power of a prism

From the equation:

$$d\lambda/dx = \frac{1/f}{d\theta/d\lambda} \tag{2.10}$$

which applies when the plate tilt is zero, it follows that:

$$dx/d\lambda = f d\theta/d\lambda \tag{2.11}$$

where f is the focal length of the camera lens. For a finite change:

$$\Delta x/\Delta\lambda = f\Delta\theta/\Delta\lambda \tag{2.12}$$

and therefore:

$$\Delta x = f\Delta\theta \tag{2.13}$$

It would appear, therefore, that by making the focal length of the camera lens large enough if would be possible to separate two spectral lines of very similar wavelength, however small the angular separation between them happened to be. Such would be the case if the image of the slit in the focal plane of the instrument could be reduced to an infinitely narrow line. In practice, however, the image of the slit is a diffraction pattern of finite size depending on the dimensions of the optical system and the wavelength of the light employed. For example, a camera lens with longer focal length will cause an increase in separation of the principal diffraction maxima of two close lines, but this increase is offset by a corresponding increase in the width of the diffraction maxima of these lines.

To appreciate the intensity distribution of the diffraction pattern an example will be considered using monochromatic light as source, and for the sake of simplicity the actual dispersing element will be omitted.

With reference to Fig 2.6, if AB is the limiting aperture, the diffraction pattern will be that of this aperture and will depend on its size only, since λ and f are constants. AB may be regarded as a diaphragm which restricts the size of the bundle of rays, and in a spectrograph its position would be occupied by the dispersing system (a prism); for monochromatic light the diaphragm AB and the prism would produce the same effect.

In Fig. 2.6 monochromatic light passes through an infinitely narrow vertical slit at S, then falls on the collimating lens, L_1, which makes the light parallel and directs it on to AB. An image of AB is formed by the lens, L_2, in the focal plane PP′ (plane in which the photographic plate is located in a spectrograph). When

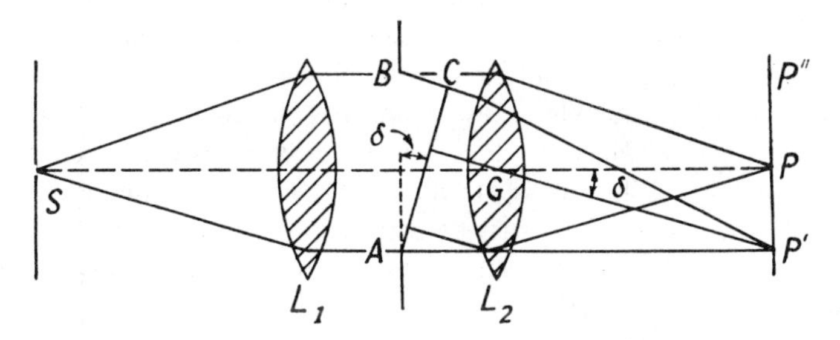

Fig. 2.6 Diffraction of a wave-front by a rectangular aperture (after Sawyer, [2.1]).

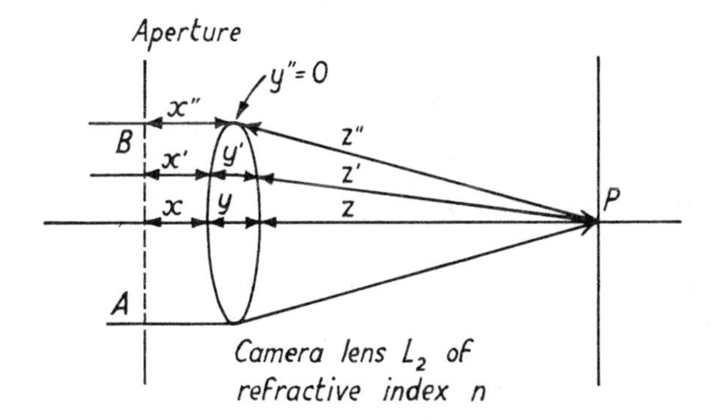

Fig. 2.7 Illustration of Fermat's principle.

the wave-front from the aperture AB (Fig. 2.6) passes through the lens L_2 and is focused at P, no phase difference in the light is introduced, since by Fermat's principle the length of any path from the aperture AB is the same. Expressed in terms of the path lengths in Fig. 2.7 this becomes:

$$x + ny + z = x' + ny' + z' = x'' + z'' = \text{constant} \tag{2.14}$$

where n is the refractive index of the lens material. Hence, at P the light waves arrive in phase, and their intensities add together and give a resultant intensity which is a maximum. In addition, according to the Huyghens principle, diffracted wave-fronts may spread in all directions; such a front brought to focus at P′ will now be considered. It may be seen from Fig. 2.6 that the light passing through the upper half of the aperture (AB) must travel further than that passing through the lower half; the path difference between the rays passing through the bottom and top of the aperture is equal to the distance BC. As before the lens will not introduce any further phase shifts. If the path difference BC is equal to an even number of half wavelengths, then the intensity at a point such as P′ will be a minimum, since for every point in the lower half of AC there will be a corresponding point in the upper half whose distance from P′ differs by $\lambda/2$

and whose contributions at P′ destructively interfere. Thus, the two rays would be exactly out of phase and would cancel each other out. If, on the other hand, the path difference BC is an odd number of half wavelengths, the intensity will give a maximum. For example, if BC = $3\lambda/2$, the bundles of light from AC may be considered to be divided into three equal parts where two of these parts will cancel each other out, but the third will produce a weak maximum. If BC = $5\lambda/2$, the diffracted wave-front may be regarded as being divided into five parts, where disturbances due to two will destroy another two, while the fifth part will provide light for a weak maximum.

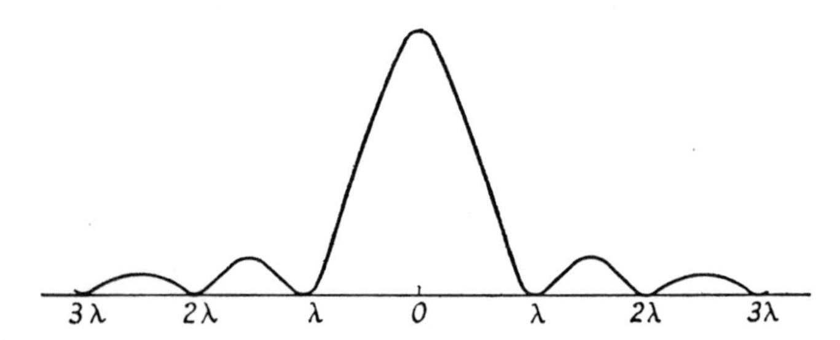

Fig. 2.8 The intensity distribution of the diffraction pattern for different path differences.

The intensity distribution of the diffraction pattern for these different path differences, BC, is represented in Fig. 2.8. It may be seen from the figure that the light intensity is a maximum when the path difference is equal to an odd number of half wavelengths, i.e. when:

$$\mathrm{BC} = (n + \tfrac{1}{2})\lambda \tag{2.15}$$

and the intensity is a minimum when the path difference is an even number of half wavelengths, or when:

$$\mathrm{BC} = n\lambda \tag{2.16}$$

where n is integral and equals 1, 2, 3.†

Since the triangles ABC and GPP′ in Fig. 2.6 are similar, then if f is the focal length of the lens, L_2, and d is the width of the aperture, AB, the distance x from P of successive maxima is such that:

$$\sin\delta = (n + \tfrac{1}{2})\lambda/d \tag{2.17}$$

or approximately for small values of δ:

$$x = (n + \tfrac{1}{2})f\lambda/d \tag{2.18}$$

while the distance of successive minima from P can likewise be shown to be:

† This symbol should not be confused with the one for refractive index.

$$x = nf\lambda/d \tag{2.19}$$

The most intense maximum is known as the zero-order fringe or central maximum, and its width is $2f\lambda/d$.

If two wavelengths are present in the incident radiation, then two sets of diffraction patterns of the aperture result. If no dispersing element is present, then the zero-order fringes would be coincident since they do not depend on a path difference. However, when a prism is present in the light path, different wavelengths suffer differing dispersions, and the zero-order fringes are not coincident. Should the wavelengths of two spectral lines be close together, the diffraction patterns of the images of the aperture will overlap as shown in Fig. 2.9 giving a resultant intensity indicated by the dotted line. Two overlapping lines may be recognized only if the two maxima appear as separate peaks with a distinct minimum between. A criterion was proposed by Lord Rayleigh by which it can be

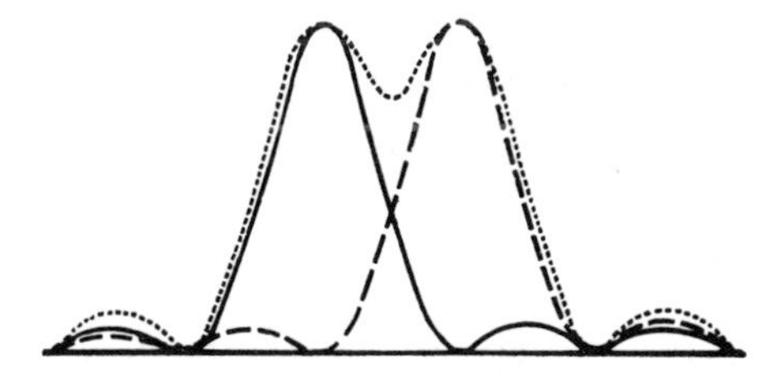

Fig. 2.9 Two overlapping diffraction patterns (full and broken lines) and their resultant intensity (dotted line) [2.1].

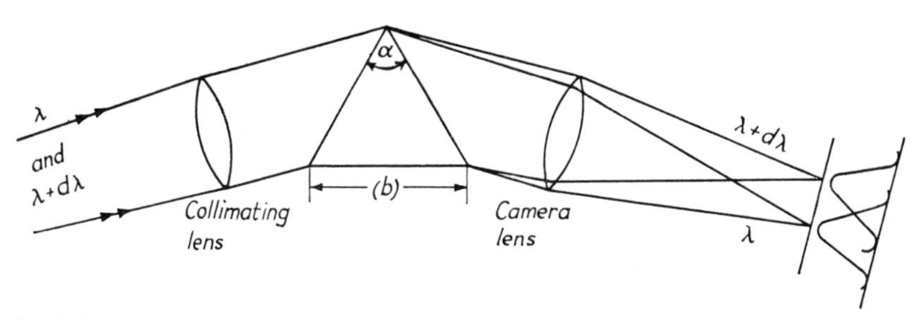

Fig. 2.10 Resolution of light by a prism.

adjudged whether two lines of wavelengths λ and $(\lambda + d\lambda)$ are separated, that is resolved. The criterion is that two wavelengths λ and $(\lambda + d\lambda)$ can be regarded to be just resolved when the central maximum of the diffraction pattern of the line of wavelength λ falls on the first minimum of the diffraction of the line of wavelength $(\lambda + d\lambda)$. This is the case represented in Fig. 2.9 where each of the lines has equal intensity. However, when the dip between the two central maxima is no longer recognizable, then the two lines remain unresolved. The ability of a spectrograph to separate two close lines of wavelength λ and $(\lambda + d\lambda)$ is called its resolving power R and is expressed as the ratio of the wavelength observed to the smallest difference between two wavelengths which can be distinguished as two lines. This ratio:

$$R = \lambda/d\lambda \tag{2.20}$$

varies widely for different types of instruments. When the whole of the prism face is employed, the resolving power may readily be shown [e.g. 2.1] to be:

$$R = b \,.\, dn/d\lambda \tag{2.21}$$

where b is the length of the base of the prism (see Fig. 2.10) and $dn/d\lambda$ is the characteristic dispersion of the material of the prism. Should only a part of the prism face be used, then b is the effective thickness and is the difference in thickness of the prism traversed by the two extreme rays.†

In practice the resolution achieved depends on several factors, and a few of the more important ones are: (i) mode of illumination of the slit; (ii) slit width; (iii) adjustment of the optics; (iv) contrast power of the photographic plate; (v) relative intensities and form of the two adjacent spectral lines to be resolved.

Selection of most suitable slit width

In practice there is no absolute rule for the choice of slit width, a compromise having to be made as regards intensity and resolution. In the ideal case with an infinitely narrow slit width and no diffraction there would be perfect resolution. However, when finite slit widths are employed and diffraction effects are considered, the resolving power is limited.

The dependence of the intensity of the central diffraction maximum I_c (in arbitrary units) on the slit width may be deduced from Fig. 2.11. Along the abscissa a function of the slit width is plotted and is $\pi BD/2\lambda f$, where B is the slit width, λ the wavelength employed, and f the focal length of the collimator lens of diameter D. It will be seen from the figure that if a very narrow slit is used and gradually increased, the intensity of the central maximum rises rapidly to a peak value. However, on further increase of the slit width beyond this peak value the intensity of the central maximum is modified only slightly. The half-width of the absorption $\Delta\nu$ is also plotted in Fig. 2.11 where $\Delta\nu$ may be defined as half the width (in cm^{-1}) at the position where the intensity of the line falls from its maximum to one-half of that value. It follows directly from the figure that beyond the peak intensity A, the central maximum intensity value is altered only slightly when the slit width is increased, whereas the half-width of the line

† It is interesting to note that whereas the dispersive power of a prism instrument is independent of the prism size (apart, of course, from the angle α) the resolving power depends on the effective thickness of the prism traversed by the beam.

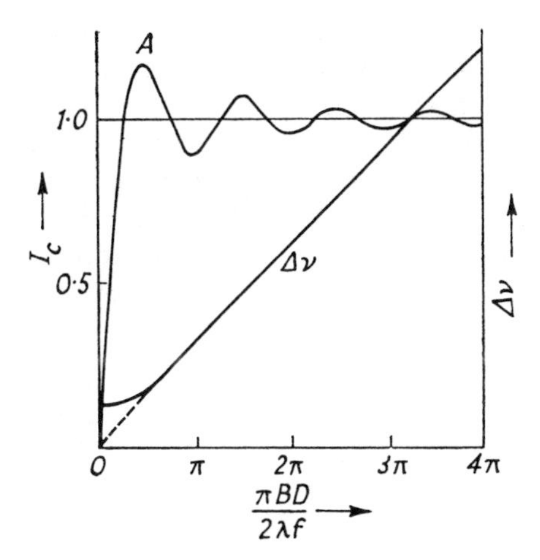

Fig. 2.11 The variation of the central maximum, I_c, against a function of the slit width, $\pi BD/2\lambda f$. The half-width of the line, $\Delta\nu$, is also plotted against this function of the slit width. A non-coherent mode of illumination of the slit was employed [2.6].

increases linearly with the slit width. From a spectroscopic viewpoint maximum resolution is required without a large increase in the exposure time, while the half-width of the line should be a minimum. These criteria may best be achieved for point A in Fig. 2.11 where:

$$\pi BD/2\lambda f = \pi/2 \tag{2.22}$$

that is when the value of the slit width B is given by:

$$B = \lambda f/D \tag{2.23}$$

Such a slit width leads to a good central diffraction maximum with negligible loss of resolving power.

The case just considered was for a non-coherent mode of illumination and was evaluated by van Cittert [2.6]. He has also calculated the effect of employing a coherent mode of illumination where all the rays from the slit are in phase. In this case the maximum corresponding to point A in Fig. 2.11 occurs when $\pi BD/2\lambda f = \pi$. Thus, for a coherent mode of illumination the most desirable slit width is $2\lambda f/D$, that is double the width of the noncoherent case. In general, in spectroscopic work, a range of wavelength is involved, and the radiation is most likely to be a mixture of coherent and non-coherent modes of illumination. Thus, the most suitable slit width will range in value between $\lambda f/D$ and $2\lambda f/D$. In practice some value just less than $\lambda f/D$ may be tried as the slit width. The slit width is then gradually increased until the best condition is found. If the intensity of the illumination is observed on a white card on the far side of the slit, then when the best slit width is obtained, a sudden increase in brightness ensues. If the slit width is increased beyond this, then the brightness increases only very slowly. To discern the effect of slit width on resolution and spectral intensity a number of exposures have to be carried out at different slit widths. In fundamental research this is often worthwhile. In practice slit widths greater than the calculated value are often advantageous. This becomes desirable as a result of imperfections in the optical system. Two conditions which call for wide slit widths are: (a) when the intensity of the incident radiation is very low, e.g. in some infrared work; (b) when high-speed recording is required, and it might be permissible to sacrifice resolving power in favour of speed.

Photographic speed of spectrograph

The photographic speed of a spectrograph regulates the time required for a given amount of blackening of a photographic plate. It can be expressed in what is known as the *F-number* which is defined by:

$$F = f/D \tag{2.24}$$

where f and D are the focal length and diameter of the camera lens, respectively. The quantity $(D/f)^2$ determines the solid angle subtended at the focus of the camera lens by the aperture and is, therefore, a measure of the amount of light entering. A small F-number indicates a high speed, and to increase the focal length without also increasing the diameter of the camera lens means a reduction in the speed of the instrument. For astrochemistry work involving weak sources speed is essential, and lenses with small F-numbers are used. However, dispersion is sacrificed with such instruments. In the Hilger large-aperture two-prism glass spectrograph (see Fig. 2.12) two interchangeable cameras are available, one for preliminary survey work with $F = 1.5$ giving a reciprocal linear dispersion of 64 Å/mm at 4358 Å, and the second for more detailed examinations with $F = 5.7$ and a reciprocal linear dispersion of 16 Å/mm at 4358 Å.

2.2.3 Types of prism instruments

Medium quartz and glass spectrograph

A spectrograph which is adaptable to many kinds of work where speed and high dispersion are not required is the medium quartz and glass type of spectrograph. The instrument is able to take either glass or quartz refracting components, the latter having a greater range (2000 to 10 000 Å) than the former (3700 to 8000 Å). The length of the spectrum in the case of the quartz instrument is 22 cm, while the glass instrument has a spectrum length of 14 cm. The glass instrument has a greater angular dispersion than the quartz, giving therefore a smaller reciprocal linear dispersion (Å/mm). A medium quartz spectrograph is illustrated in Fig. 2.13.

Basically the instrument consists of a slit, collimating lens, prism, a two-component camera lens, and a photographic plate. The slit is a narrow vertical

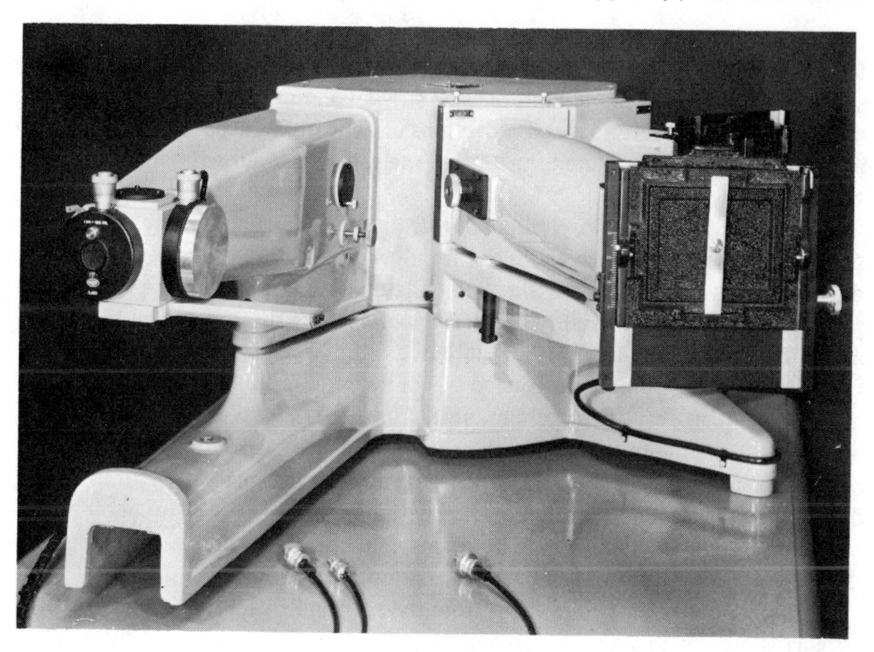

Fig. 2.12 Hilger two-prism glass spectrograph. (Courtesy of Hilger and Watts Ltd.).

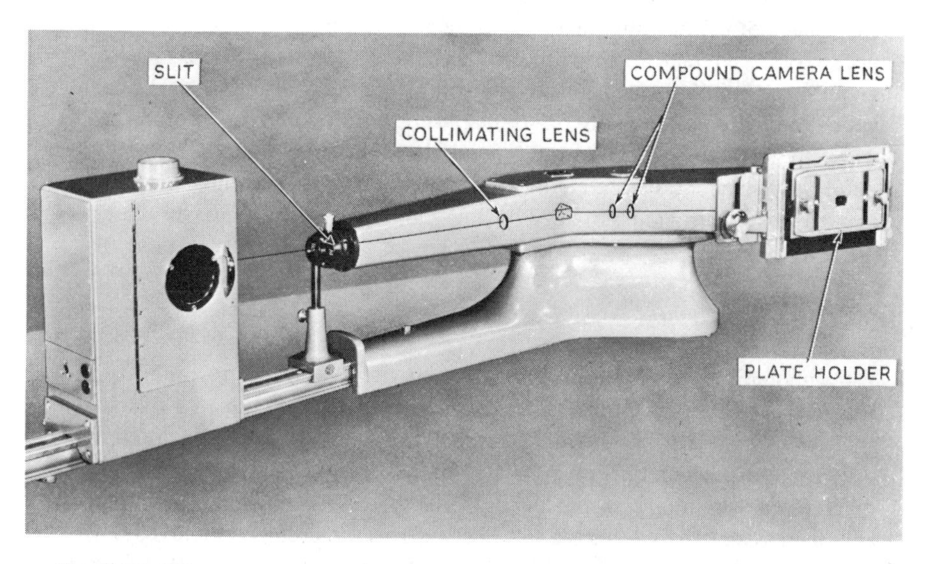

Fig. 2.13 Hilger medium quartz spectrograph showing the optical arrangement. (Courtesy of Hilger and Watts Ltd.).

rectangular aperture through which the light to the spectrograph passes. Since the image of the slit after dispersion constitutes the spectral line, it is of importance that the slit edges should not only be straight and parallel to give a clean-cut image but also sharp to avoid reflections of light from the edges. On an instrument which is to be used for routine analysis a fixed slit width may be satisfactory.

For research work, however, and on instruments of high dispersion, variable slits are provided so that the width can be adjusted to suit the type of work and the wavelength being used. The slits open bilaterally so that the centre of the spectral line is fixed for all slit widths. The width of the slit is varied by adjustment of a micrometer screw with a calibrated drum head.

The usual 60° prism, if constructed of quartz, requires to be made of two halves since quartz crystal is a doubly-refracting material and also possesses the property of rotating the plane of polarization of plane-polarized light even if the beam is parallel with the optical axis. Quartz, however, occurs in two forms, rotating the plane of polarization of the light in opposite senses. Thus, a 60° prism can be constructed of two 30° prisms, one from each variety of quartz, the second prism introducing a compensating rotatory effect.

The collimating lens renders light from the slit parallel and hence eliminates astigmatism in the prism system, while the camera lens brings to focus the beams of light of different wavelengths emerging from the prism at different angles. Images of the slit are produced in the focal plane, and in this position a photographic plate is fixed in a suitable plate holder to record the spectrum. The size of the photographic plate is 2.5 by 10 cm and several exposures may be taken on one and the same plate by vertical adjustment of the plate holder. Since the focal plane of the medium quartz instrument is flat, no adjustment of focus is necessary for any part of the spectrum. The medium quartz and glass spectrographs have built into them a wavelength (or if desired, a wavenumber) scale which can be printed directly on to the photographic plate. The error in the wavelength reading may be of the order of 100 Å at 7000 Å and falls to 1 Å at 2200 Å. A much more accurate determination of wavelength which could be applied to a spectrogram obtained on a medium quartz spectrograph is described on p. 48–49

Littrow spectrograph

When good dispersion is required the large quartz or glass spectrograph is available (see Fig. 2.14). The focal length of the camera lens of such an instrument is of the order of 170 cm. It can be appreciated that if the same construction were used for such an instrument as is employed for the medium type of spectrograph (focal length 60 cm), the length of the spectrograph would be considerably increased; to avoid this the so-called Littrow mounting is employed as is shown in Fig. 2.15. The range of the quartz instrument is from 1910 to 8000 Å, and the glass from 3700 to 12 000 Å, the spectrum lengths being 75 and 47 cm,

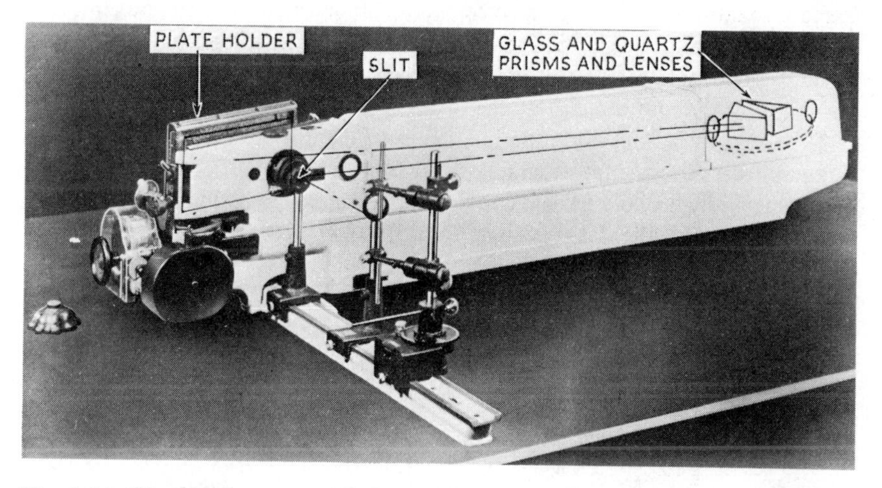

Fig. 2.14 Hilger large quartz and glass spectrograph showing optical arrangement. (Courtesy of Hilger and Watts Ltd.).

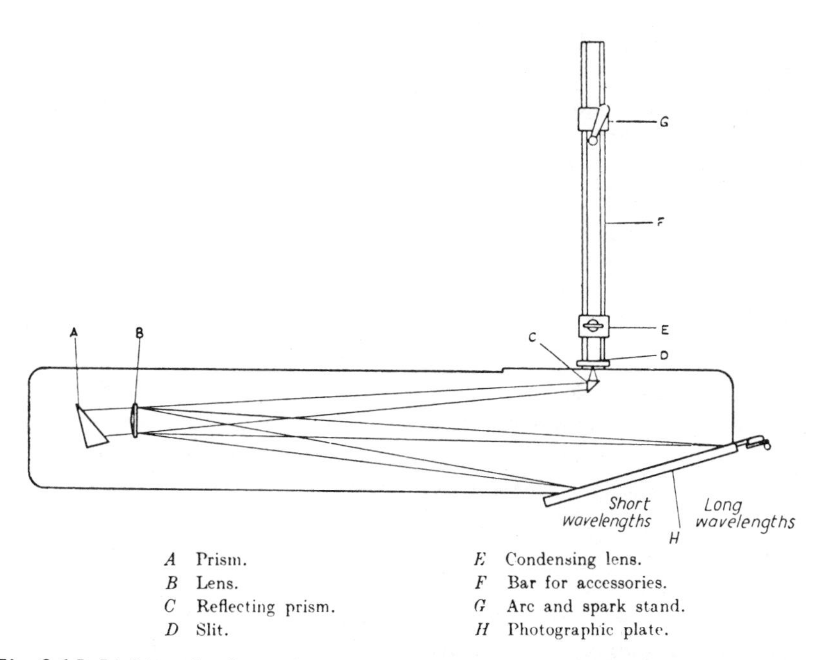

Fig. 2.15 Light path of a typical large quartz spectrograph. (Courtesy of Hilger and Watts Ltd.).

respectively. A 2.5 by 10 cm photographic plate is used, as was the case with the medium type of instrument, although with the large quartz type of spectrograph three exposures are necessary to cover the whole spectral range. One of the attractive features of this large spectrograph with quartz optical components is that it gives good dispersion in the ultraviolet region. The dispersion of the glass assembly is greater than that of the quartz, and in the visible region the glass prism produces about three times the dispersion of the quartz prism.

The operation of the slit in the Littrow instrument is the same as that for the medium quartz or glass spectrograph. Light from the slit is reflected by a small 90° prism through a collimating lens to a 30° prism which is aluminized on its rear face so that the light is reflected. Thus, the light traverses the prism twice and is brought to a focus on a photographic plate by means of a lens which acts both as a collimating and a camera lens. These points are illustrated in Fig. 2.15.

Reflected light from the surface of the lens is more troublesome in this type of instrument than in the medium quartz type, though the reflection can be cut down materially by blooming the lens. Blooming consists of depositing under an evacuated condition a transparent substance.

As may be observed in Fig. 2.14, glass or quartz refracting components may be employed merely by rotating the prism lens assembly on a turntable. To cover the whole spectrum of the large glass and quartz instruments a change from one region to another requires rotation of the aluminized prism, re-focusing of the lens, and a change in the angle of tilt of the photographic plate. All these adjustments are made from the plate-holder end of the instrument by a single control. For the quartz instrument the wavelength range 1910 to 8000 Å is spread over 75 cm of photographic plate, and for a given setting any 25 cm may be selected. In addition, a millimetre scale is so mounted that it can be imposed on the spectrogram. The plate carrier can be vertically adjusted so that many exposures can be taken on one and the same plate. A comparison of various features of five Hilger prism instruments is given in Table 2.1.

2.2.4 Grating spectra

The diffraction grating may be employed instead of the prism as a dispersion element and has the following points to recommend its use: (a) it may be used below 1200 Å and where no suitable transparent prism materials are available. In addition, gratings are frequently used in the region between 1200 Å and 40 μ where it is desirable to have the features mentioned in (b) below. (b) Generally a grating has a greater dispersion and resolving power than a prism. (c) The grating yields a linear spectrum whereas a prismatic spectrum is non-linear. This is labour-saving when the wavelengths of a number of lines are to be measured.

A diffraction grating may be regarded as being composed of a large number of parallel, equidistant, and narrow slits side-by-side made by ruling lines on a suitable surface with a diamond point. The surface can either be transparent or

Table 2.1 A comparison of various features of some Hilger prism instruments. (Courtesy of Hilger and Watts Ltd.).

	Prism		*Lens*			
Instrument	*Angle* (α)	*Height and length of face* (*mm*)	*Focal length* (f) (*cm*)	*Aperture* (D) (*mm*)	*Size of plate*	*Spectrum range and length* (x)
Large quartz	30°	56 × 94	170	75	25 × 10 cm	1910–8000 Å = 75 cm
Large glass quartz as above	26°	56 × 94	170	75	9 × 24 cm	3700–12 000 Å = 47 cm
Medium quartz Flat field	60°	41 × 65	60	51	25 × 10 cm	2000–10 000 Å = 22 cm
Medium glass	60°	41 × 65	60	51	25 × 10 cm	3700–8000 Å = 14 cm
Large aperture 2 prism glass	63°	86 × 164 86 × 130	13.2	89	11 × 8 cm	3900–8000 Å = 2.9 cm
Large aperture 2 prism glass	63°	86 × 164 86 × 130	45	101		3900–7000 Å = 10 cm

opaque to the radiation used. In the former case the grating is a *transmission grating* and in the latter a *reflection grating*. The fundamental theory is the same for both types. A reflection grating is superior to a transmission grating, however, in that, if the lines are ruled on a concave mirror, then the need for employing focusing and collimating lenses can be eliminated.

If α is the angle of incidence of the light of wavelength λ onto the grating and β is the emergence angle, where both the angles are measured with respect to the normal, then the diffraction by the grating is taken into account by the formula:

$$\pm n\lambda = \frac{A}{N}(\sin\alpha + \sin\beta) \tag{2.25}$$

A is the linear aperture of the grating and is the distance from the first ruled line to the last, and N is the number of lines ruled on the grating. n is known as the order of the spectrum and may take values $\pm 1, \pm 2, \ldots$. For the successive values of n successive images are formed. The image corresponding to $n = \pm 1$ is termed the first order and for $n = \pm 2$ the second order, and so on. Equation (2.25) gives the angles of the diffracted images for given angles of incidence. For a fixed order, images of different wavelengths will be formed at different values of β, but it follows from Equation (2.25) that overlapping of spectra of different orders may occur. Thus, for constant values of α and β the value of the right-hand side of Equation (2.25) is fixed, and therefore $n\lambda$ = constant. This constant value may be achieved for combinations such as the following:

$$n \,.\, \lambda \quad 1 \,.\, \lambda \quad 2 \,.\, \lambda/2 \quad 3 \,.\, \lambda/3$$

For example, for a given value of β the spectra of 12 000 Å in the first order, 6000 Å in the second order, and 4000 Å in the third order would be observed at the same position on the photographic plate. It is frequently possible to separate these overlapping orders by the following means.

(a) The use of a fore-prism which provides a narrow range of wavelength and acts as a monochromator for the grating.

(b) The use of photographic plates with limited spectral sensitivity.

(c) The use of certain cut-off filters which though transparent to the desired wavelengths are opaque to the sub-multiples of these wavelengths. Examples of such filters used in the infrared region are thin films of PbS, Ag_2S, and Te deposited on a material which is transparent by itself. Semiconductor filters such as Si, Ge, indium arsenide and antimonide provide a very effective discrimination against visible and short wave infrared radiation. Subsequent to optical blooming transmission at the peak wavelengths (2 μ for Si, 3 μ for Ge, 5 μ for indium arsenide, and 10 μ for indium antimonide) is between 70 and 90 per cent. A second most promising type of filter is made from the coloured alkali halide crystals [2.3].

Dispersion by a grating

Since

$$n\lambda = \frac{A}{N}(\sin \alpha + \sin \beta) \qquad (2.26)$$

and if α is arranged to be constant, then it follows that the dispersive power of the grating is given by:

$$d\beta/d\lambda = nN/A \cos \beta \qquad (2.27)$$

Thus, the dispersive power varies directly with the order and also with the number of grating lines per unit width of grating (N/A).

If the spectrum is viewed normal to the grating, $\beta = 0$, $\cos \beta = 1$, and:

$$d\beta/d\lambda = \frac{nN}{A} = \text{constant} \qquad (2.28)$$

for a given order and grating, and the dispersion has a minimum value. For this type of spectrum it follows from Equation (2.28) that $\Delta\beta$ is approximately proportional to $\Delta\lambda$ and that the wavelength is a linear function of the angle of emergence. In addition, it also follows from Equation (2.28) that for a spectrogram observed at small values of β the dispersive power is constant.

Resolving power of a grating

The resolving power of a grating may be shown to be:

$$R = \lambda/d\lambda = nN \qquad (2.29)$$

where n is the order and N the number of lines in the ruled surface. Thus, to resolve, for example, the sodium *D*-lines at an average wavelength of 5893 Å with $\Delta\lambda = 6$ Å the resolving power required would be:

$$\lambda/\Delta\lambda = 5893/6 = \text{approximately } 1000$$

Hence a grating with 1000 ruled lines would be adequate to resolve the sodium *D*-lines in the first order. It is interesting to note that it follows from Equation (2.29) that the resolving power of a grating is independent of the wavelength or the spacing of the lines in the grating.

Light distribution in a grating spectrum

In an ideal grating consisting of alternate equidistant opaque and transparent strips the distribution of light should be uniform on each side of the normal to the grating, and the intensity of successive orders should decrease in a regular manner. Such, however, is not the case, since the rulings have in addition to finite width a definite shape, usually taking the form of a flat trough, the sides of which make unequal angles with the vertical. This results in an irregular distribution of light on the two sides of the normal and also within the orders on one given side. The efficiency of a grating ruled in the ordinary manner with a diamond point cannot be predicted but must be determined experimentally.

It has already been indicated that one disadvantage of the diffraction grating is that it disperses the incident energy over a large number of orders. This may be overcome in the visible region by increasing the fineness of the groove spacing. As the wavelength being studied becomes longer, however, it is desirable to use coarser gratings; for example in the infrared, gratings are used ranging from about 6000 lines/cm down to as low as 10 lines/cm.

This waste of energy by diffraction into a variety of orders can be serious when, for example, in the infrared the total amount of energy is small in any case. In the production of modern gratings this problem can to some extent be overcome by shaping the tip of the ruling diamond so that the contours of the grooves are such as to concentrate the diffracted energy into a definite direction. This is called *blazing* the grating, and the angle the groove makes with the vertical is known as the *blaze angle.*

2.2.5 Grating mountings

A grating may be mounted in a variety of ways three of which are shown in Fig. 2.16(a), 2.16(b) and 2.16(c). Equation 2.26 holds for a concave grating as Rowland first showed. The slit, grating, and diffracted spectrum all lie on a circle, the *Rowland circle*, whose diameter is equal to the radius of curvature of the concave grating. Both the Eagle and Paschen–Runge mountings are partly built on the Rowland circle principle while the Wadsworth arrangement is rather different. Details of each of these mountings will be briefly considered.

Eagle mounting

The Eagle mounting is similar to the Littrow mounting of a prism in that the slit S and photographic plate P are mounted close together as in Fig. 2.16(a). The grating, G, and the photographic plate, but not the slit in this case, lie on the Rowland circle, R. To change the spectral region it is necessary to rotate the grating about its vertical axis, change the distance between the grating and plate holder, and also to rotate the plate holder about the slit. These operations are usually coupled together and motor-driven, the wavelength range being read on a wavelength drum. One such instrument employs a grating with a radius of curvature of 3 m and has a reciprocal linear dispersion of 5.6 Å/mm in the first order. The grating has 15 000 lines/in and a 4 in ruled surface.

The Eagle mounting though very compact has a certain amount of astigmatism associated with it and requires very precise mechanical parts for its operation in order to achieve the highest resolving power of the grating.

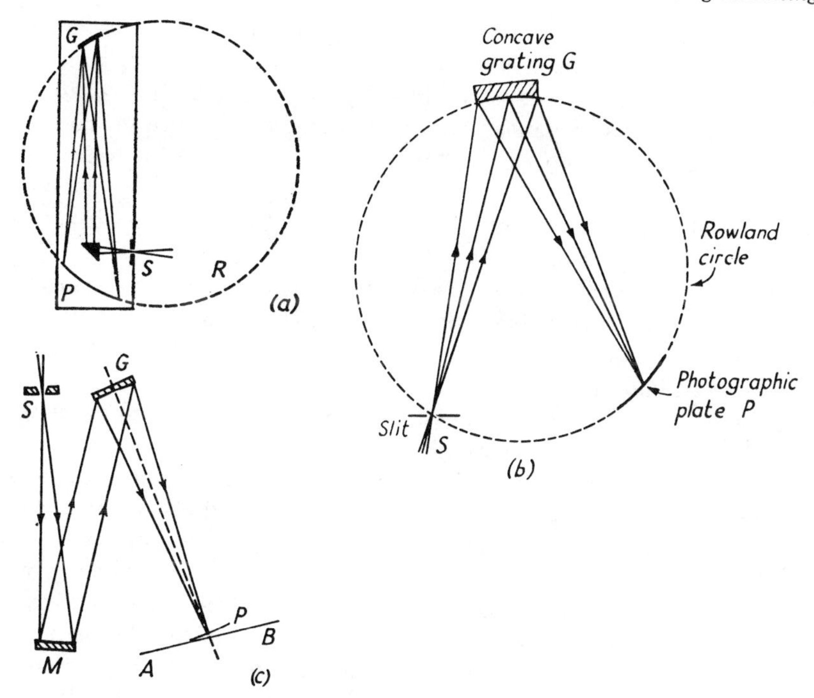

Fig. 2.16 Three arrangements for mounting a diffraction grating. (a) The Eagle mounting [2.5]. (b) The Paschen–Runge mounting. (c) The Wadsworth mounting (after Sawyer [2.1]).

Paschen–Runge mounting

A very popular mounting for large concave gratings is the Paschen–Runge one. In this case the slit, grating, and plate holder lie on the Rowland circle. One feature peculiar to this mounting is the arrangement for placing photographic plates along a large part of the Rowland circle enabling a wide spectral range to be simultaneously photographed. One minor disadvantage is the non-linear dispersion which results when operating away from the grating normal.

Wadsworth mounting

The advantage of the mounting is that it yields a stigmatic image, the beam being collimated by a concave mirror, M, before it strikes the grating, G, as indicated in Fig. 2.16(c). The plate holder, P, is mounted normal to the grating to cut down astigmatism and spherical aberration. As the grating is rotated to scan the spectrum the distance from the grating to plate holder, tilt of plate holder, and its curvature must be adjusted for each setting. The motion of the plate holder relative to the grating is automatically controlled by a cam to execute a parabolic path along AB. One instrument using this mounting has a grating with a radius of curvature of 6.4 m with 5906 lines/cm, this gives a reciprocal linear dispersion of 5 Å/mm in the first order. The Wadsworth mounting has been adapted to plane gratings. A concave mirror is arranged to reflect the radiation onto a plane transmission grating, and the radiation passes through and converges to a focus on a photographic plate.

2.2.6 Wavelength measurement

Once the desired spectrum has been obtained it is frequently required to determine the exact wavelengths of lines or band heads. The position of these is determined with respect to the positions of lines of an element whose exact wavelengths are known.

The measurement of the relative positions of the lines or bands on the plate is carried out by means of a travelling microscope under which the plate is mounted on a movable carriage. This carriage can be moved backwards or forwards by means of a screw coupled to a drum normally calibrated in divisions corresponding to 0.01 mm of travel, while an attached vernier permits estimations to 0.001 mm. The determination of wavelength is much simpler for a spectrogram resulting from a grating instrument since the position of lines on the plate has a linear dependence on wavelength. This enables linear interpolation between standard reference lines to be made. With prism spectrographs, however, there is such variation in dispersion for different wavelengths that it is necessary to have a number of standard comparison lines fairly closely dispersed between the lines whose wavelengths are to be determined and to employ a non-linear dispersion formula. For this non-linear interpolation the Hartmann formula can be used over a restricted wavelength range. Hartmann showed that the dispersion of a prism instrument could be represented over small wavelength ranges to a fair degree of approximation by the expression:

$$\lambda = \lambda_0 + \frac{c}{n - n_0} \tag{2.30}$$

λ is an unknown wavelength, λ_0 a constant having the dimensions of wavelength, c a constant valid for a range of wavelengths in the immediate neighbourhood of the unknown line, n the index of refraction of the prism material at the wavelength λ, and n_0 is a further constant. A more convenient expression of the Hartmann formula [Equation (2.30)] for the purpose of wavelength calculation is:

$$\lambda = \lambda_0 + \frac{C}{d_0 - d} \tag{2.31}$$

where d_0 is a constant and d the measured distance on the plate from an arbitrary point in the spectrum to the line of unknown wavelength.

Equation (2.31) has three unknown constants λ_0, C, and d_0. These may be determined if the separations d_1, d_2, and d_3, of three lines of known wavelength

λ_1, λ_2, and λ_3 respectively are measured with a travelling microscope. The set of three independent equations:

$$\lambda_1 = \lambda_0 + C/(d_0 - d_1) \tag{2.32}$$

$$\lambda_2 = \lambda_0 + C/(d_0 - d_2) \tag{2.33}$$

$$\lambda_3 = \lambda_0 + C/(d_0 - d_3) \tag{2.34}$$

is then solved simultaneously for the three unknown parameters. The arbitrary (fiducial) point in the spectrum referred to earlier, if convenient, may be located on one of the three lines of known wavelength. Hartmann showed that, while d_0 depends on the position of the fiducial mark C on the selected range and region of the spectrograph, λ_0 is a constant of the spectrograph and should always have the same value for a given instrument. Actually since the Hartmann formula is only an approximation, λ_0 will vary slightly with the chosen spectral range. From a knowledge of the three parameters λ_0, C, and d_0 unknown wavelengths may be calculated from Equation (2.31), if the distances (d) of these wavelengths are measured from the same fiducial point as previously employed. For reasonable accuracy the three lines of known wavelength should lie close together, and preferably the lines, band heads, or band origins, whose wavelengths are being determined, should lie within their range.

The known lines associated with the original measurement could be members of the iron spectrum which was photographed on the same plate as the unknown spectrum. From charts of the iron spectrum it should be possible to choose three convenient lines for the determination.

If the wavelengths of standard lines other than those used to calculate the constants in the Hartmann equation are evaluated and these compared with the theoretical values, it will be found that there is a periodic error throughout the spectrum. This error is due to the approximate nature of the Hartmann formula. It is necessary, then, to correct the measured values, and this can be done with the aid of a correction curve. The curve is obtained by calculating a large number of standard wavelengths for the calibration lines and plotting the difference between the wavelengths calculated from the measurements and that of the true wavelength as taken from the tables against the calculated wavelengths. Each of the calculated wavelengths is then converted into a true wavelength by means of this correction curve. Such a curve is shown in Fig. 2.17, When a corrected value of the wavelength in air has been obtained it is desirable for theoretical work to convert this value to wavelength in vacuo (λ_{vac}) by means of tables.

For most accurate wavelength determination it is desirable that the lines or band heads should be sharp and well defined. In general, lines may be determined to a greater accuracy than a few-tenths of an Ångström unit by means of a prism spectrograph. With a large diffraction grating, though, accuracies of a few-thousandths of an Ångström unit can be achieved, while for interferometer measurements on sharp lines an accuracy of 0.0001 Å can be obtained.

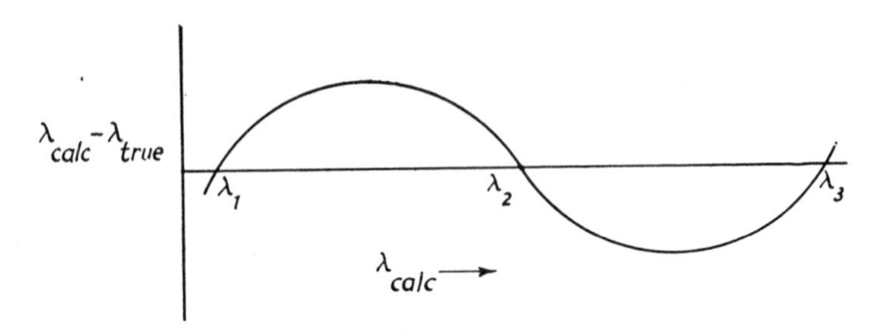

Fig. 2.17 Correction curve for wavelength determination.

3.3 ELECTRONIC EXCITATION OF DIATOMIC SPECIES

2.3.1 Introduction

If interaction between different forms of energy is neglected, then the total energy E of a molecule apart from translational energy may be regarded as the sum of the individual contributions of the rotational (E_r), vibrational (E_v), and electronic (E_e) energies where the total energy E is given by:

$$E_e + E_v + E_r = E \tag{2.35}$$

Expressed in wavenumber units this becomes:

$$T_e + G(v) + F(J) = T \tag{2.36}$$

where T_e, $G(v)$, and $F(J)$ are the term values for the electronic, vibrational and rotational energies, respectively; for example, the vibrational term value is given by:

$$G(v) = E_v/hc = (v + \tfrac{1}{2})\omega_e - (v + \tfrac{1}{2})^2 x_e\omega_e + (v + \tfrac{1}{2})^3 y_e\omega_e + \ldots \tag{2.37}$$

In a transition the wavenumbers of the spectral lines are given by the difference of the two term values in the upper (T') and lower (T'') electronic states, that is:

$$\tilde{\nu} = T' - T'' = (T'_e - T''_e) + \{G'(v') - G''(v'')\} + \{F_{v'}(J') - F_{v''}(J'')\} \tag{2.38}$$

The representation of the wavenumber ($\tilde{\nu}$) of one of the spectral lines in an electronic transition by means of term value differences is given in Fig. 2.18. Equation (2.38) has the form:

$$\tilde{\nu} = \tilde{\nu}_e + \tilde{\nu}_v + \tilde{\nu}_r \tag{2.39}$$

where the emitted or absorbed wavenumber may be regarded as the sum of an electronic, vibrational, and a rotational part. $\tilde{\nu}_v$ and $\tilde{\nu}_r$ are the wavenumbers corresponding to the vibrational and rotational energy changes, respectively, and $\tilde{\nu}_e = (E'_e - E''_e)/hc$, that is $\tilde{\nu}_e$ is the energy difference in cm^{-1} units between the minima of the upper and lower electronic states.

For the transitions between the different rotational and vibrational levels in two different electronic states $\tilde{\nu}_e$ is constant. In addition, since $\tilde{\nu}_r$ is small

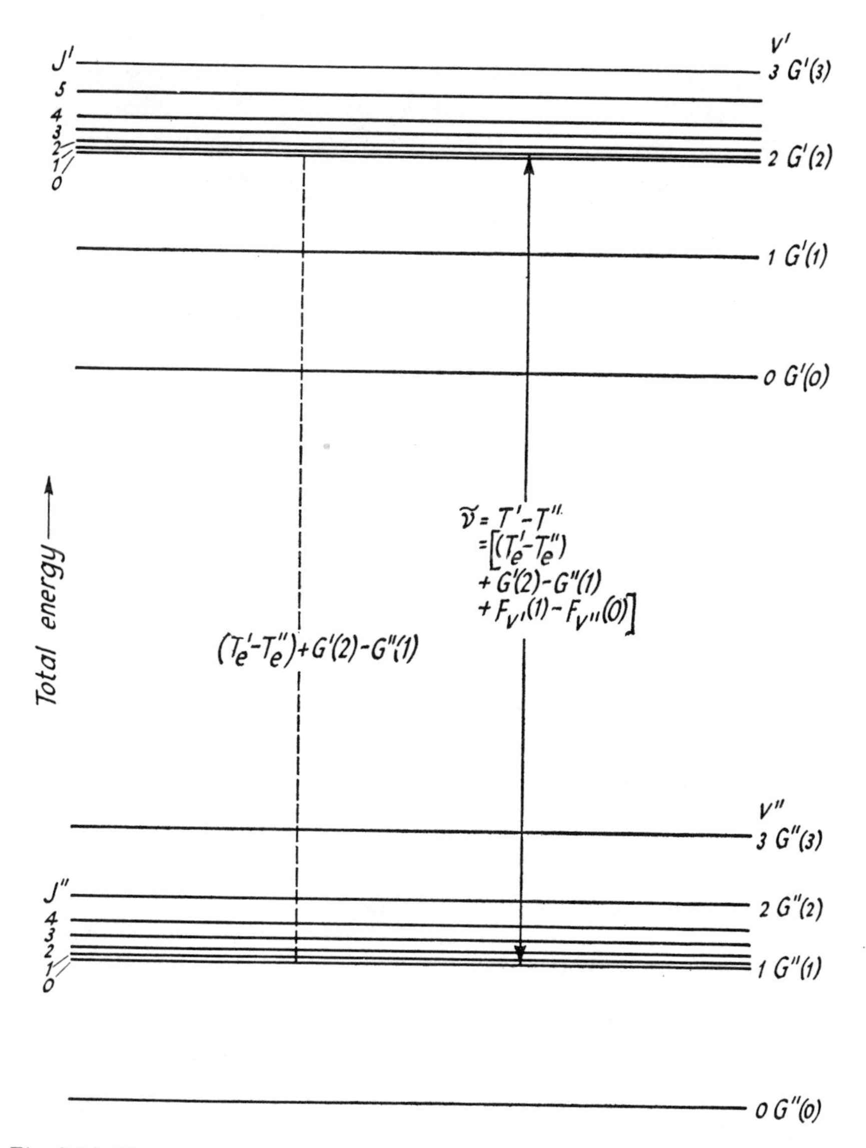

Fig. 2.18 The representation of one wavenumber ($\tilde{\nu}$) in an electronic transition by means of term value differences. The line chosen is for the $v' \leftrightarrow v''$ changes of $2 \leftrightarrow 1$ and the $J' \leftrightarrow J''$ of $1 \leftrightarrow 0$. The term difference $[(T_e' - T_e'') + G'(2) - G''(1)]$ is also indicated.

compared with $\tilde{\nu}_v$, this may in general be neglected in a vibrational analysis, and the wavenumbers of the electronic changes between the different vibrational levels are given approximately by:

$$\tilde{\nu} = \tilde{\nu}_e + [G'(v') - G''(v'')] \tag{2.40}$$

Equation (2.40) is exact only for transitions between vibrational states each of which has no rotational energy, that is for a $J' = 0$ to $J'' = 0$ transition and vice versa. On substitution for the vibrational term values from Equation (2.37) into (2.40), the following equation is obtained:

$$\tilde{\nu} = \tilde{\nu}_e + (v' + \tfrac{1}{2})\omega'_e - (v' + \tfrac{1}{2})^2 x'_e\omega'_e + (v' + \tfrac{1}{2})^3 y'_e\omega'_e + \ldots - [(v'' + \tfrac{1}{2})\omega''_e - (v'' + \tfrac{1}{2})^2 x''_e\omega''_e + (v'' + \tfrac{1}{2})^3 y''_e\omega''_e + \ldots] \tag{2.41}$$

Generally, the y'_e and y''_e terms are very small, and if these are neglected,† the following equation results.

$$\tilde{\nu} = \tilde{\nu}_e + (v' + \tfrac{1}{2})\omega'_e - (v' + \tfrac{1}{2})^2 x'_e\omega'_e - [(v'' + \tfrac{1}{2})\omega''_e - (v'' + \tfrac{1}{2})^2 x''_e\omega''_e] \tag{2.42}$$

For an electronic transition changes in v are not restricted by a selection rule, and Δv may be a positive or negative integer, though certain values will be preferred. In the case $v' = 0 \leftarrow v'' = 0$ on employment of Equation (2.42) it follows that:

$$\tilde{\nu}_{00} = \tilde{\nu}_e + \tfrac{1}{2}\omega'_e - \tfrac{1}{4}x'_e\omega'_e - \tfrac{1}{2}\omega''_e + \tfrac{1}{4}x''_e\omega''_e \tag{2.43}$$

where $\tilde{\nu}$ has been replaced by $\tilde{\nu}_{00}$. Thus, $\tilde{\nu}_{00}$ is the wavenumber of the (0, 0) band which is often one of the most intense bands in the system.

2.3.2 Vibrational analysis of band systems of diatomic molecules and radicals

In a vibrational analysis of the spectrum of a band system, a modified form of either Equation (2.41) or (2.42) is employed. The modified equation is obtained by relating the vibrational energy terms not to the minimum of the potential energy curve where $G(v) = 0$ but to the $v = 0$ vibrational level where from Equation (2.37) it follows that this vibrational term $G(0)$ is given by:

$$G(0) = \tfrac{1}{2}\omega_e - \tfrac{1}{4}x_e\omega_e + \tfrac{1}{8}y_e\omega_e + \ldots \tag{2.44}$$

The vibrational energy term values are then measured from this $G(0)$ value, which is chosen as the new zero, and the vibrational term values are represented by the symbol $G_0(v)$ where:

$$G_0(v) = \omega_0 v - x_0\omega_0 v^2 + y_0\omega_0 v^3 \tag{2.45}$$

and

$$G(v) = G_0(v) + G(0) \tag{2.46}$$

The factors involved in Equation (2.46) are represented in Fig. 2.19.

† In actual analyses the $(v + \tfrac{1}{2})^3 y_e\omega_e$ term has sometimes to be included. It should be noted that ω_e is the infinitesimal vibrational frequency and is, of course, a theoretical quantity.

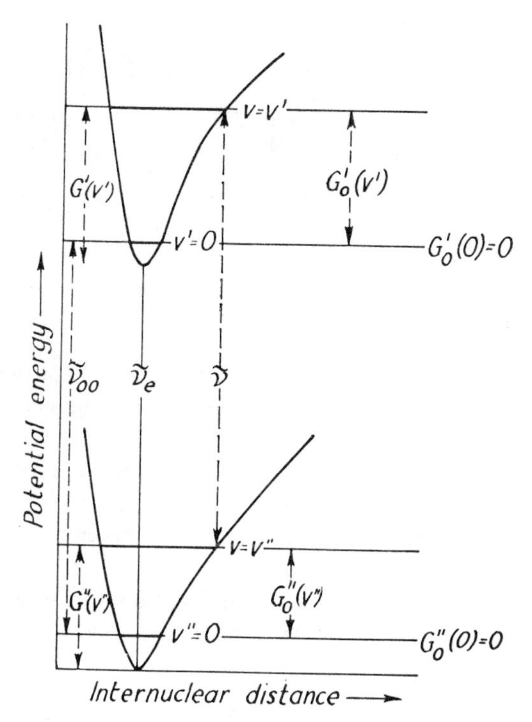

Fig. 2.19 Representation of the symbols used in the vibrational analysis of a band system.

On substitution for $G(v)$, $G_0(v)$, and $G(0)$ from Equations (2.37), (2.45), and (2.44), respectively, into Equation (2.46) and on rearrangement of the resulting equation we obtain:

$$(\omega_e - x_e\omega_e + \tfrac{3}{4}y_e\omega_e + \ldots)v - (x_e\omega_e - \tfrac{3}{2}y_e\omega_e + \ldots)v^2 + y_e\omega_e v^3 + \ldots = \omega_0 v - x_0\omega_0 v^2 + y_0\omega_0 v^3 + \ldots \quad (2.47)$$

If the coefficients of like powers of v on the left- and right-hand sides of Equation (2.47) are equated, it follows that:

$$\omega_0 = \omega_e - x_e\omega_e + \tfrac{3}{4}y_e\omega_e + \ldots \quad (2.48)$$

$$x_0\omega_0 = x_e\omega_e - \tfrac{3}{2}y_e\omega_e + \ldots \quad (2.49)$$

$$y_0\omega_0 = y_e\omega_e + \ldots \quad (2.50)$$

From the vibrational analysis the values of x_0, y_0, and ω_0 are obtained and from Equations (2.48), (2.49), and (2.50) the values of x_e, y_e, and ω_e follow. One of the main objects of the analysis is to obtain the value of ω_e since from its value the force constant of the bond may be readily calculated (see p. 60).

The equation to which the experimental data are fitted may be deduced from

Fig. 2.19 from which it may be observed that:

$$\tilde{\nu} = \tilde{\nu}_{00} + G'_0(v') - G''_0(v'') \qquad (2.52)$$

On substitution into Equation (2.51) for the values of $G_0(v)$ from Equation (2.45), we obtain the equation to which the experimental data are fitted, that is:

$$\tilde{\nu} = \tilde{\nu}_{00} + [\omega'_0 v' - x'_0\omega'_0 v'^2 + y'_0\omega'_0 v'^3 + \ldots] - [\omega''_0 v'' - x''_0\omega''_0 v''^2 + y''_0\omega''_0 v''^3 \ldots] \qquad (2.52)$$

Déslandres table

The spectroscopic procedure is to allot the values of (v', v'') to each of the bands in the band system; this is dealt with on p. 68. The wavenumbers of each of the (v', v'') bands are then arranged in what is known as a *Déslandres table*. This is illustrated in Table 2.2 where the wavenumber of each band origin is given for the purpose of representation in terms of Equation (2.51). In practice the wavenumber of each band origin experimentally observed is entered into this form of table.

The difference $G(v+1) - G(v)$, which is the wavenumber separation of two successive vibrational levels in one electronic state is represented by $\Delta G(v+\frac{1}{2})$ where:

$$\Delta G(v+\tfrac{1}{2}) = G(v+1) - G(v) = [G(0) + G_0(v+1)] - [G(0) + G_0(v)] \qquad (2.53)$$

and $\Delta G(v+\frac{1}{2})$ is known as the *first difference*. On substitution for $G_0(v)$ from Equation (2.45) into Equation (2.53) and on neglect of the cubic terms it is seen that:

$$\Delta G(v+\tfrac{1}{2}) = \omega_0 - \omega_0 x_0 - 2x_0\omega_0 v \qquad (2.54)$$

These values of $\Delta G(v+\frac{1}{2})$ may be obtained from the Déslandres table. For example, a $\Delta G'(\frac{1}{2})$ value is obtained by subtracting a $v' = 0$ value from a $v' = 1$ value vertically below it in the table. On application of this to Table 2.2 three values of $\Delta G'(\frac{1}{2})$ would result.† These three values would then be added together and divided by three, and the resulting value would be regarded as $\Delta G'(\frac{1}{2})$. By a similar procedure the values of $\Delta G'(1\frac{1}{2})$ and $\Delta G'(2\frac{1}{2})$ would follow. The values of $\Delta G''(\frac{1}{2})$ would be obtained by subtracting adjacent horizontal numbers in the $v'' = 1$ column from the ones in the $v'' = 0$ column. An arithmetic mean of the three resulting $\Delta G''(\frac{1}{2})$ values would then be regarded as the $\Delta G''(\frac{1}{2})$ value. By a similar treatment the $\Delta G''(1\frac{1}{2})$ and $\Delta G''(2\frac{1}{2})$ values would result.

If either the $\Delta G'(v+\frac{1}{2})$ or $\Delta G''(v+\frac{1}{2})$ values are plotted against v, then if the analysis was correct and it was permissible to neglect the cubic terms, it

† The value of three is an arbitrary one for the chosen Déslandres table, and in practice this would be the number of $\Delta G(\frac{1}{2})$ values which could be taken for the number of experimentally observed bands.

Table 2.2 The arrangement of the wavenumbers of the band origins in a band system in terms of their (v', v'') values; the wavenumbers inserted are derived from Equation (2.51)

v' \ v''	0	1	2	3	
0	$\tilde{\nu}_{00} + G_0'(0) - G_0''(0)$	$\tilde{\nu}_{00} + G_0'(0) - G_0''(1)$	$\tilde{\nu}_{00} + G_0'(0) - G_0''(2)$	$\tilde{\nu}_{00} + G_0'(0) - G_0''(3)$	
					$\Delta G'(\frac{1}{2})$
1	$\tilde{\nu}_{00} + G_0'(1) - G_0''(0)$	$\tilde{\nu}_{00} + G_0'(1) - G_0''(1)$	$\tilde{\nu}_{00} + G_0'(1) - G_0''(2)$	$\tilde{\nu}_{00} + G_0'(1) - G_0''(3)$	
					$\Delta G'(1\frac{1}{2})$
2	$\tilde{\nu}_{00} + G_0'(2) - G_0''(0)$	$\tilde{\nu}_{00} + G_0'(2) - G_0''(1)$	$\tilde{\nu}_{00} + G_0'(2) - G_0''(2)$	$\tilde{\nu}_{00} + G_0'(2) - G_0''(3)$	
					$\Delta G'(2\frac{1}{2})$
3	$\tilde{\nu}_{00} + G_0'(3) - G_0''(0)$	$\tilde{\nu}_{00} + G_0'(3) - G_0''(1)$	$\tilde{\nu}_{00} + G_0'(3) - G_0''(2)$	$\tilde{\nu}_{00} + G_0'(3) - G_0''(3)$	
	$\Delta G''(\frac{1}{2})$	$\Delta G''(1\frac{1}{2})$	$\Delta G''(2\frac{1}{2})$		First ↑ ← difference

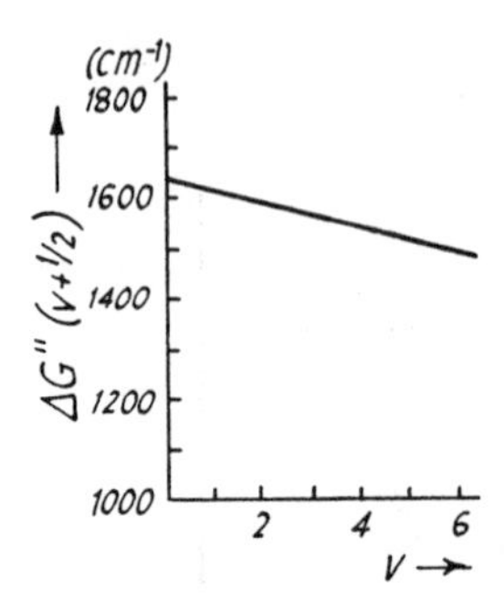

Fig. 2.20 The plot of $\Delta G''(v + \frac{1}{2})$ against v for the $X^3\Pi_u$ state of the C_2 Swan band system. The symbol X indicates this electronic state is the ground one. (After Phillips [2.7], courtesy of The Astrophysical Journal).

follows from Equation (2.54) that a straight line should be obtained, the slope of which should give the value of $-2x'_0\omega'_0$ or $-2x''_0\omega''_0$, respectively. The plot of $\Delta G''(v + \frac{1}{2})$ against v for the lower electronic energy state of the Swan emission system of C_2 is given in Fig. 2.20.

The most satisfactory check on the vibrational analysis is then to employ the determined mean first difference values and to obtain what is known as the *second difference* $\Delta^2G(v + 1)$ where this is defined by:

$$\Delta^2G(v + 1) = \Delta G(v + \tfrac{3}{2}) - \Delta G(v + \tfrac{1}{2}) \tag{2.55}$$

It follows from Equations (2.54) and (2.55) that this second difference is given by:

$$\begin{aligned}\Delta^2G(v + 1) &= [\omega_0 - \omega_0x_0 - 2x_0\omega_0(v + 1)] - [\omega_0 - \omega_0x_0 - 2x_0\omega_0v] \\ &= -2x_0\omega_0 \end{aligned} \tag{2.56}$$

Thus, if the analysis is correct, the second difference values for both the upper and lower electronic states have to give constant values of $-2x'_0\omega'_0$ and $-2x''_0\omega''_0$ respectively. If the values are not roughly constant, then the analysis is unsatisfactory. The arithmetical mean of the $\Delta^2G'(v + 1)$ values gives the value of $-2x'_0\omega'_0$, and similarly that of the $\Delta^2G''(v + 1)$ values gives the $-2x''_0\omega''_0$ values. If these $x_0\omega_0$ values are employed in conjunction with the plot of $\Delta G(v + \frac{1}{2})$ against v, then satisfactory values of x'_0, x''_0, ω'_0, and ω''_0 may be obtained. In addition, if these values are substituted in Equations (2.48) and (2.49), where the y_e terms are neglected, then values of x'_e, x''_e, ω'_e, and ω''_e result.

This method of vibrational analysis will now be applied to the C_2 molecule. This molecule is readily produced in a vigorous form of discharge through aromatic hydrocarbons. Part of the spectrogram for the C_2 molecule is given in Fig. 2.22 for the Swan band system which is the transition $A^3\Pi_g \rightarrow X^3\Pi_u$, that is an emission spectrum from the excited electronic state $A^3\Pi_g$ to the ground state $X^3\Pi_u$. The potential curves for these two electronic states are given in

Fig. 2.25(a). The band origin data for the Swan band system are arranged in Table 2.3 in the form of a Déslandres table. The $\Delta' G(v+\frac{1}{2})$ and $\Delta'' G(v+\frac{1}{2})$ values have been inserted together with the mean of the first differences and the values of the second differences. From these second difference values the values of ω_0' and ω_0'' may be evaluated as follows. The mean of the second difference $\Delta^2 G'(v+1)$ is:

$$(38.2 + 38.9 + 38.7)/3 = 38.6 = 2x_0'\omega_0'$$

and for $\Delta^2 G''(v+1)$:

$$(23.1 + 23.7 + 23.2 + 23.3 + 23.4)/5 = 23.34 = 2x_0''\omega_0''$$

Thus, $x_0'\omega_0'$ and $x_0''\omega_0''$ are equal to 19.3 and 11.67, respectively. Equation (2.54) for the $v = 0$ level becomes:

$$\Delta G(0+\tfrac{1}{2}) = \omega_0 - \omega_0 x_0 - 2x_0\omega_0 \times 0 \tag{2.57}$$

and from the Déslandres table it may be noted that the mean values of $\Delta G'(\frac{1}{2})$ are equal to 1753.8 and that of $\Delta G''(\frac{1}{2}) = 1618.2$. It follows thus from Equation (2.57) that:

$$1753.8 = \omega_0' - 19.3$$

$$\omega_0' = 1773.5$$

and that:

$$1618.2 = \omega_0'' - 11.67$$

$$\omega_0'' = 1629.87$$

Since the $\tilde{\nu}_{00}$ value is 19 373.9 cm^{-1} it follows by substitution of this value and the values for x_0', ω_0', x_0'', and ω_0'' into Equation (2.52) that the equation which represents the wavenumbers of the band origin of each of the bands is:

$$\tilde{\nu} = 19\,373.9 + (1773.5v' - 19.3v'^2) - (1629.87v'' - 11.67v''^2) \tag{2.58}$$

where the $y_0\omega_0$ terms are neglected.

The vibrational analysis just considered for the Swan bands of C_2 is based on band origin data. If, however, the dispersion of the spectrum is not sufficient for the band origins to be determined and only band head data are available, the procedure is slightly more empirical.† The band head data for the Swan bands of C_2 are listed in Table 2.4.

In this case the equation which expresses the wavenumbers of the heads of the bands with reasonable accurracy is:

$$\tilde{\nu} = 19\,355 + (1770v' - 20v'^2) - (1625v'' - 11.5v''^2) \tag{2.59}$$

It will be noted that the constants in this equation differ slightly from those in Equation (2.58).

† Since the head of the Q-branch lies closer to the band origin position than that of the P- or R- branches, it is desirable that the wavenumber of the Q-head should be employed in the Déslandres table in preference to that of the P or R.

Table 2.3 The Déslandres table for the Swan band system of C_2 where the electronic transition is $A^3\Pi_g \rightarrow X^3\Pi_u$

v' \ v''	0	$\Delta G''(\frac{1}{2})$	1	$\Delta G''(1\frac{1}{2})$	2	$\Delta G''(2\frac{1}{2})$	3	$\Delta G''(3\frac{1}{2})$	4	$\Delta G''(4\frac{1}{2})$	5	$\Delta G''(5\frac{1}{2})$	6	$\Delta G'(v+\frac{1}{2})$	$\Delta^2 G'(v+1)$
0	19 373.9	1618.2	17 755.7	1594.9	16 160.8	1571.5	14 589.3								
$\Delta G'(\frac{1}{2})$	(1754.0)		(1754.0)		(1753.2)		(1754.0)							1753.8	
1	21 127.9	1618.2	19 509.7	1595.7	17 914.0	1570.7	16 343.3	1548.2	14 795.1						38.2
$\Delta G'(1\frac{1}{2})$	(1715.4)		(1715.4)		(1716.2)		(1715.4)		(1715.4)					1715.6	
2	22 843.3	1618.2	21 225.1	1594.9	19 630.2	1571.5	18 058.7	1548.2	16 510.5	1524.8	14 985.7				38.9
$\Delta G'(2\frac{1}{2})$			(1676.7)		(1676.7)		(1676.7)		(1676.7)		(1676.5)			1676.7	
3			22 901.8	1594.9	21 306.9	1571.5	19 735.4	1548.2	18 187.2	1525.0	16 662.2	1501.4	15 160.8		38.7
$\Delta G'(3\frac{1}{2})$					(1638.0)		(1637.9)		(1638.9)		(1638.1)		(1638.0)	(1638.0)	
4					22 944.9	1571.6	21 373.3	1548.2	19 825.1	1524.8	18 300.3	1501.5	(16 798.8)		
$\Delta G''(v+\frac{1}{2})$		1618.2		1595.1		1571.4		1548.2		1524.9		1501.5		↑ ←Mean of first differences	
$\Delta^2 G''(v+1)$			23.1		23.7		23.2		23.3		23.4			←Mean of ↑ second differences	

Table 2.4 Déslandres table for the Swan bands of C_2 based on band heads

v' \ v''	0		1		2		3		4		5		6		7		
0	19 355	1615	17 740	1593	16 147												
	1749		1750		1752												$\Delta G'(\frac{1}{2})$
1	21 104	1614	19 490	1591	17 899	1569	16 330										
	1708		1712		1712		1713										$\Delta G'(\frac{1}{2})$
2	22 812	1610	21 202	1591	19 611	1568	18 043	1545	16 498	1526	14 972						
			1668		1671				1673		1677						$\Delta G'(2\frac{1}{2})$
3			22 870	1588	21 282				18 171	1522	16 649	1500	15 149				
					1620						1627		1629				$\Delta G'(3\frac{1}{2})$
4					22 902	1562	21 340				18 276	1498	16 778	1477	15 301		
		$\Delta G''(\frac{1}{2})$		$\Delta G''(1\frac{1}{2})$		$\Delta G''(2\frac{1}{2})$		$\Delta G''(3\frac{1}{2})$									↑ First ←difference

To express the band head data adequately Equation (2.42), which is a band origin formula, is modified slightly by including a small term $k(v' + \frac{1}{2}) \times (v'' + \frac{1}{2})$ and becomes:

$$\tilde{\nu} = \tilde{\nu}_e + [v' + \tfrac{1}{2})\omega_e' - (v' + \tfrac{1}{2})^2 x_e'\omega_e'] - [(v'' + \tfrac{1}{2})\omega_e'' - (v'' + \tfrac{1}{2})^2 x_e''\omega_e''] - k(v' + \tfrac{1}{2})(v'' + \tfrac{1}{2}) \quad (2.60)$$

The difference in wavenumbers between two consecutive members $(v' + 1)$ and v' in a column of the Déslandres table based on band head data would then be given by inserting these values into Equation (2.60) and is:

$$\omega_e' - 2(v' + 1)x_e'\omega_e' - k(v'' + \tfrac{1}{2}) \quad (2.61)$$

and similarly the difference between two consecutive members $(v'' + 1)$ and v'' in a row would be:

$$\omega_e'' - 2(v'' + 1)x_e''\omega_e'' - k(v' + \tfrac{1}{2}) \quad (2.62)$$

and by equating these to the mean values of actual differences between consecutive members of a column and row, respectively, a suitable value of k may be chosen.

The determination of the band origin wavenumber is dealt with on p. 81 where its value may be obtained from consideration of the fine structure data.

Application of vibrational analysis data

A large number of diatomic species have had their ω_e, $x_e\omega_e$, and $y_e\omega_e$ values determined by vibrational analysis. Herzberg [2.8] lists over 350 values and, in addition, a large number of their dissociation energies which have also been determined by these studies. The additional information which results from a knowledge of ω_e and x_e is:

(i) Calculation of the force constant for both the upper and lower electronic states from the equation:

$$\omega_e c = \frac{1}{2\pi}\sqrt{\frac{k}{\mu}} \quad (2.63)$$

where μ is the reduced mass of the molecule.

(ii) To assist in the statistical calculation of thermodynamic functions such as entropy and free energy (Chapter 7 of Vol. 2).

(iii) Calculation of the value of α in the Morse equation:

$$\alpha = \sqrt{(8\pi^2 \mu x_e \omega_e c/h)} \quad (2.64)$$

(iv) The value of the constants in the vibrational energy equation are thus known, and the following equation:

$$E_v = (v + \tfrac{1}{2})hc\omega_e - (v + \tfrac{1}{2})^2 hcx_e\omega_e + (v + \tfrac{1}{2})^3 hcy_e\omega_e + \ldots \quad (2.65)$$

may be employed then to calculate the vibrational energy for any value of v. In addition, this equation may be used to obtain a very approximate value for the dissociation energy (Chapter 3).

Progressions and sequences

So far it has been assumed that the (v', v'') values for a band can be allotted. The allocation of these values is, of course, an essential step in the analysis, and at least a provisional allotment of (v', v'') values has to be made before the Désландres table can be formed. This allotment of (v', v'') values is greatly assisted by picking out what are known as *progressions* and *sequences*.

It has been seen that the wavenumber differences between the adjacent (v', v'') values in any row or column are nearly constant for that particular row or column. The bands in a particular column where v'' is constant and v' varies are called *v' progressions*; similarly, those in a row where v' is constant and v'' progressively varies are termed *v'' progressions*. Progressions with constant v' appear as bands whose wavenumber separations decrease towards longer wavelengths, while those with constant v'' have decreasing separations towards shorter wavelengths. These points may be deduced from Fig. 2.21, bearing in mind that as v increases the spacing of the vibrational energy levels decreases and the progressions illustrated by the energy level diagrams also shown in Fig. 2.21.

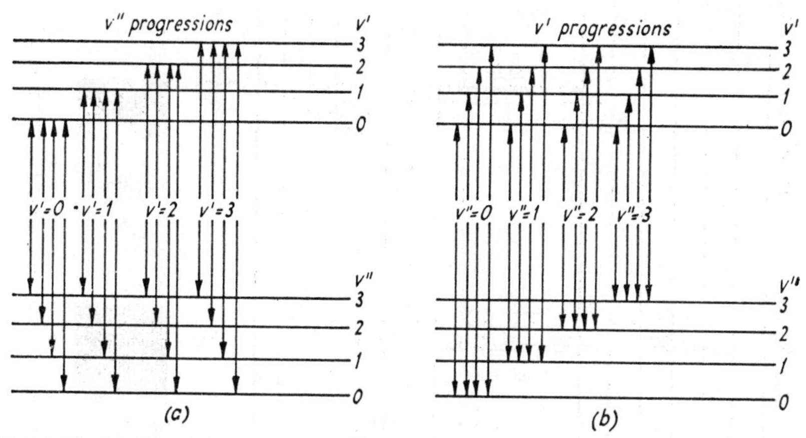

Fig. 2.21 (a) Transitions in some v'' progressions; (b) transitions in some v' progressions.

Bands which fall in diagonal rows of the Déslandres array have a constant value of $v' - v''$. These groups of bands are characteristic in the spectrum and called *sequences*; e.g. the (0, 0). (1, 1), (2, 2), and (0, 1), (1, 2), (2, 3) bands in Table 2.5 given for C_2 form two types of sequences. The wavenumbers of the bands in a sequence do not differ considerably; thus, a group of bands close together, of gradually changing intensity, would suggest a sequence. Some of the C_2 Swan bands may be observed in Fig. 2.22 where it will be noted that there are six separate groups of bands. Each of these groups is a sequence. In Table 2.5 the wavelength of the first band in each sequence is listed, and the bands in that particular sequence are identified in terms of their (v', v'') values. Thus, each of

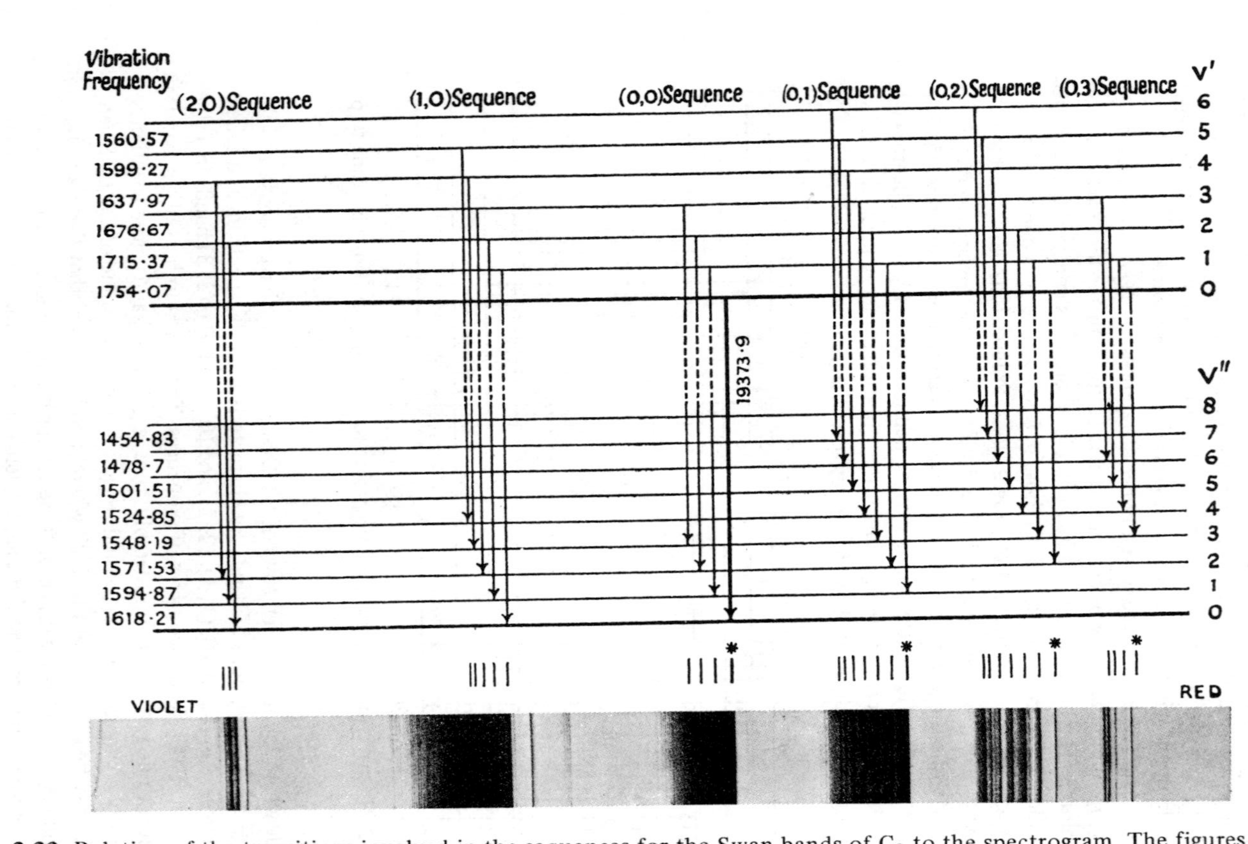

Fig. 2.22 Relation of the transitions involved in the sequences for the Swan bands of C_2 to the spectrogram. The figures quoted are in cm^{-1} units. (After Johnson [2.9]).

Table 2.5 Six sequences in the Swan bands of C_2

4383 Å	*4737 Å*	*5165 Å*	*5636 Å*	*6191 Å*	*6677 Å*
2–0	1–0	0–0	0–1	0–2	0–3
3–1	2–1	1–1	1–2	1–3	1–4
4–2	3–2	2–2	2–3	2–4	2–5
	4–3	3–3	3–4	3–5	3–6
	5–4		4–5	4–6	
			5–6	5–7	
			6–7	6–8	

the vertical groups in Table 2.5 is a sequence, Progressions may be picked out of the horizontal groups; for example, two progressions are the 2–0, 1–0, 0–0, and the 3–3, 3–4, 3–5, 3–6 bands. In the C_2 Swan band spectrogram the bands in a progression, unlike those in sequences, are widely separated and may be readily distinguished from one another (e.g. see Table 2.6). The transitions between the (v', v'') values in the electronic states involved in the sequences are related to the spectrogram in Fig. 2.22.

Table 2.6 Intensities of some of the Swan bands of C_2

Wavelength	*Intensity*	(v', v'')	
5636	8	0, 1	One sequence
5585	8	1, 2	
5541	6	2. 3	
5165	10	0, 0	One v' progression
4737	9	1, 0	
4383	2	2, 0	

Intensity distribution in a band system

One of the most important factors in deciding which v' and v'' values have to be allotted to a particular band is the intensity of the band concerned with respect to the other bands in a given system. In Fig. 2.22 it will be observed that the intensities of some of the Swan bands differ considerably. Initially the main factors will be examined, which determine (a) the intensity of a band, and (b) which particular bands might be expected from transitions between two electronic states whose potential energy curves and vibrational levels have been previously determined.

The time of an electronic transition is of the order of 10^{-16} s and the time for a vibration is about 10^{-14} s, that is relatively the vibration period is long compared with the time taken for an electronic transition. Thus, an electronic transition takes place so rapidly that immediately after the transition the internuclear distance is very nearly the same as before the transition. In other words, transitions occur vertically upwards or downwards between the two potential energy curves.

The probability of a transition between two vibrational levels in different electronic states for a common internuclear distance is dependent on the overlap integral (see Vol. 2, p. 348) value:

$$\int \Psi_{v'}\Psi_{v''}\mathrm{d}r$$

where $\Psi_{v'}$ is the eigenfunction of the higher vibrational level, $\Psi_{v''}$ that for the lower one. The value of Ψ_v for a particular vibrational level is dependent on the internuclear distance (r). The greater the value of the overlap integral the more probable it is that the transition will take place.

In Fig. 2.23 both the potential energy curves have their minima at equal internuclear distances. The variation of the eigenfunction Ψ with internuclear distance for the $v' = 0$ and $v'' = 0$ levels is given by the dotted lines in Fig. 2.24.

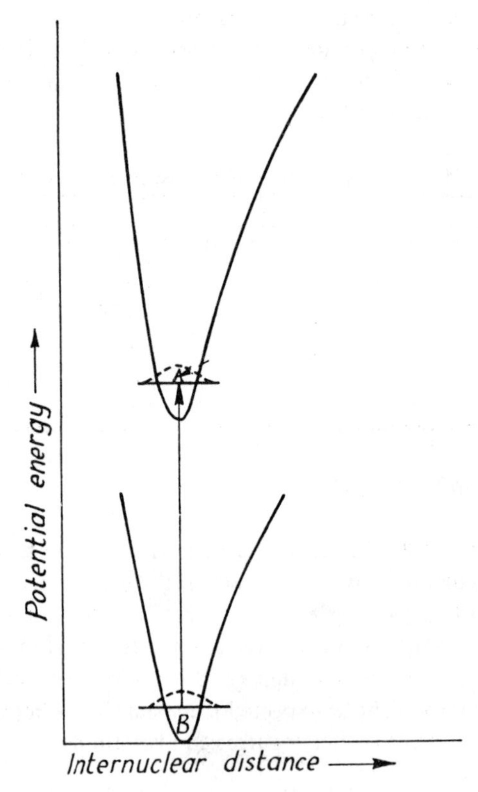

Fig. 2.23 Electronic absorption transition from the $v'' = 0$ to the $v' = 0$ transition

The dependence of the eigenfunction (Ψ_v) on the vibrational level concerned may be seen for rubidium hydride in Fig. 2.24. In order to decide which of the absorption bands might be obtained in a transition from the $v'' = 0$ level for

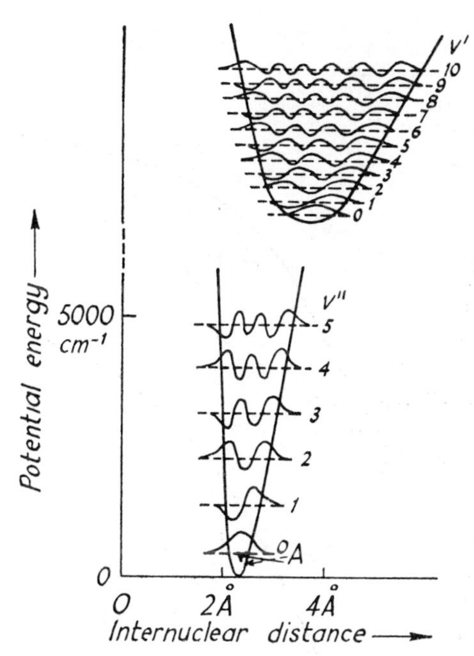

Fig. 2.24 The dependence of the eigenfunction on the vibrational levels for the upper and lower electronic states of rubidium hydride. (After Gaydon [2.10]).

rubidium hydride, a line should be drawn from the point A parallel to the ordinate axis – really applying the Franck–Condon principle – and where this line intersects the terminal maximum or minimum† of any of the wavefunctions a band might be expected. In this case the absorption bands expected would be the (7, 0), (8, 0), (9, 0). However, the electronic transitions for rubidium hydride are too complex for a detailed consideration here. Instead, we shall now examine the relevant potential energy curves for the emission transition in the Swan bands of C_2 and apply (i) the Franck–Condon principle, and (iii) the fact that the greater the product of $\Psi_{v'}\Psi_{v''}$ then the more probable is the transition, to gain some indication as to which of the bands would be expected to be the most intense.

In Fig. 2.25(a) are given the potential energy curves together with the first three vibrational levels for the two electronic states of C_2 the transitions between which result in the main Swan bands. Normally these would not be known until the vibrational analysis had been completed, but to illustrate the

† In the higher vibrational levels the two end maxima (or minima) are the largest, and in between these the contributions of the smaller and narrower maxima and minima very approximately cancel one another out when the overlap integral is evaluated. Hence for these higher vibrational levels to a rough approximation at least only the terminal loops need be considered.

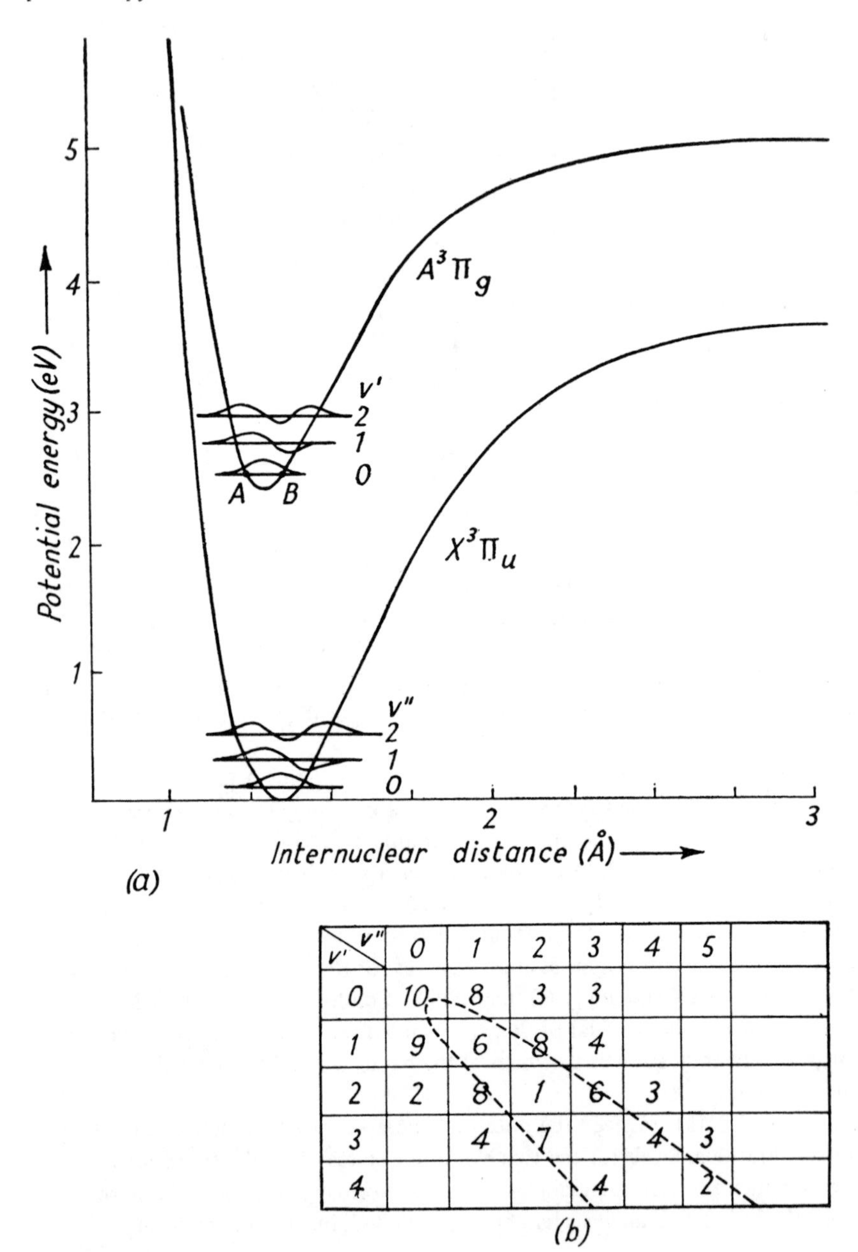

v' \ v''	0	1	2	3	4	5	
0	10	8	3	3			
1	9	6	8	4			
2	2	8	1	6	3		
3		4	7		4	3	
4				4		2	

(b)

Fig. 2.25 (a) Potential energy curves† and eigenfunction plot, (b) intensity distribution in the Swan bands of C_2 and the Condon parabola.

† The ground state is a $^3\Pi_u$ state while the excited one is a $^3\Pi_g$. In addition, there are other $^3\Pi_g$-states and this one is characterized by placing the letter A

dependence of band intensities on the (v', v'') values we shall assume that this information is available.

Since the transitions are emission ones, then the most intense bands would be expected to have a $v' = 0$ value as this might be expected to be the most heavily populated level. The most probably internuclear distance in this $v' = 0$ level is at the mid-point of AB where the wavefunction $(\Psi_{v'})$ has a maximum value, and it is from this value of the internuclear distance that we apply the Franck–Condon principle and draw a line parallel to the ordinate axis.

This line nearly intersects the maximum of the $v'' = 1$ level. Thus, the transition would take place from a highly probable internuclear distance to another probable one, and the product of the wavefunctions would have a high value, and an intense band might be expected. The intensities of the C_2 Swan bands are represented in Fig. 2.25(b), where the intensities of the bands have been placed on a relative scale by giving the most intense band an intensity of 10 and the least intense that of 1. Thus, the range of intensity is arbitrarily fixed between 1 and 10. The next most highly populated level might be expected to be the $v' = 1$, and when similar considerations are applied, the (1, 1) and (1, 2) bands might be expected; these are, in fact, obtained. From the $v'' = 2$ level the most intense bands should be the (2, 1) which again proves to be the case. However, the fact that the (0, 0) band is the most intense is not readily accounted for. In general, it is not always certain as to which v'–v'' transitions will appear and what the approximate order of their intensities will be. We have entirely neglected the effects of the intermediate maxima and minima of the eigenfunctions and have only indirectly considered the fact that many molecules have internuclear distances other than the most probable value. This range of r values in the $v' = 0$ level may lead to v'' progressions in emission transitions, while in absorption transitions from the $v'' = 0$ level a number of bands of a v' progression may result. The number of bands obtained in these progressions depends on the relative positions of the potential energy curves and the steepness of the curve to which the transition is taking place. In fact, the object has been to gain some qualitative insight into the basic factors involved. For more

in front. Thus, the excited state in the Swan band system is an $A^3\Pi_g$-state while the ground state is $X^3\Pi_u$. The letter X is commonly employed to signify that the electronic state concerned is the ground state. For molecules several excited electronic states may exist, and in the case of CuH the following are known:

$$^1\Sigma^+,\ ^1\Sigma^+,\ ^1\Pi,\ ^1\Pi$$

The ground state is $X^1\Sigma^+$. The excited electronic states are characterized by placing a letter in front of the symbol. For example, the excited electronic states of CuH are represented by $A^1\Sigma^+$, $B^1\Sigma^+$, $C^1\Pi$, and $D^1\Pi$, where the $\tilde{\nu}_{00}$ values for the

$$A^1\Sigma^+ \leftrightarrow X^1\Sigma^+,\quad B^1\Sigma^+ \leftrightarrow X^1\Sigma^+,\quad C^1\Pi \leftrightarrow X^1\Sigma^+,\quad D^1\Pi \leftrightarrow X^1\Sigma^+$$

transitions are, respectively, 23 311.1, 26 281.7, 27 107, and 44 651.2 cm^{-1}.

quantitative treatment transition probabilities have to be considered. Evaluation of the transition probability for any band is a most complex procedure, and it may be shown that the transition probability for an emission transition is proportional to:

$$\tilde{\nu}^4 \left(\int_0^\infty \Psi_{v'} M \Psi_{v''} \mathrm{d}r \right)^2$$

where $\tilde{\nu}$ is the emitted wavenumber, $\Psi_{v'}$ and $\Psi_{v''}$ are the wavefunctions for the upper and lower vibrational energy states, M is the electric dipole moment, and r is the internuclear distance. M is a variable electric dipole moment and may be subdivided into the components M_x, M_y, and M_z along the x, y, and z axes respectively. For instance, M_x is the component of the electric dipole moment of the system along the x axis, and:

$$M_x = \Sigma_j e_j x_j$$

where e_j is the charge and x_j the coordinate of the jth particle and the summation is taken over all the particles of the system. For a diatomic molecule where the nuclei and several electrons have to be considered M can be expressed in a $r-r_e$ series:

$$M = M_0 + \epsilon(r - r_e) + \ldots$$

ϵ is a constant, r_e the equilibrium internuclear distance, and M_0 the value of the electric dipole moment when the internuclear distance has the equilibrium value. If M does not vary too rapidly with r, in some cases it may be treated as constant, and then the evaluation of the transition probability is simplified.

If the most intense bands as represented in Fig. 2.25(b) are joined together by a dotted line, a parabolic curve whose axis is the principal diagonal is obtained. This is called the *Condon parabola*. The width of the Condon parabola depends on the internuclear separations in the upper and lower electronic states. When the separations of the nuclei are very different in the upper and lower states an open parabola is obtained; should they be appreciably different, the parabola is very open, and the (0, 0) and neighbouring bands are not observed. This Condon parabola is a most important feature in verifying the correct assignment of the v' and v'' values to a particular band as the most intense of them should lie on the Condon parabola. In fact, if in the above analysis a band of intensity 8 for a (3, 0) transition had been observed and this was correct, then some of the other allotted (v', v'') values would be incorrect.

Assignment of v' and v'' values

In the treatment so far when v' and v'' have been correlated it has been assumed that the two potential energy curves have been known. This is rarely the case. In practice the v' and v'' values have first to be allotted by observing various points about the spectrum itself. Such a procedure is not for the student but for the practising spectroscopist, whose previous experience and intuition greatly assist in achieving this end. However, a few of the factors which help in the allotment of values will now be considered.

The most intense sequences are noted. In the case of the C_2 molecule for the Swan bands the most intense sequences are the (1, 0) and (0, 0) ones (see p. 62). One complication is that sequences sometimes overlap. The C_2 molecule is not a case in point. Other factors which assist in the allotment of v' and v'' values are:

(i) An attempt is made to pick out the progressions. This may be assisted by long exposures when the appearance of weak additional bands may help in the identification of a progression. In emission the transitions from $v' = 0$ might be expected. For C_2 the (0, 0), (0, 1), and (0, 2) bands of such a progression are observed. For a band system in absorption at ordinary temperatures, however, then transitions from $v'' = 0$ might be expected since the molecules are mainly in the $v'' = 0$ level.

(ii) The wavenumbers of the bands heads have to be arranged in a Déslandres table so that: (a) The first difference between adjacent rows is very approximately constant and similarly so for adjacent columns. (b) The second difference should be almost constant. It was shown on p. 56 that this is $-2x_0\omega_0$, and depending on which difference is being considered this would be either $-2x'_e\omega'_e$ or $-2x''_e\omega''_e$. If this value is not nearly constant, then a third difference should be taken, and this would be $6y_0\omega_0$.† (c) The v', v'' values of the most intense bands should lie on the Condon parabola in the Déslandres table itself; otherwise, the analysis is incorrect.

(iii) In general, for the diatomic molecules and radicals so far reported in the literature, at least three or four band systems have been obtained. Hence, in the analysis of new band systems of such species it is likely that one of the previously analysed electronic states whose vibrational data will be available (i.e. ω_e, x_e, and y_e values) is one of the electronic states in the new transition. The analysis for the assignment of v' and v'' values may then be considerably assisted, since some of the frequency differences may be directly related to the known vibrational levels. This may thus assist the picking out of a progression involving such levels.

2.3.3 Rotational structure of electronic bands of diatomic molecules and radicals

As yet we have been mainly concerned with bands at poor dispersion where the vibrational analysis could be carried out after measurement of the position of the band heads. At better dispersion and resolution the rotational structure of such bands may be observed. From the measurement of the position of the lines in this fine structure of the band and by allotting the correct J' *and* J'' values to each of these lines, the $B_{v'}$ and $B_{v''}$ values may be determined. This type of study

† By taking these second and third differences, values of $x_e\omega_e$ and $y_e\omega_e$ values are obtained.

will be the next objective, and the approach will be made by consideration of the equations which may be employed to express the wavenumbers of the spectral lines in terms of J values.

It was indicated on p. 50 that the emitted or absorbed wavenumber in an electronic transition is given by:

$$\tilde{\nu} = \tilde{\nu}_e + \tilde{\nu}_v + \tilde{\nu}_r \tag{2.66}$$

For one particular band in an electronic transition both $\tilde{\nu}_e$ and $\tilde{\nu}_v$ are constant and may be replaced by ν_0 where ν_0 is termed the *band origin* or *zero line*. The wavenumbers of the spectral lines are then given by:

$$\tilde{\nu} = \tilde{\nu}_0 + \tilde{\nu}_r \tag{2.67}$$

where $\tilde{\nu}_r$ is the wavenumber for the difference in the two rotational term values for the particular spectral line. Thus, the equation may be modified to:

$$\tilde{\nu} = \tilde{\nu}_0 + [F_{v'}(J') - F_{v''}(J'')] \tag{2.68}$$

where the rotational term values are given by the formula:

$$F_v(J) = \frac{E_r}{hc} = B_v J(J+1) - D_v J^2(J+1)^2 + \ldots \tag{2.69}$$

The wavenumbers of the lines in a particular band are then represented by:

$$\begin{aligned}\tilde{\nu} = \tilde{\nu}_0 &+ [B_{v'}J'(J'+1) - D_{v'}J'^2(J'+1)^2 + \ldots] \\ &- [B_{v''}J''(J''+1) - D_{v''}J''^2(J''+1)^2 + \ldots]\end{aligned} \tag{2.70}$$

For an electronic transition, the upper and lower energy states may have different orbital angular momenta about the internuclear axis where this momentum is $\Lambda h/2\pi$. When $\Lambda = 0$, that is for a Σ-state (see p. 30), the selection rule for J is:

$$\Delta J = \pm 1 \tag{2.71}$$

but when $\Lambda \neq 0$ for either or both the electronic states between which the transition takes place the selection rule becomes:

$$\Delta J = 0, \pm 1 \tag{2.72}$$

Thus, in the case of a $^1\Sigma - {}^1\Sigma$ transition only P- and R-branches would be expected corresponding to $J' - J'' = -1$ and $+1$, respectively. However, in a $^1\Pi - {}^1\Sigma$ transition, since $\Lambda = 1$ for a Π-state, then P-, Q-, and R-branches would be expected. The wavenumbers of the R-, Q-, and P-branches are:

$$\tilde{\nu} = \tilde{\nu}_0 + F_{v'}(J+1) - F_{v''}(J) = R(J) \tag{2.73}$$

$$\tilde{\nu} = \tilde{\nu}_0 + F_{v'}(J) - F_{v''}(J) = Q(J) \tag{2.74}$$

$$\tilde{\nu} = \tilde{\nu}_0 + F_{v'}(J-1) - F_{v''}(J) = P(J) \tag{2.75}$$

where $R(J)$, $Q(J)$, and $P(J)$ are employed to represent the wavenumbers of the lines in the R-, Q-, and P-branches, respectively, for a particular value of J.

The centrifugal constants $D_{v'}$ and $D_{v''}$ are normally very small. In fact:

$$D_v = 4B_v^3/\omega^2 \dagger \qquad (2.76)$$

where $\omega \gg B_v$, and D_v is generally something of the order of $10^{-5} B_v$. The case will now be considered where this D_v term is neglected. On substitution of Equation (2.69) into Equation (2.73), (2.74), and (2.75) assuming $D_v = 0$, then by expression of the J' values in terms of J'' by means of $J' = J'' - 1$ for a P-branch, $J' = J''$ for a Q-branch, and $J' - J'' = +1$ for an R-branch, the following equations result for the P-, Q-, and R-branches:

$$\tilde{\nu} = \tilde{\nu}_0 - (B_{v'} + B_{v''})J + (B_{v'} - B_{v''})J^2 = P(J) \qquad (2.77)$$

$$\tilde{\nu} = \tilde{\nu}_0 + (B_{v'} - B_{v''})J + (B_{v'} - B_{v''})J^2 = Q(J) \qquad (2.78)$$

$$\tilde{\nu} = \tilde{\nu}_0 + 2B_{v'} + (3B_{v'} - B_{v''})J + (B_{v'} - B_{v''})J^2 = R(J) \qquad (2.79)$$

These equations which give the wavenumbers of the lines of the P-, Q-, and R-branches have exactly the same form as those derived for rotation–vibration transitions (Vol. 2). In addition, as was the case for rotational–vibrational transitions, the P- and R-branches may be represented by the equation:

$$\tilde{\nu} = \tilde{\nu}_0 + (B_{v'} + B_{v''})m + (B_{v'} - B_{v''})m^2 \qquad (2.80)$$

where $m = -J = -J''$ for the P-branch and $m = J + 1 = J'' + 1$ for the R-branch.

As regards an electronic spectrum $B_{v'}$ and $B_{v''}$ refer not only to different vibrational energy levels but also to different electronic states and depend on how one potential energy curve lies with respect to the other. In fact, sometimes these B-values may be very different from one another.‡

If there is no Q-branch (as for a $^1\Sigma$–$^1\Sigma$ transition), no line appears in the position of $\tilde{\nu}_0$ which is the band origin. A $^1\Sigma$–$^1\Sigma$ transition is illustrated diagrammatically in Fig. 2.26 between the vibrational levels $v' = 0$ and $v'' = 0$. For a $^1\Pi$–$^1\Sigma$ transition, however, a Q-branch is permitted in addition to the P- and Q-branches, and in addition the $P(1)$ line is absent.

Band head and Fortrat parabola

$B_{v'} > B_{v''}$. If the internuclear distance in the upper electronic state is smaller than that in the lower, then, since B_v varies inversely as r^2, the term $(B_{v'} - B_{v''})$ will be positive. By reference to the following formula:

† This equation is derived in Herzberg, [2.8].

‡ For a rotation–vibration infrared or Raman change the alteration in the moment of inertia is small in passing from one vibrational level to the next in the same electronic state. However, in an electronic transition there may be an appreciable change in the moment of inertia; hence, the spacing of the lines may differ markedly between the two types of spectra.

$$\tilde{\nu} = \tilde{\nu}_0 + (B_{v'} + B_{v''})m + (B_{v'} - B_{v''})m^2 \quad (2.81)$$

it may be noted that in the case of a P-branch, where $m = -J$, the term in m^2, as m increases, will eventually outweigh the term in m, and consequently $\tilde{\nu}$ at first decreases and then increases. If sufficient J values are available, the P-branch consists of lines which grow closer together then gradually turn back on themselves, and their spacing then begins to increase. The frequency at which the lines begin to turn back is known as the band head. A band where $\tilde{\nu}$ decreases

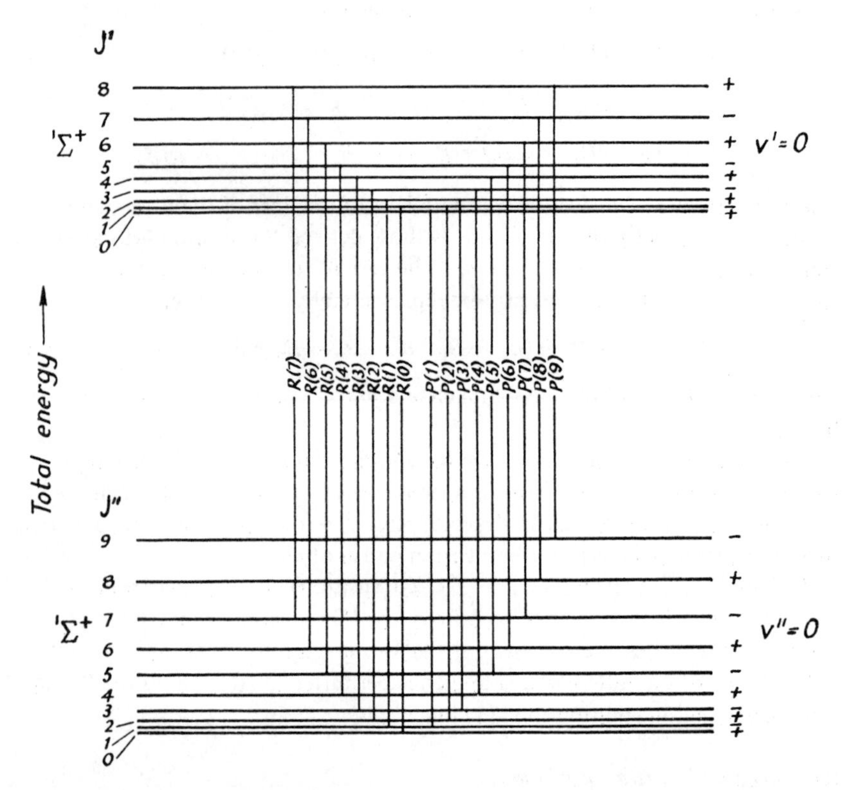

Fig. 2.26 Energy level diagram illustrating the rotational transitions of the (0, 0) band of a $^1\Sigma^+–^1\Sigma^+$ transition.

and then increases is termed *degraded towards the violet*, that is to shorter wavelengths. The P-branch represented in Fig. 2.27 is such a case. In general, the wavenumber at which the band head occurs depends on the difference between the $B_{v'}$ and $B_{v''}$ values, and normally the band head may easily be recognized and measured directly. This is to be contrasted with the band origin, that is the $J' \leftrightarrow J''$ transition, where both the J values are zero. The band origin has, in fact, to be determined from the analysis of the rotational branches. As regards the R-branch for this case of $B_{v'} > B_{v''}$ it follows from Equation (2.81) and the fact

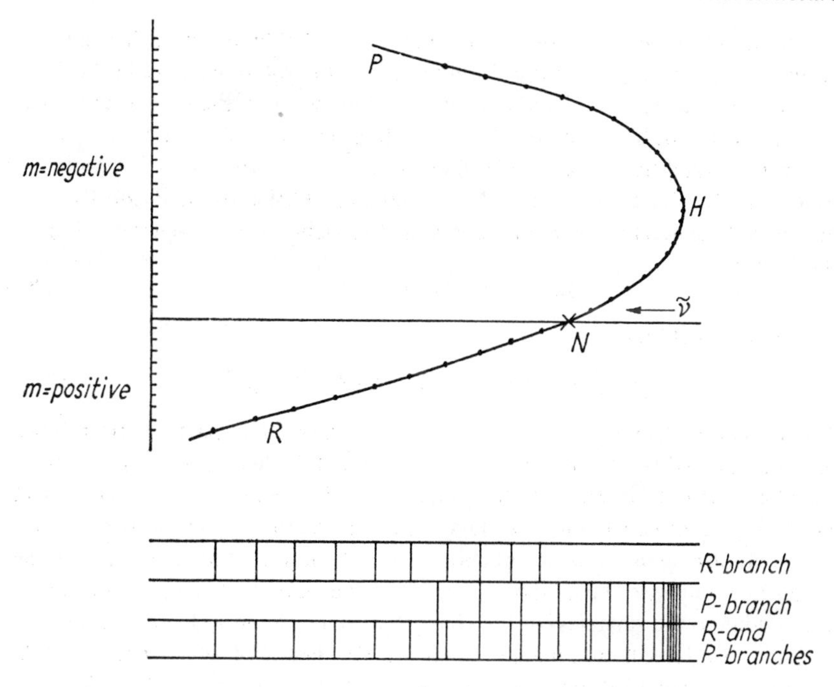

Fig. 2.27 Fortrat parabola for a $^1\Sigma$—$^1\Sigma$ transition where the band head is degraded to the violet.

that $m = J + 1$ that as m increases the wavenumbers of the lines will steadily diverge; this may also be noted in Fig. 2.27. Fortrat was the first to employ the type of representation given in Fig. 2.27 for the P- and R-branches, and this $\tilde{\nu}$–m plot is known as the *Fortrat parabola.*

The case shown in Fig. 2.27 is one which does not exhibit a Q-branch. The negative values of m, which are those of the P-branch, are drawn above the zero line, and the positive values of m, that is those of the R-branch are shown below this line. The diagram illustrates most clearly the formation of a band head in the P-branch. The parabola cuts the axis at $m = 0$ at a point N and has its vertex $\Delta\tilde{\nu}/\Delta m = 0$ at a point H which is the band head.

$B_{v'} < B_{v''}$. The condition which has to be satisfied for the formation of a head in the R-branch is that the internuclear distance in the lower electronic state has to be smaller than that in the upper, that is $(B_{v'} - B_{v''})$ must be negative. It then follows from Equation (2.81) and the fact that $m = J + 1$ for an R-branch that initially the $(B_{v'} + B_{v''})m$ term may outweigh the $(B_{v'} - B_{v''})m^2$ term and the wavenumbers will increase. However, as J increases the squared term will eventually outweigh the single power term and the wavenumbers will decrease. In this case the band is termed *degraded towards the red*, that is to longer wavelengths. Such an example is that of the R-branch in the (0, 0) band of the $A^1\Sigma^+ - X^1\Sigma^+$ transition of CuH in Fig. 2.29(a). In addition, an intensity distribution representation of the lines in the P- and R-branches of this band is given in Fig. 2.29(c), and in Fig. 2.29(b) each of the rotational lines in the P- and R-branches may be related to an appropriate point on the Fortrat parabola.

For transitions between two electronic states where in one or both of the electronic states the orbital angular momentum is not zero, that is for transitions other than $\Sigma - \Sigma$, a Q-branch is to be expected in addition to the P- and R-branches. In general the lines in the P- and R-branches may be represented by the formula:

$$\tilde{\nu} = \tilde{\nu}_0 + (B_{v'} + B_{v''})m + (B_{v'} - B_{v''})m^2 \tag{2.82}$$

but those in a Q-branch by:

$$\tilde{\nu} = \tilde{\nu}_0 + (B_{v'} - B_{v''})J'' + (B_{v'} - B_{v''})J''^2 \tag{2.83}$$

The lines in the Q-branch will be represented by a different parabola from those of the P- and R-branches because the coefficients for the second terms on the right-hand side of Equations (2.82) and (2.83) differ. Since the terms in J''^2 and m^2 in Equations (2.83) and (2.82) have the same coefficient the vertices of the two parabolae lie in the same direction. If the $B_{v'}$ and $B_{v''}$ values are of the same magnitude, then the Q parabola will intersect the wavenumber axis nearly vertically owing to the smallness of the $(B_{v'} - B_{v''})$ value. When a Q-branch is present, then two band heads are generally observed: (i) a head in P and Q when $B_{v'} > B_{v''}$ and these are degraded towards shorter wavelengths; or (ii) when $B_{v''} > B_{v'}$ a head in R and Q which are degraded to longer wavelengths. A representation of the fine structure for a $^1\Pi - {}^1\Sigma$ transition together with the Fortrat parabolae is given in Fig. 2.28 for the case where $B_{v''} > B_{v'}$. In general all the bands in a system are degraded in the same direction, since the sign of the difference $B_{v'} - B_{v''}$ does not normally change. However, if $B_{v'}$ and $B_{v''}$ are almost equal, and since the B values for the upper and lower electronic levels are dependent on the value of v' and v'', respectively, the possibility exists that $B_{v'}$ may be greater than $B_{v''}$ for one band but less than $B_{v''}$ for another band in the same band system. In fact, this is so for the electronic transition $B^2\Sigma^+ - X^2\Sigma^+$ in the system of CN where some of the bands are degraded to the violet and others to the red.

In order to determine the m value corresponding to the vertex of the Fortrat parabola for a P- or R-branch, the following equation is differentiated:

$$\tilde{\nu} = \tilde{\nu}_0 + (B_{v'} + B_{v''})m + (B_{v'} - B_{v''})m^2 \tag{2.84}$$

and we obtain:

$$\frac{\Delta\tilde{\nu}}{\Delta m} = (B_{v'} + B_{v''}) + 2m(B_{v'} - B_{v''}) \tag{2.85}$$

where $\Delta\tilde{\nu}/\Delta m = 0$ is the band head and the corresponding m value is termed m_{head}:

$$m_{\text{head}} = -\frac{(B_{v'} + B_{v''})}{2(B_{v'} - B_{v''})} \tag{2.86}$$

On substitution of this m_{head} value into Equation (2.84) we obtain:

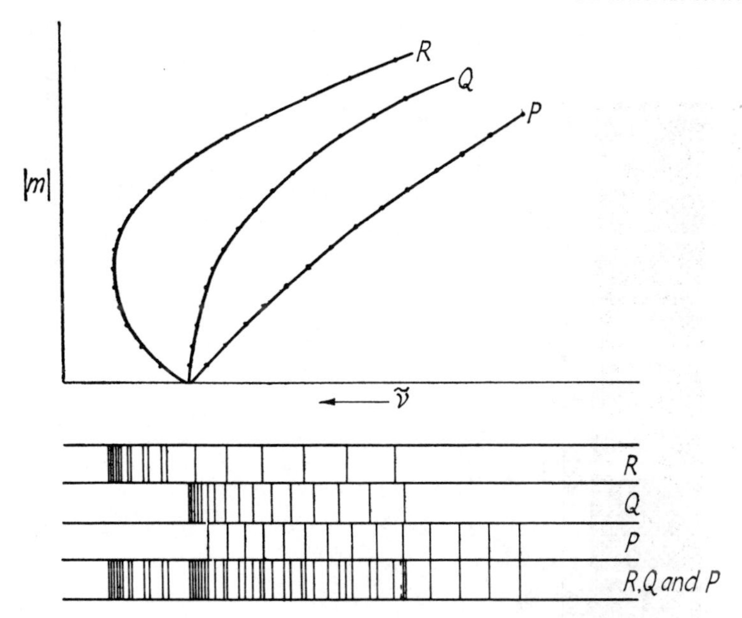

Fig. 2.28 Fortrat parabola for a $^1\Pi-^1\Sigma$ transition where the lines in R- and Q-branches are degraded to the red. $|m|$ values have been plotted on the ordinate axis. The m values for the P-branch have a negative sign. If respect were paid to the sign of m, then the P-branch parabola would be a continuation of the R-branch parabola. It should be noted that m does not take the values 0 or -1 for a $^1\Pi-^1\Sigma$ transition.

$$\tilde{\nu}_{\text{head}} - \tilde{\nu}_0 = -\frac{(B_{v'} + B_{v''})}{4(B_{v'} - B_{v''})} \tag{2.87}$$

When $B_{v'} < B_{v''}$, then it follows from Equation (2.87) that $\tilde{\nu}_{\text{head}} - \tilde{\nu}_0$ is positive, that is the head of the hand is at a higher frequency than the band origin. Hence, the band must be degraded to the red. However, if $B_{v'} > B_{v''}$ then the value is negative and the shading must be towards the violet. It can be seen from Equation (2.87) that if $B_{v'} \approx B_{v''}$, then the wavenumber separation of $\tilde{\nu}_{\text{head}}$ and $\tilde{\nu}_0$ will be very great; thus, the band head may be so far separated from the band origin that for these high m values the intensity of the lines will be so weak that the head is not observable.

Intensity distribution within a band

It has just been considered how the appearance of a band depends on the position of its constituent lines and on the relative values of $B_{v'}$ and $B_{v''}$. An additional factor which affects its appearance is the relative variation of intensities in these constituent lines and their distribution. The intensity distribution of the lines in the $^1\Sigma^+-^1\Sigma^+$ (0, 0) band of CuH under certain experimental conditions is given in Fig. 2.29(c).

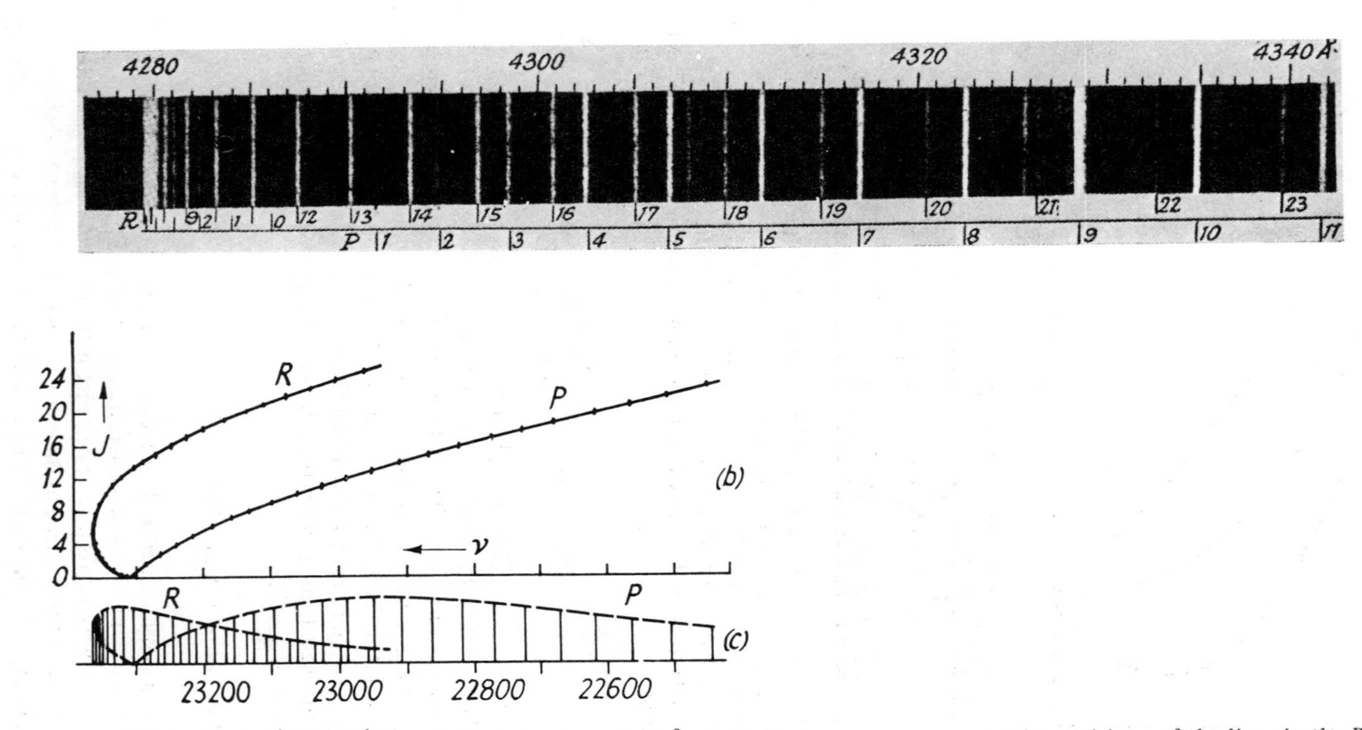

Fig. 2.29 (a) The CuH $A^1\Sigma^+ \rightarrow X^1\Sigma^+ - K$ (0, 0) band at 4280 Å obtained from an arc source. The positions of the lines in the P- and R-branches are indicated [2.11]. (b) Fortrat parabola for the P- and R-branch lines in the (0, 0) band of CuH [2.12]. (c) An intensity distribution representation of the lines in the (0, 0) band of CuH. (Courtesy of Dr. W. Jevons and the Council of the Physical Society [2.12]).

Combination relationships

For P- *and* R-*branches*. In order to determine the values of $B_{v'}$ and $B_{v''}$ most accurately and to allot the J' and J'' values unambiguously it is necessary to make use of what are known as *Combination relations*. To illustrate the combination principle Fig. 2.26 will be employed, and this involves one band of a $^1\Sigma$–$^1\Sigma$ transition. In this figure the spectral lines, whose wavenumbers are $R(2)$ and $P(4)$, have a common upper rotational level of $J = 3$. The wavenumber difference of these lines is:

$$R(2) - P(4)$$

and this may be interpreted from the figure as the difference in wavenumber between the rotational levels $J'' = 4$ and $J'' = 2$. Hence, it follows:

$$R(2) - P(4) = F_{v''}(4) - F_{v''}(2) \tag{2.88}$$

where $F_{v''}(4)$ is the term value (see p. 70) for the $J'' = 4$ level. In general, for two such lines involving a common upper rotational level J, the difference of wavenumbers of the two lines would be:

$$R(J-1) - P(J+1) = F_{v''}(J+1) - F_{v''}(J-1) = \Delta_2 F''(J) \tag{2.89}$$

where $\Delta_2 F''(J)$ is the wavenumber difference of one of the lower $^1\Sigma$-state rotational energy levels from the second lower down. This equation could also be deduced from Equation (2.73) and (2.75) as follows. Since:

$$R(J) = \tilde{\nu}_0 + F_{v'}(J+1) - F_{v''}(J) \tag{2.90}$$

then

$$R(J-1) = \tilde{\nu}_0 + F_{v'}(J) - F_{v''}(J-1) \tag{2.91}$$

and similarly

$$P(J+1) = \tilde{\nu}_0 + F_{v'}(J) - F_{v''}(J+1) \tag{2.92}$$

On subtraction of Equation (2.89) from Equation (2.91) we obtain:

$$R(J-1) - P(J+1) = F_{v''}(J+1) - F_{v''}(J-1) \tag{2.93}$$

It may be readily shown that the difference between the wavenumbers of two lines in electronic transitions with a common lower rotational level (J) equals the separation of one of the upper electronic state rotational levels from the one next but one above it. This may be shown by means of the figure or by means of Equations (2.73) and (2.75) as follows:

$$\begin{aligned} R(J) - P(J) &= [\tilde{\nu}_0 + F_{v'}(J+1) - F_{v''}(J)] - [\tilde{\nu}_0 + F_{v'}(J-1) - F_{v''}(J)] \\ &= F_{v'}(J+1) - F_{v'}(J-1) = \Delta_2 F'(J) \end{aligned} \tag{2.94}$$

where $\Delta_2 F'(J)$ is the separation in wavenumbers between these two rotational terms of the same vibrational level of the upper electronic state.

On substitution for the appropriate rotational term values from the equation:

$$F_v(J) = B_v J(J+1) - D_v J^2(J+1)^2 \tag{2.95}$$

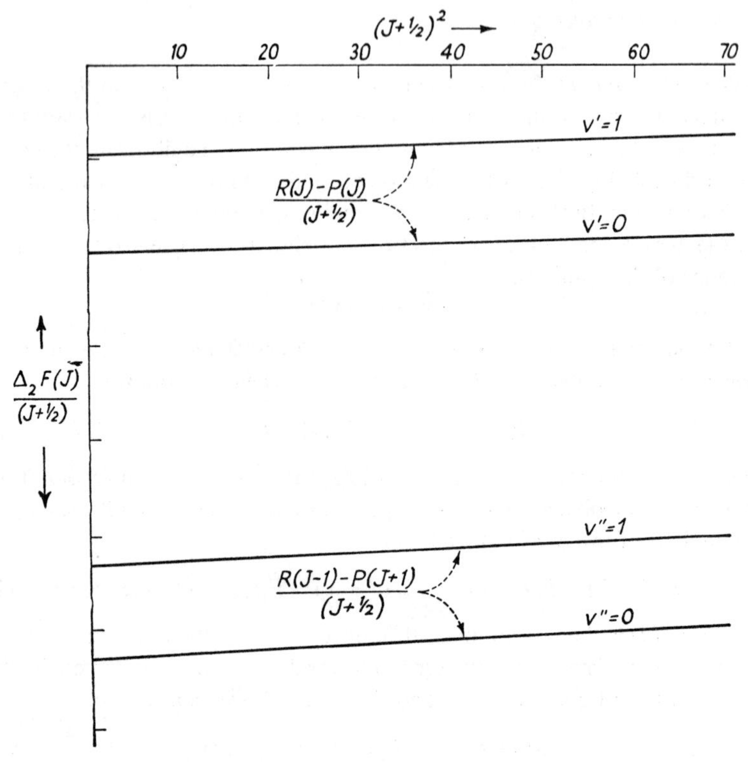

Fig. 2.30 Plot of $[R(J-1)-P(J+1)]/(J+\frac{1}{2})$ against $(J+\frac{1}{2})^2$ for the $v''=0$. $v''=1$ vibrational levels and the plot of $[R(J)-P(J)]/(J+\frac{1}{2})$ against $(J+\frac{1}{2})^2$ for the $v'=0$ and $v'=1$ levels. The data were obtained from the (0, 0) and (0, 1) bands of CuH for the $A^1\Sigma^+ \rightarrow X^1\Sigma^+$ transition [2.12].

into Equations (2.94) and (2.93) we obtain:

$$R(J)-P(J) = (4B_{v'}-6D_{v'})(J+\tfrac{1}{2})-8D_{v'}(J+\tfrac{1}{2})^3 \tag{2.96}$$

$$R(J-1)-P(J+1) = (4B_{v''}-6D_{v''})(J+\tfrac{1}{2})-8D_{v''}(J+\tfrac{1}{2})^3 \tag{2.97}$$

Equations (2.97) and (2.96) may be employed to determine the $B_{v'}$, $B_{v''}$, and $D_{v'}$ and $D_{v''}$ values. One of the graphical methods by which this is achieved is to plot $[R(J)-P(J)]/(J+\frac{1}{2})$ against $(J+\frac{1}{2})^2$ when a straight line should be obtained which has a slope of $-8D_{v'}$ and an intercept of $4B_{v'}-6D_{v'}$. In a similar way the plot of $[R(J-1)-P(J+1)]/(J+\frac{1}{2})$ against $(J+\frac{1}{2})^2$ gives a straight line of slope $-8D_{v''}$ and an intercept value of $(4B_{v''}-6D_{v''})$. Thus, in this way the B and D values may be obtained. In Table 2.7 the $R(J)-P(J)/(J+\frac{1}{2})$ and $[R(J-1)-P(J+1)]/(J+\frac{1}{2})$ data are given for the (0, 0) and (0, 1) bands of CuH at the wavelengths of 4280 Å and 4650 Å respectively, in the $A^1\Sigma^+ \rightarrow X^1\Sigma^+$ transition. The plot of $[R(J)-P(J)]/(J+\frac{1}{2})$ and $[R(J-1)-P(J+1)]/(J+\frac{1}{2})$ against $(J+\frac{1}{2})^2$ is given in Fig. 2.30. The rotational structure of the 4280 Å (0, 0) band is given in Fig. 2.29(a) for this $A^1\Sigma^+ \rightarrow X^1\Sigma^+$ transition.

Table 2.7 Part of the rotational data for the (0, 0) and (0, 1) bands of CuH for the $^1\Sigma^+ \rightarrow {}^1\Sigma^+$ transition

Band	*J*	*Wavenumber of observed lines*		$\Delta_2F(J)$ values			
		$P(J)$	$R(J)$	$R(J)-P(J)$	$\frac{R(J)-P(J)}{(J+\frac{1}{2})}$	$R(J-1)-P(J+1)$	$\frac{R(J-1)-P(J+1)}{(J+\frac{1}{2})}$
	0		23 324.68				
	1	23 295.47	23 335.95	40.48	26.986	46.89	31.260
	2	23 277.79	23 345.06	67.27	26.908	78.09	31.236
(0, 0)	3	23 257.86	23 352.18	94.32	26.948	109.11	31.176
head	4	23 235.95	23 357.04	121.09	26.908	140.24	31.164
at	5	23 211.94	23 359.70	147.76	26.865	171.19	31.124
4280 Å	6	23 185.85	23 360.26	174.41	26.832	201.99	31.075
	7	23 157.71	23 358.45	200.74	232.72	232.72	31.028
	8	23 127.54	23 354.64	227.10	26.717	263.06	30.94
	0		21 458.39				
	1	21 429.97	21 470.26	40.29	26.860	45.41	30.273
	2	21 412.98	21 480.55	67.57	27.028	75.60	30.240
(0, 1)	3	21 394.66	21 488.95	94.29	26.940	105.94	30.268
head	4	21 374.61	21 496.61	121.40	26.976	135.72	30.160
at	5	21 353.23	21 501.31	148.08	26.924	166.26	30.228
4650 Å	6	21 329.75	21 504.57	174.82	26.895	195.96	30.147
	7	21 305.35	21 506.52	201.17	26.824	225.38	30.057
	8	21 279.19	21 506.52	227.33	26.744	254.80	29.976

From the procedure outlined the values of $B_{v'}$, $B_{v''}$, $D_{v'}$, and $D_{v''}$ are obtained.. The correctness of the analysis may be checked by calculating the values of ω_e' and ω_e'' from the formula:

$$\omega_e = \sqrt{(4B_e^3/D_e)} \tag{2.98}$$

and comparing the values of ω_e with the ones obtained from the vibrational analysis (see p. 56).

For P-, Q- *and* R-*branches*. The combination relations considered so far have been for P- and R-branches. When a Q-branch is observed further combination relationships exist in addition to those already given involving the P- and R-branches, since a common upper or lower state rotational level may be common to three spectral lines. In an exactly similar way to which the combination relations for P- and R-branches were developed, it may be deduced from Equations (2.73), (2.74), and (2.75) that:

$$R(J) - Q(J) = F_{v'}(J+1) - F_{v'}(J) = \Delta_1 F'(J) \tag{2.99}$$

$$R(J) - Q(J+1) = F_{v''}(J+1) - F_{v''}(J) = \Delta_1 F''(J) \tag{2.100}$$

$$Q(J) - P(J+1) = F_{v''}(J+1) - F_{v''}(J) = \Delta_1 F''(J) \tag{2.101}$$

$$Q(J+1) - P(J+1) = F_{v'}(J+1) - F_{v'}(J) = \Delta_1 F'(J) \tag{2.102}$$

Since the right-hand sides of Equations (2.99) and (2.102) are equal it follows that:

$$R(J) - Q(J) = Q(J+1) - P(J+1) \tag{2.103}$$

and, in addition, from Equations (2.100) and (2.101) that:

$$R(J) - Q(J+1) = Q(J) - P(J+1) \tag{2.104}$$

If two bands have the same upper vibrational state, the combination differences (2.99) and (2.102) must be satisfied for each J value while for the same lower vibrational level the combination relationships (2.100) and (2.101) must be satisfied. A thorough test of the successful analysis of a band then follows, if the combination relationships (2.103) and (2.104) hold.

In a similar way as for $^1\Sigma$–$^1\Sigma$ transitions the equation:

$$F_v(J) = B_v(J+1) - D_v J^2(J+1)^2 \tag{2.105}$$

may be employed to substitute for the $F(J+1)$ and $F(J)$ terms in Equations (2.99) to (2.102) and the resulting equations may be used as before to determine $B_{v'}$, $B_{v''}$, $D_{v'}$, and $D_{v''}$ values.

Determination of B_e, D_e, and r_e

If the rotational analysis has been carried out on a number of bands in a band system, then the dependence of $B_{v'}$, $B_{v''}$, $D_{v'}$, and $D_{v''}$ on the value of the vibrational quantum number in the upper and lower electronic states may be followed. In general, their dependence on v is adequately represented by equations of the type:

$$B_v = B_e - \alpha_e(v + \tfrac{1}{2}) \quad (2.106)$$

$$D_v = D_e + \beta_e(v + \tfrac{1}{2}) \quad (2.107)$$

where B_e and D_e are the extrapolated values of B_v and D_v correspond to the hypothetical vibrationless state. From the plot of B_v against v the constant α_e may be determined, and from Equation (2.106) it then follows that the value of B_e is given by:

$$B_e = B_0 + \tfrac{1}{2}\alpha_e \quad (2.108)$$

where B_0 is the value of the rotational constant for the $v = 0$ vibrational level. Since

$$B_e = \frac{h}{8\pi^2 cI_e} \quad (1.109)$$

$$I_e = \frac{m_1 m_2}{(m_1 + m_2)} r_e^2 \quad (2.110)$$

then the internuclear distance (r_e) at the minimum of the potential energy curve may be determined from the analysis. The electronic spectrum method is unique in that it yields r and r_e values not only for the ground state but for excited electronic states as well. The values of r_e are usually quoted to within 0.01 to 0.001 Å. Such r_e values have been obtained for a very large number of diatomic molecules and radicals. In fact, some of the species recorded may be quite unfamiliar, e.g. AgO, AlH, AsN, BF, CaBr, CS, HS, SiH, VO, ZnH, and $(ZnH)^+$.

It is most important to note that the detailed analysis of the rotational structure of the electronic bands not only yields r_e values but if, in addition, it gives what appears to be a reasonable value of r_e, then this is proof that the correct molecular species was postulated.

Determination of band origin

In order to obtain accurate vibrational constants for a molecule it is necessary to determine the value of $\tilde{\nu}_0$, the wavenumber of the band origin. In fact, as was indicated when the vibrational analysis was considered, for the best analysis of the vibrational frequencies by means of the Déslandres table, it is desirable to employ band origin data and not band head.

The determination of $\tilde{\nu}_0$ may be accomplished by a combination relationship. For a band with only a P- and R-branch present, on employment of Equations (2.77) and (2.79) we obtain:

$$R(J-1) + P(J) = 2\tilde{\nu}_0 + 2(B_{v'} - B_{v''})J^2 \quad (2.111)$$

By plotting $R(J-1) + P(J)$, which is obtained from the fi e structure analysis, against J^2, a straight line is obtained whose intercept on the ordinate axis gives the value $2\tilde{\nu}_0$.

Allotment of J values to spectral lines

So far it has been assumed that the branches of a band may be picked out and that the J values have been allotted. If the fine structure of the band is completely resolved, and the zero line position (band origin) is apparent, then the allotment of J values to a particular band is readily carried out. An example of this type of band may be observed for the (0, 0) band of CN at 3883 Å of the $B^2\Sigma^+ – X^2\Sigma^+$ violet band system at low temperatures. Only P- and R- branches are present. The first line on the short wavelength side of the zero gap is the $J'' = 0$ line of the R-branch, and the first line on the longer wavelength side of the zero gap is the $J'' = 1$ line of the P-branch. Thus, the numbering of the P- and R-branches is readily achieved for a Σ type transition where the zero line position is apparent as a gap between the two branches.

The assignment of the m values and consequently J values of a particular series of lines in a branch may be checked as follows. The lines of a P- and R-branch of a band may be, at least approximately, represented by the formula:

$$\tilde{\nu} = c + dm + em^2 \qquad (2.112)$$

where $m = J + 1$ for an R-branch and $m = -J$ for a P-branch. It follows that:

$$\Delta\tilde{\nu}/\Delta m = d + 2em \qquad (2.113)$$

and

$$\Delta^2\tilde{\nu}/\Delta m^2 = 2e \qquad (2.114)$$

where e is a constant. Hence, the second difference between the wavenumbers of consecutive m values should have a constant value for all the members of the branch. This point is illustrated in Table 2.8 where the first and second difference between the wavenumbers of consecutive lines in each of the P- and R-branches for the (0, 0) band of the $^1\Sigma^+ – ^1\Sigma^+$ transition of CuH are given in columns 3 and 4, respectively. The second difference values are sufficiently constant for the assignment of the m values to be regarded as correct.

When, however, the fine structure of adjacent bands overlaps, it may become difficult to pick out the lines in the branches of a particular band. One of the procedures is then to attempt to allot the m values of a few of the lines which, from their wavenumber separation, look as though they are members of the same branch. The first and second differences for these lines are formed, and from these differences it becomes possible to calculate the anticipated wavenumbers of additional lines on each side of these lines. In this manner a number of lines may be allotted m values and designated to a particular branch. To check this m allocation in a particular branch the first difference of wavenumber values for consecutive m values is taken and plotted against m. From formula (2.113) it follows that these wavenumber differences for a particular P- or R-branch should lie on a straight line.† This type of plot may be employed to determine whether

† Since formula (2.112) does not normally now hold exactly for higher m values, slightly curved lines usually result.

Table 2.8 The first and second differences between consecutive lines in part of the P- and R-branches in the ${}^1\Sigma^+ - {}^1\Sigma^+$ (0, 0) band of CuH at 4280 Å

m	*Observed wave-number* (cm^{-1})	*First difference*	*Second difference*
P-branch			
−10	23 061.27		
		34.12	
−9	23 095.39		1.97
		32.15	
−8	23 127.54		1.98
		30.17	
−7	23 157.71		2.03
		28.14	
−6	23 185.85		2.05
		26.09	
−5	23 211.94		2.08
		24.01	
−4	23 235.95		1.90
		21.91	
−3	23 257.86		1.98
		19.93	
−2	23 277.79		2.25
		17.68	
−1	23 295.47		
R-branch			
1	23 324.68		
		11.27	
2	23 335.95		2.16
		9.11	
3	23 345.06		1.99
		7.12	
4	23 352.18		2.26
		4.86	
5	23 357.04		2.20
		2.66	
6	23 359.70		2.10
		0.56	
7	23 360.26		2.37
		−1.81	
8	23 358.45		2.00
		−3.81	
9	23 354.64		2.26
		−6.07	
10	23 348.57		2.24
		−8.31	
11	23 340.26		

the lines allocated to particular P- and R-branches belong solely to these two branches.† If this is the case, the points should group themselves along two lines of different slope. If any further branches are present, these should group themselves along other lines. In addition, if scattered points occur which cannot be identified with these P- and R-lines, these may be attributed to spectral lines in other bands. If a $^1\Pi-^1\Sigma$ transition is considered, the lowest rotational level in a Π-state is $J = 1$ and two rotational lines are missing, one at $\tilde{\nu}_0$ and another corresponding to the first line of the P-branch. In addition, for this type of transition a Q-branch is to be expected.

The numbering of the lines in the P-, Q-, and R-branches of the 2430 Å $C^2\Pi \rightarrow X^2\Sigma^+$ band of MgH is indicated in Fig. 2.31 where the Q-branch overlaps the R-branch and the P-branch is separated from these branches by a zero gap. When the Q-branch overlaps either the P- or R-branches, the lines of the Q-branch may be fairly readily distinguished by the spectroscopist from those of the other branch, since the spacing of the lines in the two branches is normally quite different.

A case where overlap of the P- and R-branches occurs is in the 4280 Å $A^1\Sigma^+ \rightarrow X^1\Sigma^+$ band of CuH which is given in Fig. 2.29(a) and in which no Q-branch is observed. From the appearance of the lines near the head of the band at 4280 Å and the gradually increased line spacing between 4283 and 4290 Å the branch responsible for the band head may be readily identified. Furthermore, the faint lines at about 4291 Å and 4295 Å, which interrupt the spacing pattern of the 4285, 4287, 4290, and 4297 Å lines, suggest the commencement of a new branch. In this manner the lines in the two branches may be separated, and the spectroscopist is then in a position to attempt a trial numbering of the lines in each of the branches. The correctness of the allocated J values may be checked by considering another band of the same electronic band system. If, for example, the same upper vibrational state is involved for both of these bands, and then if the J numbering is correct in both of these bands, the combination difference:

$$R(J) - P(J) = \Delta_2 F'(J)$$

must be in good agreement for every J value. If the two bands have the same lower vibrational energy state, the $\Delta_2 F''(J)$ values must be in good agreement for every J value. This approach provides a most sensitive means of detecting whether the correct J values have been allotted. In general, the J values first allotted will be incorrect, and in that case the numbering has to be continually adjusted until the $\Delta_2 F'(J)$ and $\Delta_2 F''(J)$ criteria are satisfied.‡

† The same procedure may be employed if there are P-, Q-, and R-branches present in the band. In this case three straight lines would be obtained.

‡ These $\Delta_2 F'(J)$ and $\Delta_2 F''(J'')$ criteria apply not only to bands with the same lower or upper vibrational level of one electronic band system but also to bands with a common lower or upper vibrational level in two different band systems. In fact, these criteria may be employed to check whether or not a particular electronic state is common to each of the band systems.

Determination of the nuclear spin quantum number from band spectra

For a diatomic molecule whose nuclei posses a nuclear spin quantum number I_1 and I_2, respectively, the total nuclear spin quantum number T of the molecule is given by:

$$T = I_1 + I_2,\ I_1 + I_2 - 1, \ldots, |I_2 - I_2|$$

When $I_1 = I_2 = \frac{1}{2}$, T has the values 1 and 0. The value of 1 corresponds to parallel nuclear spins and antisymmetric rotational levels, while that of 0 corresponds to antiparallel spins and symmetric levels. In the presence of a magnetic field each state with a given T value is split into $(2T + 1)$ levels of slightly different energy which are characterized by the quantum number (M_T) of the component of $\sqrt{[T(T + 1)]}\, h/2\pi$ in the direction of the applied field. In the absence of a magnetic field a $(2T + 1)$ degeneracy in T exists, and the total statistical weight for a given J value is $(2T + 1)(2J + 1)$.† When $T = 1$ the statistical weight for a particular J value is $3(2J + 1)$ while for $T = 0$ the statistical weight is $1(2J + 1)$. It follows, therefore, that the antisymmetric levels are three times more frequent than the symmetric.

Since, according to the selection rule, transitions are only possible between two symmetric or two antisymmetric states‡ and the latter has a statistical weight three times that of the former, it follows that the intensities of alternate rotational lines in a band of the electronic spectrum will be in the ratio of 3 : 1. The rotational lines in the band spectrum of $^1H_2$ have been observed to have just this 3 : 1 intensity alternation, the lines with odd J values being more intense than those with even J values. It may be concluded, therefore, that the spin of the proton is $\frac{1}{2}$.

In the general case the intensity ratios of adjacent strong to weak lines of a homonuclear molecule, where each nucleus has a nuclear spin quantum number I, is given by the ratio:

$$(I + 1)/I$$

When $I = 0$, as for $^{16}O_2$, $(I + 1)/I = \infty$ and alternate rotational lines are absent, and when I is large, $(I + 1)/I$ tends to unity.

By photometric measurements of the relative intensities of a rotational line in a band of an electronic transition with respect to an adjacent rotational line in the same branch the alternation intensity may be obtained§ and the nuclear

† The $(2J + 1)$ factor arises since each rotational level is $(2J + 1)$-fold degenerate in the absence of a magnetic field.

‡ Owing to a small interaction between the magnetic moment of a spinning nucleus and the remainder of the molecule a small probability of a transition. between symmetric and antisymmetric rotational levels does exist.

§ Allowance must be made for the variation in the Boltzmann factor with increasing J.

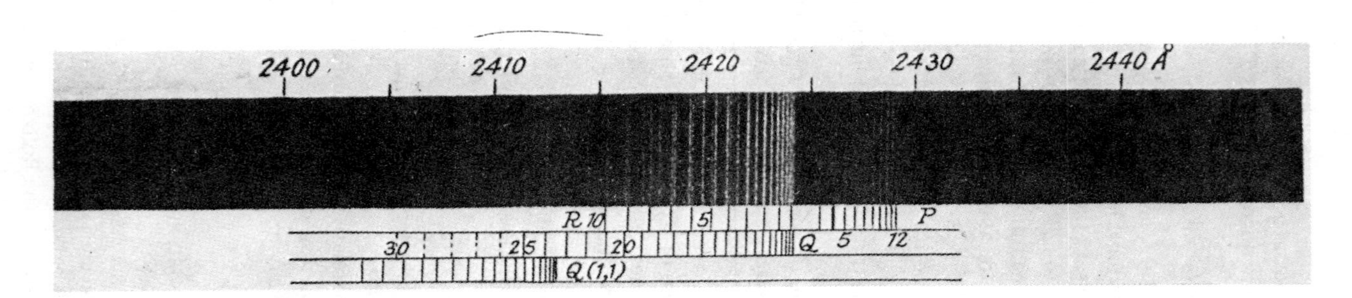

Fig. 2.31 The 2430 Å $C^2\Pi - X^2\Sigma^-$ (0, 0) band of MgH taken by a large quartz spectrograph. This emission band was obtained from a magnesium arc in hydrogen (after Gaydon [2.10]).

Fig. 2.32 Part of the P- and R- branches in the 3914 Å (0, 0) band in the $B^2\Sigma_g^-$ transition of the N_2^+ negative system where alternating intensities may be observed in consecutive rotational lines in each branch (after Jevons [2.12] courtesy of W.H.J. Childs).

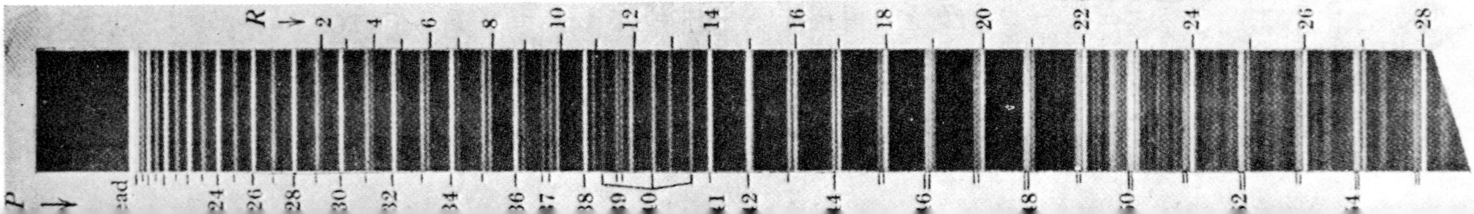

spin quantum number determined. For example, in Fig. 2.32 in the P-branch of the (0, 0) band in the electronic transition $B^2\Sigma_u^+ \rightarrow X^2\Sigma_g^+$ of the N_2^+ negative system the intensity ratio is 2:1. Hence, it follows that the nuclear spin quantum number of ^{14}N is 1. In Table 2.9 the observed intensity alternation was used to calculate the nuclear spin quantum numbers of homonuclear molecules.

Table 2.9 Values of nuclear spin quantum numbers derived from observed intensity alternations

Molecule	*Observed intensity alternation*	I
$^1H_2$	3:1	$\frac{1}{2}$
$^7Li_2$	5:3	$\frac{3}{2}$
$^{12}C_2$	∞	0
$^{14}N_2$	2:1	1
$^{16}O_2$	∞	0
CH≡CH	3:1	$\frac{1}{2}$
$^{19}F_2$	3:1	$\frac{1}{2}$

The acetylene molecule has been examined and a rotational intensity alternation of 3:1 has been obtained in the near infrared rotation–vibration bands. This ratio of 3:1 gives a value of $I = \frac{1}{2}$ which must be due to the proton, since ^{12}C has no nuclear spin. For a homonuclear molecule such as $^{16}O_2$, where every alternate rotational line is missing, it is certain proof that the nuclear spin quantum number is zero.

2.3.4 Effect of isotopes on the spectra of diatomic molecules

The isotope effect has already been considered several times in various types of spectra. The use of isotopes is, in fact, a general procedure in microwave and infrared spectra for the determination of internuclear distances, and bond angles in certain simple polyatomic molecules. The technique is also employed in Raman spectra for the determination of internuclear distances. The isotope effect, which is reflected in the spacing of the lines, in a band is dependent on the inverse value of the moment of inertia, and for a diatomic molecule on:

$$(m_A + m_B)/m_A m_B r_{AB}^2$$

where m_A and m_B are the masses of the two atoms, and r_{AB} is their internuclear distance. The spacing of the lines in a band is thus dependent on the value of:

$$(m_A + m_B)/m_A m_B$$

and this is very dependent on the relative magnitudes of the masses of the atoms A and B. To illustrate this the following cases will be considered:

(a) When $m_A = 1$ and $m_B = 200$, then $(m_A + m_B)/m_A m_B = 201/200$, and if m_A is now altered to the value of 2, then $(m_A + m_B)/m_A m_B = 202/400$.

(b) When $m_A = 200$ and $m_B = 202$, then $(m_A + m_B)/m_A m_B = 402/40\,400$, and if m_A is now altered to 201, then $(m_A + m_B)/m_A m_B = 401/40\,602$.

In case (a), where a light atom is joined to a heavy one, when the the isotopic mass of the light atom was increased by the 1 value of $(m_A + m_B)/m_A m_B$ was almost halved, whereas in (b) there was only a very small change. In case (a) the isotopic effect would be readily observed, and for a given $v'-v''$ transition of the two isotopic molecules the band would be readily separated even at only moder-- ate dispersion. However, in case (b) the separation would be difficult to achieve even at the highest resolution. In general, when the ratio of m_B/m_A differs appreciably from unity, then the better the chance of separating the lines in a given band due to the different isotope. Of course, in the examples taken the problem has been simplified, no account having been taken of the r_{AB}^2 term, the value of which may differ appreciably if an electronic transition is being considered. However, for the infrared and Raman rotation–vibration bands the r_{AB}^2 term is less important, while for infrared, Raman, and microwave rotational lines the change in the value of r_{AB} is negligible.

As an example of the isotope effect in an electronic transition the spectrum of ^{10}BO and ^{11}BO may be seen in Fig. 2.33(a) and (b). This spectrogram is for β-bands in the $B^2\Sigma^+ \rightarrow X^2\Sigma^+$ electronic transition of ^{10}BO and ^{11}BO. Two sets of isotopic bands for the $1 \rightarrow 3$, $0 \rightarrow 2$, and $2 \rightarrow 3$ transitions are readily discernible. In Fig. 2.33(b) the $2 \rightarrow 6$ band is given at a much greater dispersion than that employed in Fig. 2.33(a), and the rotational structure of the two isotopic bands has been completely separated. In addition, the line intensities within this band are given in Fig. 2.33(c) in the microphotometer trace. Thus, in the case of BO the bands due to both the boron isotopes were fairly readily separated and were examined apart.

The introduction of an isotope into a molecule leaves essentially unchanged all the properties of the molecule associated with the electronic structure. In fact, the force constant of the bond is unchanged, and the same potential energy curve may be used to represent the energies of both isotopic molecules. The substitution, however, does affect those properties of the molecule which depend on the reduced mass and which are: (a) the internuclear distances in the various vibrational levels; (b) the moment of inertia. The change referred to in (a) can be readily appreciated by reference to $H^{35}Cl$ and $D^{35}Cl$. The heavier molecule DCl, for a given value of v, will have a lower vibrational frequency than the lighter molecule HCl. The dependence of the equilibrium vibrational frequency ω_e on the mass of the two atoms is given by the equation:

$$\omega_e c = \frac{1}{2\pi}\sqrt{k\,\frac{(m_1 + m_2)}{m_1 m_2}} \qquad (2.115)$$

where k is the force constant and has the same value for both isotopes. The dependence of the zero-point energy on ω_e is given by Equation (2.65) when

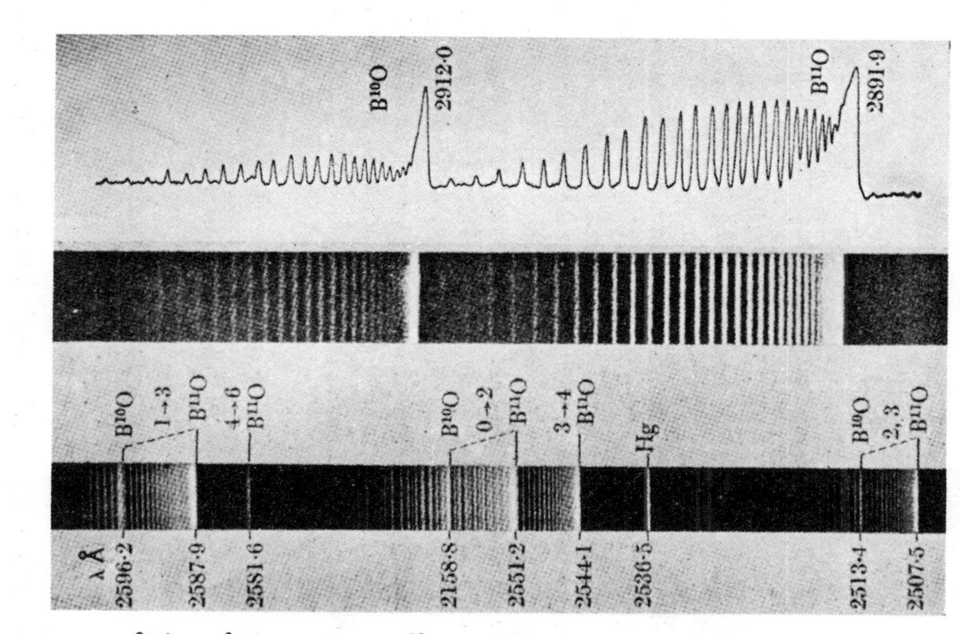

Fig. 2.33 The β-bands for the $B^2\Sigma^+ \rightarrow X^2\Sigma^+$ transition in ^{10}BO and ^{11}BO in active nitrogen. In spectrogram (a) may be observed the $1 \rightarrow 3$, $0 \rightarrow 2$, $2 \rightarrow 3$ bands for both species. In (b) the $2 \rightarrow 6$ band is given at much greater dispersion while (c) gives the microphotometer trace of the same band. (After Jevons [2.12] courtesy of: (a) R.S. Mulliken, (b) A. Elliott).

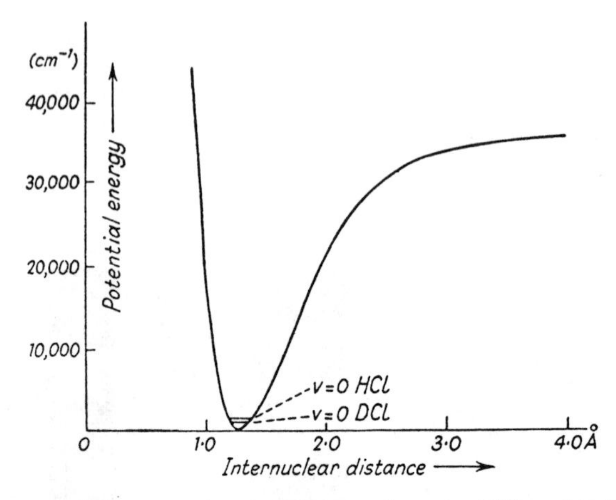

Fig. 2.34 The zero-point energies and the r_0 values for HCl and DCl.

$v = 0$, and if the x_e and y_e terms in this equation are neglected, the zero-point energy is $\frac{1}{2}hc\omega_e$, and from this and Equation (2.115) it follows that DCl will have a smaller zero-point energy than HCl. Thus, the internuclear separations in the ground vibrational levels are different, but the internuclear distance r_e, however, which corresponds to the hypothetical state of zero vibrational energy, is the same for both isotopic forms of the molecule. The dependence of the zero-point energy on the isotopic mass in HCl and DCl is represented in Fig. 2.34.

The influence of the isotope effect on the rotation–vibration spectrum of $H^{35}Cl$ and $H^{37}Cl$ may be seen in a paper by Meyer and Levin [2.13].

The isotopic mass difference does not affect the $\tilde{\nu}_e$ value between two electronic states, and if it is assumed that the vibration is simple harmonic, then for band origin wavenumbers:

$$\tilde{\nu} = \tilde{\nu}_e + \omega_e'(v' + \tfrac{1}{2}) - \omega_e''(v'' + \tfrac{1}{2}) \tag{2.116}$$

For the isotopic molecule, since the force constants are identical, then:

$$_i\omega_e = \rho\omega_e \tag{2.117}$$

where

$$\rho = (\mu/_i\mu)^{\frac{1}{2}}$$

and

$$_i\tilde{\nu} = \tilde{\nu}_e + \rho\omega_e'(v' + \tfrac{1}{2}) - \rho\omega_e''(v'' + \tfrac{1}{2}) \tag{2.118}$$

$$\Delta_i\tilde{\nu} = \tilde{\nu} - {_i\tilde{\nu}} = (v' + \tfrac{1}{2})(\omega_e' - \rho\omega_e') - (v'' + \tfrac{1}{2})(\omega_e'' - \rho\omega_e'') \tag{2.119}$$

$$\Delta_i\tilde{\nu} = (v' + \tfrac{1}{2})(1 - \rho)\omega_e' - (v'' + \tfrac{1}{2})(1 - \rho)\omega_e'' \tag{2.120}$$

$$\Delta_i\tilde{\nu} = (1 - \rho)[(v' + \tfrac{1}{2})\omega_e' - (v'' + \tfrac{1}{2})\omega_e''] \tag{2.121}$$

Thus, the isotopic shift is greater the more the value of $(1 - \rho)$ deviates from unity. The rare oxygen isotope of mass 18 was the first to be discovered from electronic spectra study. In the solar spectrum the oxygen molecule ($^{16}O^{16}O$) has an intense (0, 0) band at 7596 Å containing thirteen lines and is called the A-band. Near this occurs a very weak band called the A′ band which contains twenty-six lines. Both the A and A′ bands appear to have the same structure, but the A′ band cannot be fitted into the Déslandres table of the $^{16}O^{16}O$ bands, and hence this very weak band could not be attributed to $^{16}O^{16}O$. The explanation of this very weak band was provided by Giauque and Johnston [2.14] in 1929 who attributed the A band to the $^{16}O^{16}O$ molecule and the weaker A′ band to the $^{16}O^{18}O$ molecule. According to Equation (2.121) for the (0, 0) band, the isotope shift of the origin of the band should be:

$$\Delta_i\tilde{\nu} = \tfrac{1}{2}(1-\rho)(\omega_e' - \omega_e'') \qquad (2.122)$$

For the $^{16}O^{16}O$ molecule the ω_e' and ω_e'' values have been accurately determined as 1432.69 and 1580.36 cm^{-1} respectively, and therefore:

$$(\omega_e' - \omega_e'') = 1432.69 - 1580.36 = -147.67\ \mathrm{cm}^{-1}$$

On substitution for this value into Equation (2.122), and on the determination of ρ from ${}_i\mu$ for the $^{16}O^{18}O$, and μ for the $^{16}O^{16}O$ molecule, then a $\Delta_i\nu$ value of -2.1 cm^{-1} is obtained. The experimentally observed band origin shift is $\Delta_i\nu = -2.067$ cm^{-1} which is in good agreement with the over-simplified theoretical considerations. The negative sign implies that the shift due to the ^{18}O isotope has taken place towards shorter wavelengths. This close agreement was taken as proof of the existence of the ^{18}O isotope. A third band very close to the A′ band, whose intensity was five times weaker than that of the A′ band, was found shortly after the discovery of the A′ band, and this, the A″ band, was shown to be due to the $^{16}O^{17}O$ molecule, and the existence of ^{17}O was thus established. The fact that for the $^{16}O^{18}O$ molecule the A′ band contains twice as many rotational lines as the A band is due to the unsymmetrical nature of the $^{16}O^{18}O$ molecule. Owing to the symmetry of the $^{16}O^{16}O$ molecule, certain transitions are forbidden by the selection rules resulting in every other rotational line being missing. In the $^{16}O^{18}O$ molecule as in $^{16}O^{17}O$ because of the asymmetry no such restrictions are imposed by the selection rules, and double the number of rotational lines is observed. This is dealt with more fully on p. 20.

2.3.5 Further applications of isotope effect

For identification purposes

From a vibrational analysis it is occasionally difficult to be certain of the diatomic species emitting a particular band system. On comparison of the experimentally determined value of ρ with that of the calculated one it is occasionally possible to confirm or reject the identity of a proposed emitter. One such case was the identification of the emitter of some bands produced in a discharge containing BCl_3 in active nitrogen. The emitter was first thought to be BN; however, the experimentally determined value of ρ taking into account the other boron isotopic species present was 1.0291, whereas the calculated value of ρ for BN was 1.0276. This value of ρ is, in fact, in excellent agreement with that to be anticipated for ^{10}BO and ^{11}BO the calculated value of which is 1.0292. Further investigation showed that the presence of a trace of oxygen was essential in this discharge for the production of these bands. Hence, there was no doubt that BO and not BN was the emitter.

The band spectra of a diatomic species enable a distinction to be made between certain isotopic species which could not otherwise be achieved with absolute certainty even by means of a mass spectrograph. For example, in the case of silicon, whose main isotope is ^{28}Si, a mass spectrograph could not distinguish between ^{30}Si and $^{28}SiH_2$ or ^{29}SiH produced from a volatile silicon hydride. Mulliken, however, from a study of the band spectrum of SiN definitely proved the presence of the ^{30}Si isotope.

Abundance ratio

The bands in the spectra from different isotopic species differ not only in their relative displacements but also in their relative intensities. The intensity ratio of the corresponding $v' \leftarrow 0$ bands in an absorption spectrum of isotopic molecules gives the abundance ratio of the molecules. For example, in the electronic absorption spectrum of Br_2 the intensity ratio of the isotopic species:

$$^{79}Br{-}^{79}Br,\ ^{79}Br{-}^{81}Br,\ ^{81}Br{-}^{81}Br \text{ is approximately } 1:2:1.$$

It follows, therefore, that the two isotopes of bromine are almost equally abundant, and, in fact, this is borne out by the atomic weight of bromine which is 79.916.

2.4 ELECTRONIC SPECTRA OF DIATOMIC SPECIES IN FLAMES

2.4.1 General considerations

One of the important areas for the study of both electronic absorption and emission spectra is in flames. Diatomic species play a particularly important role in flames; for example, C_2, CH, and OH may be identified by both their absorption and emission electronic spectra. The fundamental aim in this type of work is to determine which species are reacting together in the combustion process, and spectroscopy helps to identify and to estimate quantitatively the intermediate species and compounds involved and to formulate the reactions taking place.

When combustion occurs electromagnetic radiation is emitted. Analysis of the band spectra of flames has been responsible for the identification of the presence of the radicals such as CH and OH in normal combustion processes, and flash photolysis studies have been made on the radicals present in $H_2 + O_2$ and

$C_2H_2 + O_2$ flames. From the electromagnetic radiation emitted it may be possible to examine either (a) the visible and ultraviolet emission spectra, or (b) the infrared emission spectra, and although both types of emissions have been studied, the greater proportion of the work has been done in the visible and ultraviolet regions. Most of the infrared work appears to have been of an applied nature, e.g. to find out why gases behind a flame front still glow after the flame has passed by. The treatment given here will be confined to visible and ultraviolet emission, that is to electronic transitions in the molecules and radicals present in flames. In fact, the object will not be to understand the processes occurring in different types of flames, on which so much work has been directed, but only the more direct spectroscopic information which emerges.

One example which may be considered to illustrate the procedure of the identification of the emitter in a flame is the work on the band spectrum of stannous oxide. These bands were first observed in the oxy-coal-gas flame spectra of tin sulphide and tin chloride, and then later by Mahanti [2.15] in the arc spectrum of metallic tin on carbon electrodes. The feasible emitters of these bands were considered to be SnH, Sn_2, or SnO. Analysis of the bands showed that SnH could not be the emitter. In addition, Mahanti considered that it was improbable that Sn_2 was the emitter since no alternation of intensity in the lines of a given branch in a band could be observed even at high dispersion (Fig. 2.35).

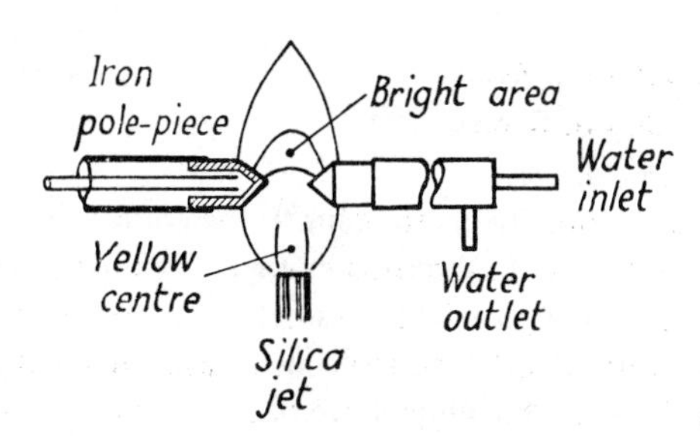

Fig. 2.35 Apparatus for the study of the emission spectrum of a flame with an arc across it. (After Connelly [2.16] courtesy of the Council of the Physical Society).

Such alternations are to be expected for a symmetrical diatomic emitter (see p. 20). This work was extended by Connelly [2.16] who employed the type of arc illustrated in Fig. 2.35.

In practice the arc was enclosed in a chamber and could be maintained in gases other than air. An arc was struck across the flame by means of an iron cup electrode which contained tin. The other electrode was a hollow copper rod which was cooled right up to the tip by flowing water. When the arc was struck the tin became molten, and when air was used in the arc chamber the characteristic bands were readily obtained. These bands were also obtained by a hydrogen flame burning in a silica jet as in Fig. 2.35. The hydrogen was taken from a cylinder, then passed through a Woulffe's bottle containing anhydrous stannic chloride. A picture of the flame is given in Fig. 2.36. The flame was from hydrogen containing stannic chloride vapour while the electrodes had an oxide impurity.

The arc was fed by a transformer applying about 0.2 A at 5000 V. The wavelengths of the bands from the SnO spectra were arranged in a Déslandres table, where the v' values ranged from 0 to 8 and those of v'' from 0 to 9; the most intense bands lay on a Condon parabola. Connelly obtained the values of x_e'' and ω_e'' and the dissociation energy of the molecule in the ground state from the analysis, as well as x_e and ω_e values for one of the excited electronic states.

Mahanti had claimed three band systems for the SnO in emission, known as A, B, and C systems. Some doubt remained as to which of these systems involved a transition to the ground state of the molecule. This was settled by the work of Connelly when he took an absorption spectrum of the SnO in a flame by directing through it a continuous radiation obtained from burning magnesium. The absorption spectrum of SnO is given in Fig. 2.37(b) and is to be compared with the emission bands in Fig. 2.37(a). The bands in Fig. 2.37(a) correspond in wavelength to some of those in Fig. 2.37(b), and these were the bands which Mahanti had attributed to band system A. The presence of only the A system in absorption indicates that the lower electronic state in this system must be the ground state.

2.4.2 Information derived from emission studies of flames

The work on flames not only helps to identify the emitter but yields information on the vibrational constants, dissociation energies, and the electronic states involved in the transition. Many such studies have been made on different types of flames, and a number of radicals and molecules have been identified. Some of the radicals such as BrO and IO which have been recognized by means of the band spectra emitted by flames are not readily detectable from other sources. The BrO and IO radicals were identified by means of the band system produced from methyl bromide and methyl iodide flames, respectively.

Various types of flames have been studied. Organic compounds such as hydrocarbons and ethers when burning in air produce similar spectra. With the normal bunsen type of flame the inner cone from such gaseous mixtures yields electronic band spectra from which C_2, CH, and OH may be identified. The outer cone of the flame consists mainly of OH and CO emission. Quite a number of different types of inorganic flames have been examined, and one case was the spectra from an S_2 flame where emission bands of SO have been detected. The equation suggested to account for the SO bands is:

$$S_2 + O_2 \rightarrow S_2O_2 \rightarrow 2SO$$

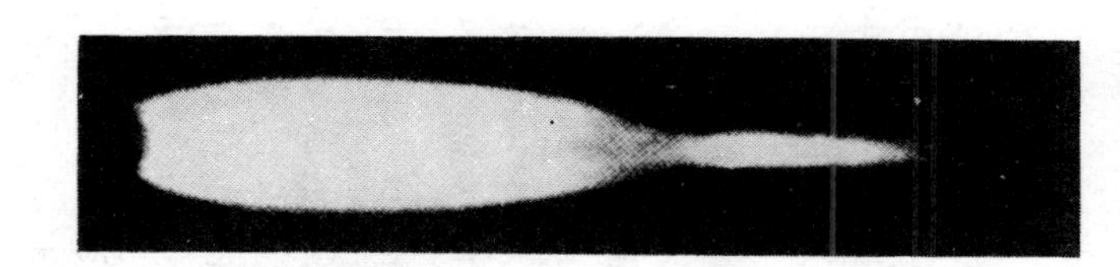

Fig. 2.36 The flame of hydrogen containing stannic chloride vapour. Iron and copper electrodes were employed containing oxide impurity. The inner region of the flame was yellow and the outer mantle a bright blue. (After Connelly [2.16], courtesy of the Council of the Physical Society).

Fig. 2.37 (a) Emission spectrum of SnO obtained from a flame (b) The corresponding absorption spectrum obtained by directing continuous radiation through the same flame. The reference spectrum at the head of the figure is that of the iron arc. (After Connelly [2.16], courtesy of the Council of the Physical Society).

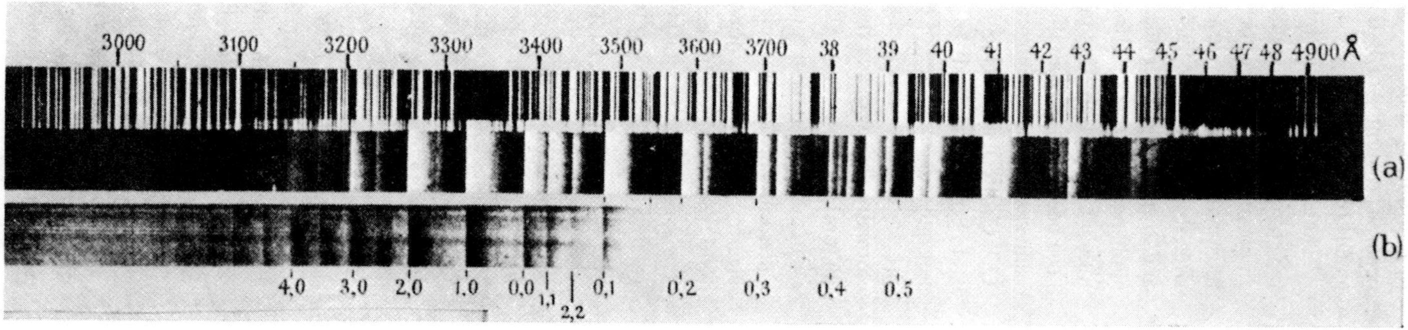

In addition, S_2 emission bands are observed, and SO_2 absorption bands also occur. The burning of H_2S produces all these features and, besides this, shows an OH band.

There is no new principle in the procedure of spectrographically recording the emitted electronic radiation; in fact, the same medium quartz spectrograph employed for studies of certain electronic transitions (see p. 36) is suitable for most purposes. Since, however, flames are usually relatively weak emitters, a wide slit is employed. Such studies may reveal the v' and v'' values in the upper and lower electronic levels, and a vibrational analysis may be performed in a similar way to that described on p. 56. An equation may be derived for the wave-number of the transition in terms of v' and v'' and the x_e and ω_e values which have been derived from the analysis. A few of the radicals and molecules whose band systems have been studied in electronic emission include CN, CO, NH, NO, CCl, CS, FeO, and NH_2.

In general, the study of the electronic emission spectra during combustion not only suggests some of the chemical reactions occurring during the oxidation process but also identifies some of the intermediate radicals. Band spectra are normally observed, and the most abundant information has been obtained on diatomic molecules and radicals.

In addition to band spectra, most flames emit a certain amount of continuous radiation which may result from the recombination of positive ions with free electrons. Sometimes, however, a strong continuum is obtained; this is particularly true of flames containing halogens and of halogens reacting with hydrogen. The cause of such a continuum has been ascribed to the reaction of normal and excited halogen atoms to form the halogen molecule, e.g. in the case of bromine:

$$Br(^2P_{1/2}) + Br(^2P_{3/2}) \rightarrow Br_2 + \text{continuous radiation}$$

A large fraction of the radiation emitted by flames may by attributed to *chemiluminescence*. In such a process the chemiluminescence results from the formation of a radical in an excited electronic state by means of a chemical reaction. One case which has been explained this way is the intense emission from the hydroxyl radical in hydrocarbon flames where the reaction may be:

$$CH + O_2 \rightarrow CO + OH^*$$

Line spectra are only detected from impurities in the flame (e.g. Na) since the atoms which could be present in flames, such as H, O, and C, require a greater excitation energy than can normally be supplied by means of a flame; an exception to this is the 2478 Å line of carbon. The exact cause of the electronic excitation of atoms and molecules in flames is as yet not certain. It is considered that although flames contain free electrons—these being the means of achieving excitation in discharges—their number and kinetic energy is insufficient to account for the intensity of the observed emission spectra. The excited electronic state in the emitter could be produced by a collision process involving some particle (atom, molecule, or ion) which itself possesses either electronic excitation energy or is in an excited vibrational state. A further possibility is that the emitter may become activated by being the third body present in combination of free atoms or radicals. Such low-energy processes can only produce excitation from the ground state to low-lying electronic levels in the radical or molecule which is to become the emitter. In fact, for all the electronic flame emission spectra observed the transition has always been from a low-lying excited state to the ground state itself. In molecules such as N_2, where there is no electronic state close to the ground one, no such emission spectrum is observed in flames.

2.4.3 Spectroscopic determination of the temperature of a flame

Generally, the rotational fine structure of the bands has not been examined. One of the reasons for this is that the intensity of the spectra is sometimes too weak for a diffraction grating to be used. In some cases the radicals present in flames have been studied exhaustively by other means. In a few cases the rotational fine structure has been studied with a view to determining what is known as the effective rotational temperature. Normal methods of temperature measurement are often inadequate for assessing the thermal energy present in flames. A more satisfactory means is to consider the energy of the molecules as being composed of rotational, vibrational, and translational contributions, and so obtain a corresponding rotational, vibrational, and translational temperature for the molecules. The rotational temperature is obtained by resolving the rotational fine structure, determining the rotational constant, and measuring the intensities of the lines in a particular band. The intensity of a single line of the rotational fine structure in a band is given by:

$$I = C_{em} P \tilde{\nu}^4 \exp\left(-E_r/kT\right) \qquad (2.123)$$

where I is the intensity of the line; C_{em} is a constant and is the same for all the rotational lines in the band; P is a transition probability and may be evaluated; $\tilde{\nu}$ is the wavenumber of the line; k is the Boltzmann constant; E_r is the value of the rotational energy for the upper state and is obtained from the rotational analysis of the band spectrum; T is the effective rotational absolute temperature.

A plot is made of $(\log I - \log P\tilde{\nu}^4)$ against E_r, and if a straight line is obtained its slope is $-1/2.303\ kT$, from which T, the effective rotational temperature, may be evaluated. This was done for the radical OH in hydrocarbon flames at atmospheric pressure which produces well-spaced lines in each band, and the effective rotational temperature in the flames studied was about 5700 C. The effective rotational temperature of CH was also evaluated from the rotational structure of one of its bands, and this was found to have a rotational temperature similar to that of the expected flame temperature.

By a similar type of study of the intensities of v' or v'' progressions an effective vibrational temperature may be determined; the value of the effective rotational temperature only coincides with this when there is thermodynamic equilibrium. This is also true of the effective translational temperature which is determined by measuring the Doppler broadening of spectrum lines. This broadening results from the random thermal motion of the hot emitter. When the molecule is moving towards the spectrograph, the spectrum line is displaced very slightly towards the violet, whereas when moving away the displacement is towards the red. Thus, when the process is considered for a number of molecules moving in various directions, the resultant effect is that the line is broadened. The line breadth due to the Doppler broadening varies as the square root of the temperature.

For a detailed account of combustion, the various types of flames, their theory and electronic spectra, and their infrared emission and flame temperature, books by Gaydon [2.17, 2.20, 2.21] should be consulted; there is also a review [2.18] article on the visible and ultraviolet emission spectra of flames by the same author, and a most detailed book by Gaydon and Wolfhard [2.19]. The various Combustion Symposia should also be examined.

2.4.4 Some more recent studies in the spectroscopy of flames

Gaydon [2.19, 2.20] has recently reviewed the experimental techniques employed in the absorption spectroscopy of flames. In addition, he has discussed the absorption spectra in the following types: (a) pre-mixed flames (C_2, CH, OH, etc.); (b) diffusion flames where a pyrolysis continuum precedes soot formation; (c) cool flames from aldehydes.

Gaydon indicates the considerable difficulties in identifying and quantitatively estimating the intermediate radicals and compounds in combustion reactions by studying the absorption spectra of flames.

One study on pre-mixed flames was made by Bulewicz, Padley, and Smith [2.22] who examined a low-pressure acetylene flame at various mixture strengths. Through absorption and emission studies they identified C_2, CH, and OH and formulated the reaction:

$$C_2(X^3\Pi_u) + OH(X^2\Pi) \rightarrow CH(A^2\Delta) + CO(X^1\Sigma^+)$$

where the CH radicals are produced in an excited electronic state.

Another interesting study on pre-mixed flames is described by Jessen and Gaydon [2.23, 2.24] for the luminous mantle above rather rich oxyacetylene flames and they detected C_3. In addition they found C_2 and CH in absorption and estimated their concentrations for various mixture strengths and at various positions in the flame; they showed that in the reaction zone the C_2 and CH concentrations were much above the equilibrium values whereas above the reaction zone they corresponded to them.

The study of transient species in explosive mixtures has been made by flash photolysis where the flash ignition is brought about by an intense source of light.

For example, in oxyacetylene explosions, sensitized with NO_2, after the photochemical decomposition the absorption bands of C_2, CH, OH, and C_3 are detected. Gaydon and Hurle [2.25] describe rather similar studies of ignition and pyrolysis employing a bursting-diaphragm shock-tube and flash-tube background.

2.5 MORE RECENT STUDIES OF ELECTRONIC SPECTRA

2.5.1 Introduction

The types of studies feasible in the electronic spectra of diatomic species were well developed even a quarter of a century ago. A large number of diatomic molecules, free radicals, and ions had been examined in both absorption and emission, and most precise information had been gained on vibrational frequency, force constants, interaction constants, dissociation energies, and internuclear distance for both the ground and excited electronic states. An adequate variety of absorption and emission techniques was available and the resolution of the spectrographs was excellent. Special methods for examining transient species, such as flash photolysis, the matrix technique, and mild discharges were beginning to develop. Thus even a quarter of a century ago the scope and nature of electronic spectra as applied to diatomic species seemed well laid down and many such spectroscopists turned their attention to more challenging areas such as evaluating the parameters for triatomic, tetratomic, and larger species in an attempt to appreciate fully their geometry and valence in both the ground and excited electronic states. Two of the most important applications are considered elsewhere under the headings of Astrochemistry (Chapter 6) and Electronic Spectra of Diatomic Species in Flames (p. 92–99). In the latter case one of the fundamental aims is to determine which species are reacting together in the combustion process; and spectroscopy helps to identify and to estimate quantitatively the intermediate species and compounds involved, and attempts to formulate the reactions taking place. Such studies have an interesting theoretical base and possible industrial applications. In astrochemistry the electronic spectra of diatomic species continue to be invaluable in formulating theories on various planets, stars, interstellar matter, and the nature of the Universe. Diatomic species which have been identified in such studies include CH, CN, and H_2.

Two areas in which electronic spectra are still yielding valuable and topical information will now be considered. One of the most recent books is one by Dunford [2.26].

2.5.2 Matrix studies

One development of matrix studies of trapped molecules is to mix the species to be examined with at least a hundred-fold excess on an inert gas dilutent and then to freeze and condense the mixture at a low temperature (e.g. 20 K). The sample can be then irradiated and free radicals may be produced, e.g. $A{-}B \rightarrow A\cdot + B\cdot$.

If either A· can diffuse away from B· or alternatively if there is an appreciable activation energy for the recombination process, then the free radicals A· and B· may be stabilized in sufficient concentrations for their absorption spectra to be detected. Such electronic spectra as compared with the data for a gas usually have (a) shifted band origins of their electronic transitions, (b) similar vibrational spacings, (c) much sharper bands, and (d) the difference bands do not appear.

The matrix technique is thus useful in formulating the vibrational assignment of stable molecules and identifying unstable species. Several species have been identified by this approach [2.21, 2.22, 2.33] and include CH, OH, OD, and C_2^-.

A most useful review on matrix spectra has been made by Milligan and Jacox [2.27].

2.5.3 Spectra of diatomic ions

Even a quarter of a century ago quite a number of diatomic ions had been identified by means of their electronic emission spectra [2.31], for example, N_2^+, CO^+, O_2^+, Cl_2^+, BeH^+, CH^+, OH^+, HCl^+, and HBr^+. Considerable interest has centred around the study of diatomic ions since they lead to a fuller appreciation of electronic structure and the bonding in the corresponding neutral molecules. More recent studies have been made on the following diatomic ions: H_2^+, NH^+, SiH^+, PH^+, SH^+, CN^+, NO^+, F_2^+, CS^+, NS^+, PO^+, PF^{2+}, P_2^+, PS^+, AsO^+, N_2^{2+}, and C_2^-. One study of particular interest was made by Douglas and Lutz [2.32] who examined the spectrum of SiH^+ in the visible region. The levels of SiH^+ and CH^+ are very similar and it will be noted that both these species have ground electronic states of the type $^1\Sigma^+$ involving $\sigma^2\sigma^2$ electrons while the excited electronic state is of the $^1\Pi$ type involving $\sigma^2\sigma\Pi$ electrons. The electronic states are in line with what is to be expected from the electron configuration and states of the corresponding neutral molecules SiH and CH. One important subsequent development was that Grevesse and Sauval [2.33] were then able to identify several SiH^+ lines in the solar spectrum and this was the first time a molecular ion had been identified in a stellar atmosphere.

An authoritative account of the spectra and structure of molecular ions is given by Herzberg in the 1970 Faraday Lecture. In this account he considers:

(a) Relation of the electronic structure of the diatomic ion to the corresponding neutral molecule.

(b) The methods employed to study molecular ions, including photoelectron spectroscopy (see Chapter 7).

(c) The application of the molecular ion studies to astrochemistry. For example, it is well known that in the spectrum of the aurora the N_2^+ ion is prominent; however, it has now been shown that O_2^+, NO^+, H_2O^+, and NO_2^- are present in the ionospheric layers when no aurora is present. Herzberg points out that the upper atmospheres of other planets would be expected to contain similar layers in which molecular ions play a great role. CO_{2+} has been detected in the upper atmosphere of Mars, while in the upper atmosphere of Jupiter the ions H_2^+, H_3^+, and CH_4^+ are expected since the the principal constituents of the atmosphere are H_2 and CH_4.

In comet tails the most prominent feature is CO^+. In addition, though N_2^+, CO_2^+ and CH^+ have also been observed where the ions are formed by u.v. radiation from the sun by a photoionization process.

The first ion to be detected in the interstellar medium was CH^+; N_2^+ has also been identified. It is considered likely that H_2^+, H_3^+, N_2^+, and CN^+ are also present since such species are likely from photoionization processes while the time for collision between molecules and ions is of the order of 10 to 100 years; that is once the ion is formed it will not disappear for a very long time as a result of chemical reaction.

As Herzberg points out, the subject of spectra and structure of simple molecular ions is developing rapidly; this field involves a close interaction amongst chemistry, physics, and astronomy.

REFERENCES

2.1 Sawyer, R.A., *Experimental Spectroscopy*, Chapman and Hall, London (1954).

2.2 *Molecular Spectroscopy*, Institute of Petroleum Conference (1968).

2.3 *Molecular Spectroscopy*, Institute of Petroleum Conference (1972).

2.4 Strouts, G.R.N., Gilfillan, J.H. and Wilson, H.N., *Analytical Chemistry*, Vol. 2, Oxford University Press, London (1955).

2.5 Harrison, G.R., Lord, R.C. and Loofbourow, J.R., *Practical Spectroscopy*, Prentice-Hall, New York (1948).

2.6 Van Cittert, P.H., *Z. Phys.*, **65**, 547 (1930); **69**, 298 (1931);

2.7 Phillips, J.G., *Astrophys. J.*, **108**, 434 (1948).

2.8 Herzberg, G., *Spectra of Diatomic Molecules*, Van Nostrand, New York (1950).

2.9 Johnson, R.C., *An Introduction to Molecular Spectra*, Methuen, London (1949).

2.10 Gaydon, A.G., *Dissociation Energies and Spectra of Diatomic Molecules*, Chapman and Hall, London (1968).

2.11 Schüler, H., Haber, H. and Gollnow, H., *Z. Phys.*, **111**, 484 (1939).

2.12 Jevons, W., *Report on Band-Spectra of Diatomic Molecules*, The Physical Society, London (1932).

2.13 Meyer, C.F. and Levin, A.A., *Phys. Rev.*, **34**, 44 (1929).

2.14 Giauque, W.F. and Johnston, H.L., *Nature* (Lond.), **123**, 318, 831 (1929), *J. Amer. Chem. Soc.*, **51**, 1436, 3528 (1929).

2.15 Mahanti, P.C., *Z. Phys.*, **68**, 114 (1931).

2.16 Connelly, F.C., *Proc. Phys. Soc.*, **45**, 791 (1933).

2.17 Gaydon, A.G., *Spectroscopy and Combustion Theory*, Chapman and Hall, London (1942).

2.18 Gaydon, A.G., *Quart. Rev.*, **4** 1 (1950).

2.19 Gaydon, A.G. and Wolfhard, H.G., *Flames, Their Structure, Radiation and Temperature*, 3rd ed., Chapman and Hall, London (1970).

2.20 Gaydon, A.G., *The Spectroscopy of Flames*, Chapman and Hall, London (1957).

2.21 Gaydon, A.G., *Molecular Spectroscopy 1971* (editor P. Hepple), Proceedings of the Fifth Conference on Molecular Spectroscopy, Institute of Petroleum (1972).

2.22 Bulewicz, E.M., Padley, P.J. and Smith, R.E., *Proc. Roy. Soc., A*, **325**, 129 (1970).

2.23 Jessen, P.F. and Gaydon, A.G., *Combust. Flame*, **11**, 11 (1967).

2.24 Jessen, P.F. and Gaydon, A.G., *Twelfth Combustion Symposium*, 481, (1969).

2.25 Gaydon, A.G. and Hurle, I.R., *The Shock Tube in High-Temperature Chemical Physics*, Chapman and Hall, London (1963).

2.26 Dunford, H.B., *Elements of Diatomic Molecular Spectra,* Addison-Wesley (1968).

2.27 Milligan, D.E. and Jacox, M.E., *J. Chem. Phys.,* **47**, 5146 (1967).

2.28 Tinti, D.S., *J. Chem. Phys.,* **48**, 1459 (1968).

2.29 Milligan, D.E. and Jacox, M.E., *J. Chem. Phys.,* **51**, 1952 (1969).

2.30 *Molecular Spectroscopy: Modern Research* (editors Rao, K.N. and Mathews, C.W.), Ch. 5, Academic Press (1972).

2.31 Herzberg, G., *Faraday Lecture*, 201 (1970).

2.32 Douglas, A.E. and Lutz, B.L., *Canad, J. Phys.,* **48**, 247 (1970).

2.33 Grevesse, N. and Sauval, A.J., *Astronomy and Astrophysics*, **9**, 232, (1970).

3 Dissociation energies of diatomic molecules

3.1 INTRODUCTION

The term *dissociation energy* may be appreciated by reference to potential energy internuclear distance curves. At about 0 K all molecules have no rotational energy but are merely vibrating with their zero-point energy. Thus, diatomic molecules are in the $v = 0$ vibrational level. The energy required to separate the stable molecule A – B initially in the $v = 0$ level into two unexcited atoms A and B, that is:

$$A - B \rightarrow A + B$$

is known as the dissociation energy (D), its value being represented in Fig. 3.1.

Another symbol which is often used by spectroscopists is D_e where:

$$D_e = D + G(0) \tag{3.1}$$

$G(0) = E_{v=0}/hc$ and is the value of the vibrational energy in the $v = 0$ level. As $G(0)$ is in cm^{-1} units, then D_e and D would also be in these units.

In the literature dissociation energy values are mainly to be found quoted in $kcal\,mol^{-1}$, cm^{-1}, or electron-volts (eV). The SI unit system employs $kJ\,mol^{-1}$. It is useful to have the following conversion factors: $1\ cm^{-1} = 1.23981 \times 10^{-4}$ eV/molecule $= 0.002859\ kcal\,mol^{-1} = 0.011962\ kJ\,mol^{-1}$.

The value of the dissociation energy can be found by thermochemical methods. However, its value differs very slightly from the corresponding one (D) obtained from spectroscopic data since the latter is calculated for 0 K while the former is for 298 K. It is, however, possible to convert by making certain assumptions from one to the other [3.1].

Infrared spectroscopy can be used to determine D and D_e for the ground state of the molecule, while electronic spectra can be used to determine D and D_e not only for the ground state of the molecule but also for some of the excited states as well.

In the electronic spectra approach for the determination of the dissociation energy at least one of the products of dissociation is frequently in an excited state, and in some cases both atoms are excited. In some instances the dissociation may even take place into ions.

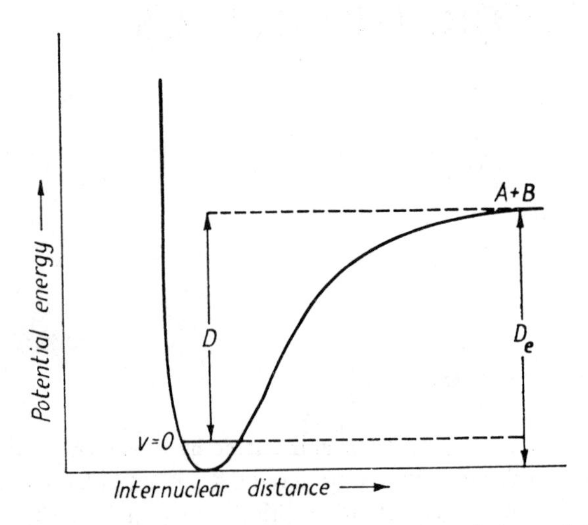

Fig. 3.1 Potential energy curve and the representation of thermal dissociation D and spectroscopic dissociation energy D_e

In order to determine the dissociation energy from spectroscopic data it is necessary either (1) to determine the lowest absorption frequency which will produce dissociation of the molecule, and to identify the electronic state of the dissociation products, or (2) to determine the energy values of as many vibrational levels for the electronic state whose dissociation energy is required.

Electronic spectra may proceed by either (1) or (2) whereas the infrared approach cannot be used to study actual dissociation and is therefore restricted to (2).

The aim in this chapter is to give a simple account of dissociation energies of diatomic molecules. For a more rigorous and fuller treatment the book by Gaydon [3.1] should be consulted.

3.2 EVALUATION OF *D* BY BAND CONVERGENCE METHOD

In Fig. 3.2 five of the potential energy curves are given for the oxygen molecule. When dissociation of the molecule takes place from any of the four lowest electronic states given in Fig. 3.2 it leads to two oxygen atoms each in a ^3P state, that is:

$$O_2 \rightarrow O(^3P) + O(^3P)$$

where both atoms would be released in their ground states. If, however, dissociation is produced in the $B^3\Sigma_u^-$ excited state, then one ^3P-state oxygen atom is produced and one in the excited ^1D state, that is:

$$O_2^* \rightarrow O(^3P) + O(^1D)$$

Dissociation products thus differ in that those resulting from the lowest electronic states have both atoms in the ^3P states whereas in the other case one is in a ^3P state and the other in a ^1D. If the energy absorbed in the $B^3\Sigma_u^- \leftarrow X^3\Sigma_g^-$ transition raises the energy of the diatomic molecule above the level ab, then the energy is sufficient to bring about dissociation of the molecule, and a continuum

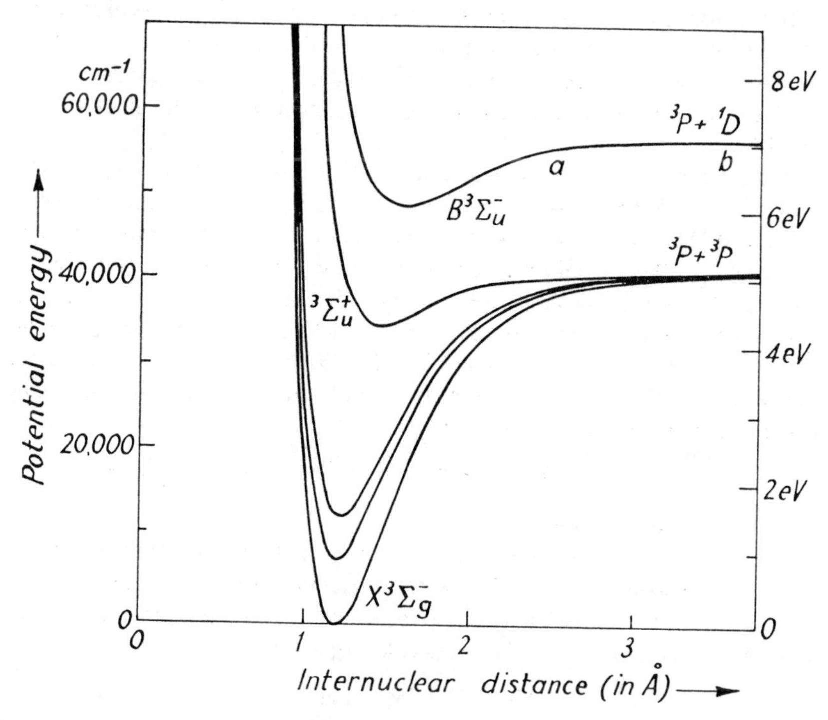

Fig. 3.2 Five potential energy curves of the oxygen molecule (*After Gaydon* [3.1])

is observed.† If $\tilde{\nu}_c$ is the lowest wavenumber at which the continuum begins where the transition takes place from the v'' level, and D'' is the dissociation

† In addition, there may be a number of bands corresponding to transitions from the ground state to definite vibrational levels in the upper state. This is the case for the electronic transition of the oxygen molecule which is now being considered.

energy in the ground electronic state then it follows from Fig. 3.2 that:

$$\tilde{\nu}_c = D'' + (^1\text{D} \leftarrow {}^3\text{P}) - [G(v'') - G(0)] \tag{3.2}$$

where $G(v) = E_v/hc$ and $[G(v'') - G(0)]$ is the difference in vibrational energy (in cm^{-1}) between the v'' and the 0 level in the ground state of the molecule. When the transition involving $\tilde{\nu}_c$ takes place from the $v'' = 0$ level then:

$$\tilde{\nu}_c = D'' + {}^1\text{D} \leftarrow {}^3\text{P} \tag{3.3}$$

where $^1\text{D} \leftarrow {}^3\text{P}$ represents the difference in energy for the two atomic states of oxygen. Hence, for this case if $\tilde{\nu}_c$ is observed and if from atomic spectra the energy corresponding to $^1\text{D} \leftarrow {}^3\text{P}$ is known, then the dissociation energy in the ground state, D'', may be determined.

In this particular O_2 electronic transition, known as the *Schumann-Runge* ($B^3\Sigma_u^- \leftarrow X^3\Sigma_g^-$), the bands converge to a readily discernible limit at 1759 Å and are then followed by a continuum. The convergence limit corresponds to 7.047 eV and is an upper limit for the dissociation energy for this electronic transition.

The ground state of the oxygen atom is a ^3P. The first excited state is a ^1D lying at 1.967 eV above the ground state, while the second and third excited states lie 4.188 and 9.1 eV above the ground state. If the $G(v'') - G(0)$ term is neglected, it follows from Equation (3.2) that the 9.1 eV is unacceptable since it would lead to a negative value for D''. Hence, the only atomic states which need be considered as dissociation products for the Schumann–Runge bands are the ^3P, ^1D, and the second excited state which is a ^1S, since these would lead to positive D'' values. The feasible atomic states from the dissociated oxygen molecule in the higher electronic state are then:

$$^3\text{P} + {}^3\text{P}, \quad {}^3\text{P} + {}^1\text{D}, \quad {}^3\text{P} + {}^1\text{S}, \quad {}^1\text{D} + {}^1\text{D}, \quad {}^1\text{D} + {}^1\text{S}, \quad {}^1\text{S} + {}^1\text{S}$$

That is there are six possibilities.

From the rotational structure of the Schumann–Runge bands the spectroscopist can deduce that the upper electronic state must be a triplet. However, on the basis of quantum mechanics it may be shown that the combination of atoms in singlet states cannot result in a molecular triplet state.† Hence:

$$^1\text{D} + {}^1\text{D}, \quad {}^1\text{D} + {}^1\text{S}, \quad \text{and} \quad {}^1\text{S} + {}^1\text{S}$$

are not feasible atomic states for the dissociation products in this excited electronic state.

The $^3\text{P} + {}^3\text{P}$ combination may be eliminated since it is known from theory that from the combination of two such atomic states it is not possible to produce a $^3\Sigma_u^-$ molecular state. Thus, the two remaining possibilities are:

† This is one of the Wigner and Witmer correlation rules [3.2]. These rules govern the permissible atomic states which may result from the dissociation of a given molecular state.

$$^3P + {}^1D \quad \text{and} \quad {}^3P + {}^1S$$

for the dissociation products in the upper electronic state. From the observed transition it is known that:

$$^1D \leftarrow {}^3P \quad \text{is } 1.967 \text{ eV}$$

and

$$^1S \leftarrow {}^3P \quad \text{is } 4.188 \text{ eV}$$

Since the convergence limit † is 7.047 eV, it follows from Equation (3.2) that two feasible dissociation energies are given at least approximately by:

$$\begin{aligned} & 7.047 - 1.967 = 5.080 \text{ eV} \\ \text{or} \quad & 7.047 - 4.188 = 2.859 \text{ eV} \end{aligned} \qquad (3.4)$$

However, the 2.859 eV value may be eliminated immediately since it is smaller than the energy of the highest observed vibrational level in the ground electronic state which lies 3.4 eV above the $v = 0$ level. Hence, the only possible value is 5.080 eV corresponding to a dissociation into $^3P + {}^1D$ oxygen atoms. This value is unambiguous since it was determined from a single accurately known dissociation limit. In the case of N_2 and NO and some other molecules, however, the spectroscopic approach has led to different values being quoted for the dissociation energy. If the dissociation energy concerned has also been obtained from thermochemistry or other independent methods, then this value may be used in identifying the atomic products in the spectroscopic dissociation process. This is done by determining which electronic state of the atom products would lead to a corresponding value of the dissociation energy in the ground state of the molecule.

The band convergence approach has also been applied to determine the dissociation energies of the halogens Cl_2, Br_2, and I_2. The ground state of the molecule is compounded of two halogen atoms each in a $^2P_{3/2}$ state. In the excited state where the convergence limit was observed the molecule was composed of one atom in the $^2P_{3/2}$ state, but the other atom in a $^2P_{1/2}$ state where the latter state has the higher energy. Other systems analysed this way are H_2 and Na_2.

Under favourable circumstances this method of band convergence may probably give the most accurate of all values for dissociation energies. This occurs when in an electronic absorption spectrum one of the v' progressions may be followed almost to where the continuum begins, so that ν_c may be obtained directly or by a negligible extrapolation of the wavenumbers of the band heads. For example, if the first band in the v' progression is the $(0, v'')$ and the wavenumber of its band head (or even better the band origin) is $\nu_{(0, v'')}$, and D' is the dissociation energy in the excited electronic state, then D' may be immediately obtained from the formula:

† The convergence limit may be determined most accurately in this case because the $(v', 0)$ progression may be followed to the beginning of the continuous absorption. This fixes the value of ν_c with great accuracy. In many other cases it is very difficult to decide exactly where the continuum begins.

$$D' = hc(\tilde{\nu}_c - \tilde{\nu}_{(0,v'')}) \tag{3.5}$$

To determine the dissociation in the ground state, however, as has already been indicated, it is necessary to be able to identify the atomic states of the dissociation products and know the value of this atomic excitation energy from the line spectrum of the relevant atoms.

3.3 EVALUATION OF D_e BY EXTRAPOLATION TO CONVERGENCE LIMITS

In many cases the bands become too faint to trace a progression up to the convergence limit or even to any appreciable convergence at all. In such cases, if sufficient bands are observed, an extrapolation method is employed such as that used by Birge and Sponer [3.3] who extrapolated the observed band head or band origin wavenumber values to the convergence limit. Their initial method has been widely used to estimate dissociation energies and is based on the vibrational energy in cm^{-1} units being represented by the equation:

$$G(v) = (v+\tfrac{1}{2})\omega_e - (v+\tfrac{1}{2})^2 x_e\omega_e + (v+\tfrac{1}{2})^3 y_e\omega_e + \ldots \tag{3.6}$$

where

$$G(v) = E_v/hc \tag{3.7}$$

If all the terms on the right-hand side of Equation (3.6) are neglected except the first two, then the mean interval between the successive levels $v+1$, and $v-1$ is given by:

$$\Delta G(v) = \tfrac{1}{2}[G(v+1) - G(v-1)] = \omega_e - 2(v+\tfrac{1}{2})x_e\omega_e \tag{3.8}$$

where $\Delta G(v)$ may be regarded as the separation between the non-existent $(v+\frac{1}{2})$ or $(v-\frac{1}{2})$ levels, or, in fact, simply as $(d/dv)[G(v)]$.

From Equation (2.54) it follows that as v increases the difference in energy between consecutive vibrational levels decreases; this is illustrated in Fig. 3.3. When $(d/dv)[G(v)] = 0$ the molecule dissociates, and if v_c is the corresponding value of the vibrational quantum number then from Equation (3.8):

$$0 = \omega_e - 2(v_c+\tfrac{1}{2})x_e\omega_e \tag{3.9}$$

Thus the value of the vibrational quantum number at which dissociation takes place is:

$$v_c = 1/2x_e - \tfrac{1}{2} \tag{3.10}$$

and when this value of v_c is substituted in Equation (3.6), then $G(v_c) = D_e$ where D_e is in cm^{-1} units. If only the first two terms on the right-hand side of Equation (3.6) are considered, then on substitution for v_c the value of D_e obtained is:

$$D_e = \omega_e/4x_e \tag{3.11}$$

Thus, from the values of ω_e and x_e (determined from infrared spectra or from electronic spectra) the spectroscopic heat of dissociation may be calculated. In the case of electronic spectra this Birge and Sponer method has been applied to data derived from the analyses of both emission and absorption band system data.

The experimental vibrational data† for the ground state of the carbon monoxide molecule may be fitted to the formula:

$$G(v) = [2167.4(v+\tfrac{1}{2}) - 12.70(v+\tfrac{1}{2})^2] \quad (3.12)$$

This formula was based on experimental data where the vibrational quantum number ranged from 0 to 25. From this formula v_c is found to be ~85, and if the dissociation energy corresponding to this v_c value is calculated from Equation (3.12), it turns out to be about 170 kJ mol^{-1} higher than what is considered to be the correct value. In fact, this procedure rarely yields the correct result, and we shall now consider the reason for this.

If, for example, three or four experimental values for $(v+1) \leftarrow v$ transitions are available, it follows from Equation (3.8) that when $\Delta G(v)$ is plotted against v a straight line ought to be obtained which could then be extrapolated to $\Delta G(v) = 0$ to give $v = v_c$. This v_c value when inserted in Equation (3.6) ought to give the dissociation energy (D_e). This procedure, which should yield the same value of D_e as the method just outlined, also rarely gives the correct value of the dissociation energy, and Gaydon suggested that a figure of about 20 per cent below that of the linear extrapolated value is generally best adopted, and this will then have a probable error of ± 20 per cent. In the case of hydrides and halides of group II elements the true dissociation energy may be even less than this 20 per cent lower value. When a few of the actual $\Delta G(v) - v$ experimental curves are viewed it becomes immediately obvious why this type of extrapolation yields an inaccurate result. In Fig. 3.3 $\Delta G(v+\tfrac{1}{2})$ is plotted against v for the ground and an excited electronic state of the oxygen molecule, and it is to be noted that curves result. It is interesting to observe that, if only the data for the first five vibrational levels had been available for the ground electronic state, it would have been possible to be misled into fitting a straight line whereas it ought to have been a curve. In general, the Birge–Sponer extrapolation gives an upper limit to the dissociation energy for covalent molecules and a lower limit to molecules where the ionic forces predominate. In the case of ionic molecules such as Na^+Cl^- no reliable estimate of the dissociation energy can be made by this Birge–Sponer type of approach since the curve approaches the v-axis asymptotically.

If only the first two terms on the right-hand side of Equation (3.6) are taken, it leads to a linear plot of $\Delta G(v)$ and $\Delta G(v+\tfrac{1}{2})$ against v. If, however, the equation:

$$G(v) = (v+\tfrac{1}{2})\omega_e - (v+\tfrac{1}{2})^2 x_e\omega_e + (v+\tfrac{1}{2})^3 y_e\omega_e \quad (3.13)$$

† These data were obtained from the analysis of the fourth positive system of CO, which is the $A^1\Pi - X^1\Sigma^+$ transition.

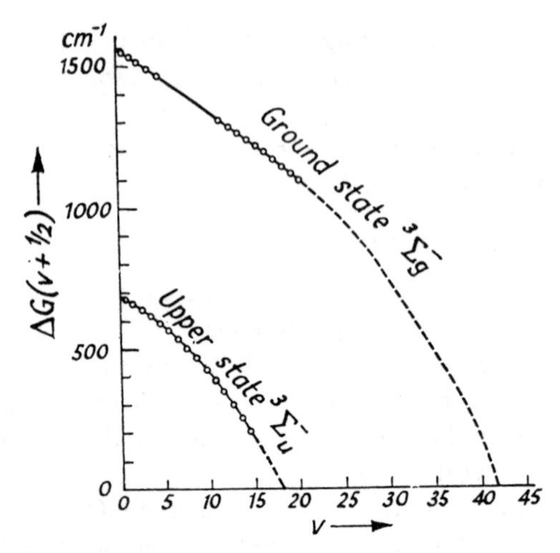

Fig. 3.3 $\Delta G(v + \frac{1}{2})$ against v for the ground and an excited state of the oxygen molecule. (*After Gaydon* [3.11]).

had been used, the $\Delta G(v)$ values would no longer show linear dependence on v However, even such an equation may prove inadequate, and, in fact, Birge showed that even a polynomial of the fourth degree failed to represent adequately the $\Delta G(v)$–v data for some diatomic molecules. In fact, he demonstrated that the complete curve can never be exactly covered by a single function but that two functions are required.

One of the best procedures when a sufficient number of bands is available is to plot the experimental $\Delta G(v)$ values against v, and, if the shape of the resulting curve seems reasonably certain, to extrapolate the resulting curve until it cuts the v-axis at the point where $v = v_c$. The dissociation energy would then be obtained by finding the area enclosed by the curve and the two axes, that is:

$$D = \int_0^{v_c} \Delta G(v)\, dv \tag{3.14}$$

3.4 EVALUATION OF D BY ATOMIC FLUORESCENCE METHOD

When a continuous or diffuse absorption spectrum is obtained for a molecule, and dissociation has taken place giving an excited atom and one in its ground state, if the excited atom is not metastable, then atomic fluorescence of this atom may accompany this photodissociation. The emitted atomic line may then be identified. The wavelength of the incident light is varied, and the lowest frequency which still produces this atomic line (owing to dissociation of the molecule) is sought. Thus, the lowest frequency of the incident light which will still cause the atom to fluoresce will lead to at least an upper limit for the dissociation energy. This process may be illustrated by considering the dissociation of sodium iodide by the absorption of light.

In its ground state sodium iodide is in the ionic form, and what takes place on the continuous absorption of light depends on the electromagnetic region in which absorption takes place. The dissociation products which occur in the different frequency regions are: (i) an unactivated Na + unactivated I; (ii) an activated Na + unactivated I; (iii) an unactivated Na + activated I. Case (ii) will now be considered. By the absorption of light an excited sodium iodide molecule $(NaI)^*$ is obtained, and this dissociates into an iodine atom in its ground state $(^2P_{3/2})$ and an activated sodium atom (3^2P); thus:

$$Na^+ + I^- \xrightarrow{hc\tilde{\nu}} (NaI)^*$$

$$(NaI)^* \rightarrow I(^2P_{3/2}) + Na(3^2P)$$

and the excited Na then emits light (the sodium D-lines) and falls to its ground state (3^2S), that is:

$$Na(3^2P) \rightarrow Na(3^2S) + hc\tilde{\nu}_0$$

where $\tilde{\nu}_0$ represents the wavenumber of the sodium D-lines. The condition sought is that the minimum wavenumber of the absorbed incident light should still produce this emission line $(\tilde{\nu}_0)$. If the wavenumber of the incident light is $\tilde{\nu}_1$, then:

$$\tilde{\nu}_1 - \tilde{\nu}_0 = D'' \tag{3.15}$$

where D'' is the dissociation energy in the ground state in cm^{-1} units.

The great merit of this method is that it is possible to be reasonably certain of at least one of the atomic states in the dissociation products. The method has been applied to the halides of the alkali metals and thallium where the results agree fairly well with those from other methods. Only a bare outline of the process has been given here; for further details a paper by Sommermeyer [3.4] should be consulted.

3.5 PREDISSOCIATION AND EVALUATION OF *D*

The phenomenon of predissociation was first observed in the electronic absorption spectrum of S_2, where for transitions from the v'' levels to low values of v' the bands appeared normal, and the rotational structure in each band consisted of sharp lines. However, for transitions from v'' to higher values of v', while the general appearance of the bands remains, some or all of the rotational fine structure of the band becomes diffuse.† This suggests that the molecule has definite vibrational energy in both the upper and lower electronic states but that above certain vibrational levels in the excited state some of the rotational energy is not quantized. Generally, predissociation may be detected when the sharp lines of a branch of a band due to the rotational fine structure suddenly end and the last few lines take on a diffuse appearance. There are several types of predissociation

† A spectrogram of CaH which illustrates this point may be observed in Gaydon, [3.1].

but two features stand out:

(a) The predissociation spectrum may be followed by a region of continuous absorption, or

(b) Bands with no diffuse structure are found on both low and high frequency sides of the predissociation bands.

One of the simplest cases of predissociation will now be explained with respect to the potential energy curves in Fig. 3.4. The molecular electronic state

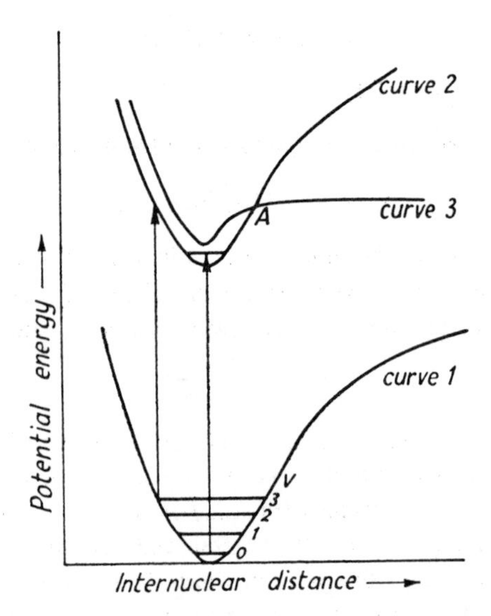

Fig. 3.4 Potential energy curves involved in the phenomenon of predissociation

quantum numbers for these three electronic states are such that the selection rules permit only transitions between curves 1 and 2. For transitions from the $v'' = 3$ level and higher, the potential energy of the excited molecule is greater than that at point A, and during its vibration in this excited state, when the internuclear distance corresponds to that at A, it has the possibility of changing from curve 2 to 3. When this transition takes place it is termed a *radiationless* one. The time for a molecular vibration is of the order of 10^{-14} s and for a molecular rotation 10^{-10} s. Hence, a molecule may vibrate several times before completing a rotation, but when the molecule is excited to curve 2 it may during its vibration transfer to curve 3 before it has had time to rotate in excited state 2. Thus, in this case the rotational energy will no longer be strictly quantized, and the rotational fine structure of the band will become diffuse. However, the vibrational energy which fixes the gross structure will still be quantized and exhibit itself in the appearance of diffuse bands. Although the energy of the molecule was insufficient to cause dissociation in state 2, it is now sufficiently great in state 3 to bring about dissociation, and the dissociation limit may be fixed within a very narrow range.

However, if in some of the transitions from potential energy states 1 to 2, high values of v' are involved, this may lead to an energy relatively much greater than that at point A. In such a case the vibrational energy of the molecule, as it passes through the point A, may be sufficiently great to prevent the transition to state 3, and when this occurs the predissociation region resulting from the v'' transitions to lower v' values will be followed by bands showing sharp rotational structure.

One supporting fact which favours the given interpretation of predissociation is that where predissociation occurs in the absorption spectrum of S_2 bands, the same bands are completely absent in the same emission system. This would be anticipated if the given theory of predissociation were correct, since emission of such bands should be precluded by spontaneous dissociation of the molecule in such v' levels.

By employing a rather similar technique to that used in the band convergence method of determining dissociation energies and by estimating the lowest frequency at which the vibrational bands become diffuse, it might be expected that it would be possible to calculate the dissociation energy. There are many complicating factors; one of them is illustrated in Fig. 3.5, where the dissociation occurs through the point P, and the radiation along PM is non-quantized, and the dissociation limit of LPM lies below the predissociation limit.

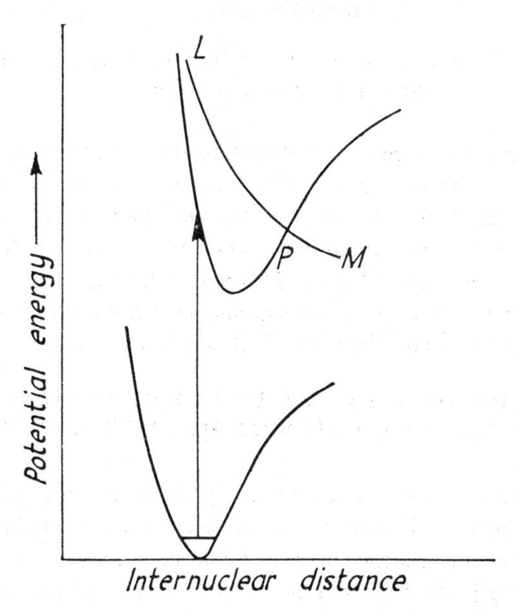

Fig. 3.5 Predissociation involving a non-quantized potential energy curve

Another complication in the predissociation determination of dissociation energies is that another type of phenomenon may give a similar appearance in the spectrum to that described to characterize predissociation. This phenomenon may occur if two potential energies nearly cross (see Fig. 3.6) and if the selection rules permit a transition between them. In such a case the energy levels may become perturbed, and the electronic spectrum from the ground state III to I may result in a radiationless transfer to II, which results in the spectrum having a similar appearance to what it would have had if the potential energy curves had actually crossed.

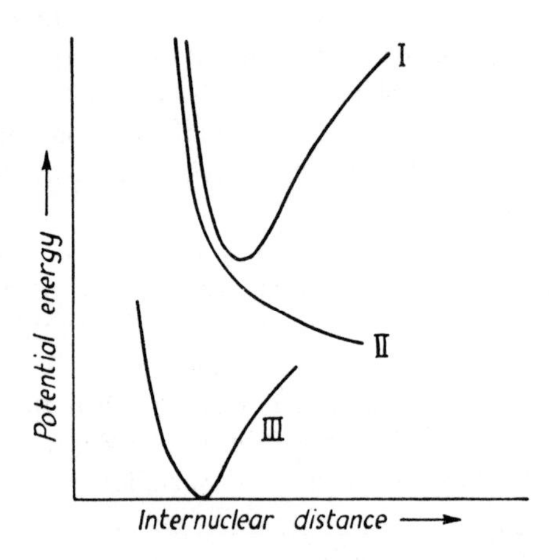

Fig. 3.6 Potential energy curves where a radiationless transfer may occur from I to II which could be mistaken for predissociation

Generally, the commencement of predissociation gives at least a maximum value of D, and in some cases a very accurate value of D may be obtained. The method has been applied to N_2, H_2, CO, NO, and several other molecules. As should be apparent from what has been indicated, the route to deduce such an accurate value is highly specialized and involves considerations of selection rules.† For the determination of dissociation energies by the predissociation method, Gaydon [3.1] and Herzberg [3.2] should be consulted.

3.6 CONCLUSIONS ON THE DETERMINATION OF DISSOCIATION ENERGIES OF DIATOMIC MOLECULES

The most prolific source of dissociation energies for diatomic molecules has been by means of electronic spectra. The infrared approach has also been employed A large number of dissociation energies of diatomic molecules and radicals is given by Herzberg [3.2]. In favourable cases the electronic spectra approach is

† Quantum mechanics has also shown that for these radiationless transitions the Λ and Σ selection rules are the same as for radiative transitions. However, the selection rule for J is $\Delta J = 0$ while, for the g and u states it now becomes $g \rightarrow g$ and $u \rightarrow u$ instead of $g \rightarrow u$ and $u \rightarrow g$.

capable of giving the most accurate dissociation energies for diatomic molecules. The band convergence method where a progression is followed to the region where dissociation commences is particularly good since the dissociation limit is fixed with reasonable certainty, and wavelengths can be measured most accurately. Except in the atomic fluorescence method, where at least one of the atomic states is identified with certainty, the identification of the excited state of the atomic products is difficult and sometimes uncertain and may lead to publication of considerably differing values for a particular dissociation energy. However, the atomic fluorescence method is very limited in application, while the band convergence method may lead to errors of ± 20 per cent. The predissociation method can sometimes yield most accurate results, but again this is limited in application. Recent procedures (see later) have permitted the determination of more accurate values in some cases.

In the treatment outlined here one possible complicating factor has been omitted, i.e. it has been assumed that there is no maximum in the potential energy curve either in the excited or ground electronic states. Herzberg interpreted the electronic spectrum of NO by assuming that a potential energy maximum was involved. Another case was that studied by Rowlinson and Barrow [3.5] who examined the electronic emission spectrum of the diatomic AlF for the $A^1\Pi \rightarrow X^1\Sigma$ system. Their results indicate that the dissociation energy in the ground state must be greater than 628 kJ mol^{-1}, and from a short extrapolation of the vibrational levels they obtained a value of 695 kJ mol^{-1}. The thermochemical value for the dissociation energy (D'') in the ground state is 613 kJ mol^{-1}. If the value of the dissociation energy from the electronic spectrum is truly appreciably greater than this value, it could most satisfactorily be explained by a maximum in the potential energy curve in the ground state. This is represented diagrammatically in Fig. 3.7.

The dissociation energy of F_2 is not in the order which might be first expected in relation to the other halogens, as is illustrated in Table 3.1. In fact,

Table 3.1 Dissociation energies of diatomic molecules

Halogen molecule	*Dissociation energy (kJ mol^{-1})*
F_2	155
Cl_2	238
Br_2	188
I_2	150

for a number of years the dissociation energy of F_2 was thought to be 264 kJ mol^{-1}. The value quoted in Table 3.1 for F_2 was obtained by Caunt and Barrow [3.6] from extensive studies on the ultraviolet absorption spectra of the gaseous

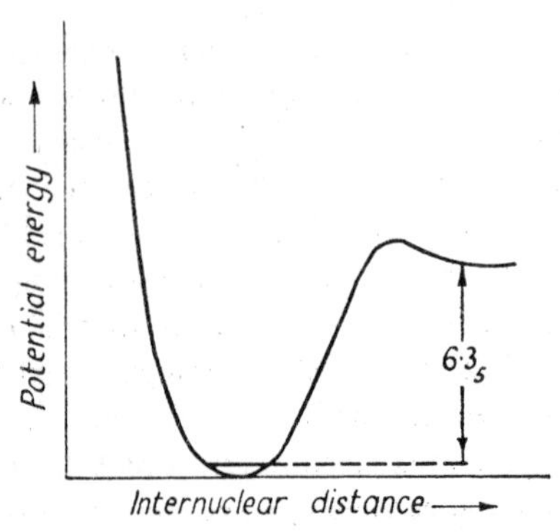

Fig. 3.7 Potential energy curve exhibiting a maximum. The dissociation energy (D) is 6.3 eV.

alkali metal fluorides, where the dissociation energy of the metallic fluoride was related to the dissociation energy of fluorine. A subsequent analysis by Stamper and Barrow [3.7] of the pressure of partly dissociated F_2 has fixed the dissociation energy as 158 kJ mol^{-1}. Caunt and Barrow [3.6] studied all the twelve gaseous potassium, rubidium, and caesium halides and deduced the upper limits of the dissociation energy for each of these substances.

Several detailed accounts on the dissociation energies of diatomic molecules exist. A review article on bond dissociation energies by Szwarc [3.8] should be consulted, and this includes the non-spectral methods such as the electron impact method, the thermochemical method, and the kinetic method, in addition to the spectral methods. For more detailed treatment there are excellent works by Cottrell [3.9], Gaydon [3.1], Jevons [3.10], and Herzberg [3.2].

Studies have been made of the dissociation energies of simple polyatomic molecules but this is beyond the scope of this chapter and for this Herzberg Vol. III should be consulted [3.11]. Such work is of considerable interest and had attracted attention even forty years ago. In fact, even in 1931 Herzberg [3.12] had studied the electronic absorption spectrum of formaldehyde and followed the rotational fine structure of a band, until it eventually merged into continuous absorption. This corresponded to the following dissociation:

$$CH_2O \rightarrow HCO + H$$

and fixed the upper limit for this dissociation energy as 439 kJ mol^{-1}.

3.7 RECENT DISSOCIATION ENERGY STUDIES FOR DIATOMIC MOLECULES

In the past twenty years a considerable amount of dissociation energy data has accumulated, new methods and procedures for determining dissociation energies have appeared, and, in particular, old data have been reassessed and more precise dissociation energies obtained. Two important incentives in gaining accurate values are: (a) for their application in thermochemistry in helping to formulate the atomic heats of formation of polyatomic molecules; (b) the dissociation energies of molecules and radicals (e.g. O_2, H_2, and OH) involved in flames are essential to determine the equilibrium constants employed in calculating the temperature of flames.

A detailed account of dissociation energies has been given by Gaydon [3.1], and for the work carried out up to 1967 this should be consulted. In the material now to be considered a few studies carried out since that date are given.

Much work has been carried out using the Birge and Sponer method to estimate the dissociation energy; many of the old values have been modified. Only very rarely can the transitions be obtained as far as the convergence limit and, as a consequence, when more accurate and detailed data become available the data are revised. However, in some cases the early data can be quite accurate. For example, in 1934 Beutler [3.13] carried out a Birge–Sponer plot [$\Delta G(v + \frac{1}{2})$ versus v] for H_2 in the ground electronic state and was able to use values of v'' up to 15 which led to a short extrapolation and yielded a dissociation energy of 431.8 kJ mol^{-1}. A recent most detailed study by Herzberg [3.14] yielded an upper limit of 432.068 kJ mol^{-1}. In this work Herzberg studied the electronic absorption spectra of H_2, HD, and D_2 at the temperature of liquid nitrogen. Dissociation limits for these molecules were found in the ultraviolet region and a vacuum grating spectrograph was employed with high resolving power. The use of low temperatures simplified the analysis since it eliminated some of the overlapping rotational structure of the bands. In the case of HD the $J'' = 0$ limit was identified and this enabled a most reliable dissociation energy to be evaluated in the ground electronic state of which the value was considered to be accurate to better than ± 0.4 cm^{-1}.

Gaydon [3.1] has made an interesting comparison of linear Birge–Sponer extrapolated values D_{lin} with those obtained in other ways. He carried this out for 72 molecules. For the comparison value (D_{true}) he selected a dissociation energy which was considered to be accurate to within 5 per cent. Such values had been determined by, for example, methods such as predissociation, convergence limits, or mass spectrometry. The average ratio for these 72 molecules of $D_{\text{true}}/D_{\text{lin}}$ is 0.85; thus there is a general tendency of the Birge–Sponer linearly extrapolated value to be too high. Beckel, Shafe, and Engelke [3.15] extended this study to 98 diatomic molecules and found the following.

(1) $D_{\text{true}}/D_{\text{lin}}$ is 0.74 ± 0.05 for homonuclear molecules and 0.75 ± 0.08 if both atoms are distinct but from the same column in the periodic table. Average ratios as a function of constituent atom periodic table separations are: 0.91 ± 0.10 if one column apart, 0.97 ± 0.06 two columns, 1.03 ± 0.12 three columns, 1.48 ± 0.16 four columns, and 1.87 ± 0.29 six columns. Thus there is an overall trend of this ratio with increasing ionicity. In addition, the variations within one class are significant and for the non-metals the ratio tends to decrease with increasing atomic size.

(2) The removal of an electron from a neutral molecule increases D_{true}/D_{lin} by ~ 25 percent.

(3) When D_i, the dissociation energy to ions, is compared with D_{lin} for the alkali halides, the average $D_i/D_{lin} = 2.50$ with almost no scatter.

Several interesting studies have been made involving the necessity of taking into account long-range forces when considering the magnitude of dissociation energies. Stwalley [3.16] took these into account when considering the dissociation energy of H_2 in the excited electronic state $B'\Sigma_u$. He showed, from new assignments of the vibrational levels near the dissociation limit in the $B'\Sigma_u$ state, that the vibrational quantum number is proportional to the one-sixth power of the binding energy, as would be expected theoretically from the long-range r^{-3} behaviour of the internuclear potential. This was used to obtain an improved dissociation limit of 1416.03 kJ mol^{-1}, corresponding to a ground state $X'\Sigma_g^+$ dissociation energy value of 432.05_1 kJ mol^{-1} which compares well with Herzberg's [3.14] upper limit of 432.04_7 kJ mol^{-1}.

LeRoy and Bernstein have determined dissociation energies and long-range potentials of diatomic molecules from the vibrational spacings of the higher (vibrational) levels. For a given diatomic species for vibrational levels near the dissociation limit (D) the attractive long-range potential, $V(R)$, is given by:

$$V(R) = D - C_n/(R^n) \tag{3.16}$$

where C_n is a long-range potential constant. They developed an expression which relates the distribution of vibrational levels near the dissociation limit to the nature of the long-range interatomic potential. They fitted experimental energies directly to their expression and, consequently, determined values of D, n, and C_n. Their procedure requires a knowledge of the relative energies and relative vibrational numbering for at least four rotationless states near the dissociation limit. Their procedure enables D to be evaluated with a much smaller uncertainty than by Birge–Sponer extrapolations [3.17].

Another interesting study involving long-range forces was carried out by LeRoy and Bernstein [3.18]. They determined the asymptotic long-range potential constants (C_n) from the vibrational spacings for:

(1) the $B^3\Pi^+$ states of Cl_2, Br_2, and I_2; and

(2) the ground $X'\Sigma_g^+$ state of Cl_2

The C_n values they obtained from the spectra agreed well with the best theoretical estimates. The analysis also yielded improved from dissociation energy values for the ground $X'\Sigma_g^+$ states of $^{35,35}Cl_2$, $^{79,79}Br_2$, $^{81,81}Br_2$, and $^{127,127}I_2$ respectively as follows: 239.21, 190.13, 190.15, and 148.82 kJ mol^{-1}.

REFERENCES

3.1 Gaydon, A.G., *Dissociation Energies and Spectra of Diatomic Molecules*, Chapman and Hall, London (1968).

3.2 Herzberg, G., *Spectra of Diatomic Molecules*, Van Nostrand, New York (1950).

3.3 Birge, R.T., and Sponer, H., *Phys. Rev.,* **28**, 259 (1926).

3.4 Sommermeyer, K., *Phys.,* **56**, 548 (1929).

3.5 Rowlinson, H.C., and Barrow, R.F., *Proc. Phys. Soc.* (Lond.), **A66**, 437, 771 (1953).

3.6 Caunt, A.D., and Barrow, R.F., *Proc. Roy. Soc.,* **A219**, 120 (1953).

3.7 Stamper, J.G., and Barrow, R.F., *Trans. Faraday Soc.,* **54**, 1592 (1958).

3.8 Szwarc, M., *Quart Rev.,* **5**, 22 (1951).

3.9 Cottrell, T.L., *The Strengths of Chemical Bonds,* Butterworths, London (1954).

3.10 Jevons, W., *Report on the Band Spectra of Diatomic Molecules,* Physical Society, London (1932).

3.11 Herzberg, G., *Spectra of Polyatomic Molecules,* Van Nostrand, New York (1968).

3.12 Herzberg, G., *Trans. Faraday Soc.,* **27**, 378 (1931).

3.13 Beutler, H., *Z. Phys. Chem.,* **B27**, 287 (1934).

3.14 Herzberg, G., *J. Molec. Spect.,* **33**, 147 (1970).

3.15 Beckel, C.L., Shafe, M., and Engelke, R., **40**, 519 (1971).

3.16 Stwalley, W.C., *Chem. Phys. Lett.,* **6**, 241 (1970).

3.17 LeRoy, R.J., and Bernstein, R.B., *J. Chem. Phys.,* **52**, 3869 (1970).

3.18 LeRoy, R.J., and Bernstein, R.B., *J. Molec. Spect.,* **37**, 109 (1971).

4 Electronic spectra of polyatomic molecules

4.1 INTRODUCTION

Chapter 2 dealt with the analysis of high resolution electronic spectra of diatomic molecules where the electronic transition also involved simultaneous changes in vibrational and rotational quantum numbers. For an N-atomic molecule there are up to three degrees of rotational freedom and $(3N-6)$ vibrational modes ($3N-5$ if the molecule is linear). Although some of these may be degenerate if the molecular symmetry is high enough, the overall appearance of the spectrum is as a result generally very complex indeed. For example, in the analysis of the 455 nm absorption band of the asymmetric molecule glyoxal many hundreds of individual rotational transitions are observed in each of the vibrational progressions. The analysis of such spectra is a complex task and the adventurous reader is directed to the book by Herzberg [4.1] and for a modern review to Innes [4.2]. Molecular geometry parameters associated with the molecule in ground and excited electronic states constitute one of the end products and these results have been used to test qualitative and quantitative theories of bonding and molecular structure. In particular the simple molecular orbital scheme developed by Walsh [4.3] some years ago, and often called Walsh's rules, have proven very useful in this respect for small molecules.

The overall appearance of the electronic spectrum of the polyatomic molecule in the gas phase is then generally rather complex in detail. However, most electronic spectra are recorded under conditions of poorer resolution or in solution or the solid state. In general the rotational fine structure disappears and quite often vibrational features are replaced by a broad envelope. Figure 4.1 shows the spectrum of benzene at about 260 nm as the vapour and in solution. Identical recording conditions were used in both cases and the loss of detail on moving to the solution phase is apparent.

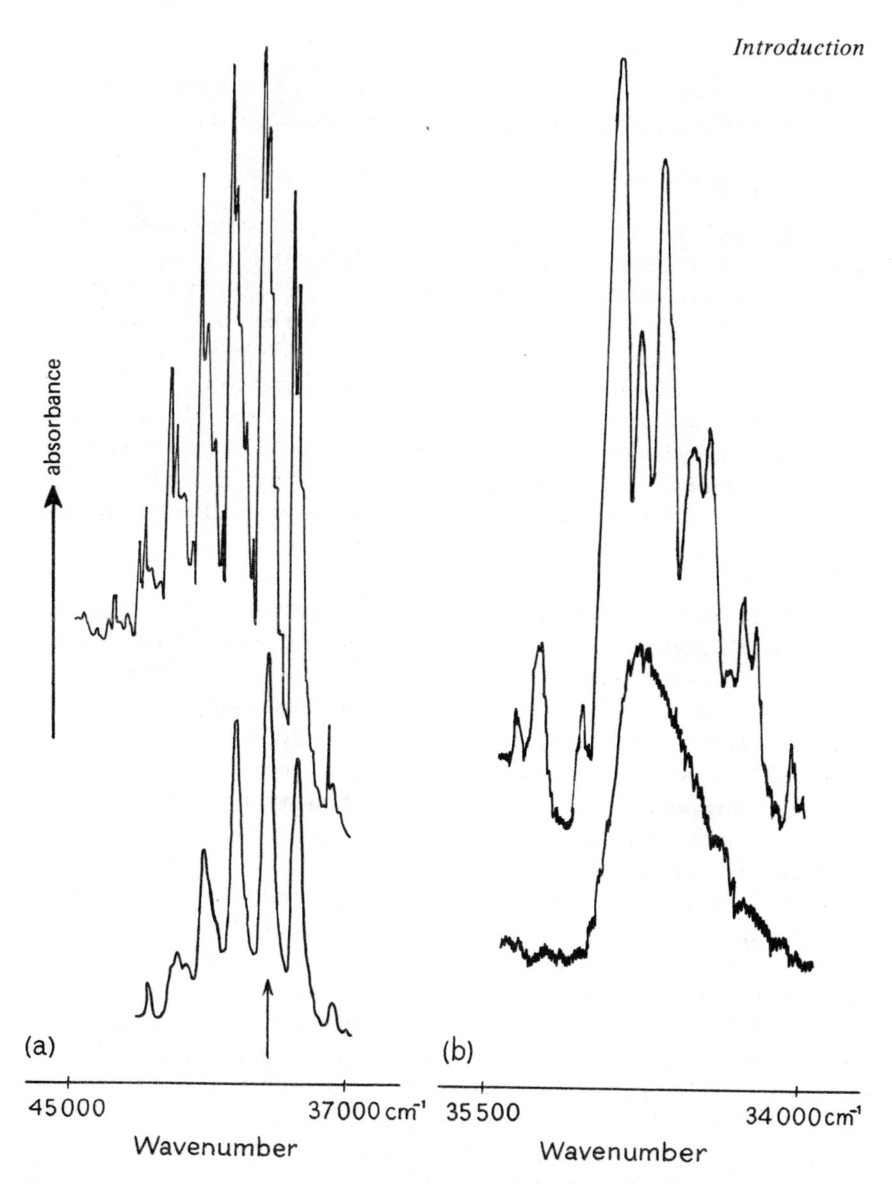

Fig. 4.1 Electronic spectrum of benzene around 250 nm. (a) Complete spectrum in the gas phase (top trace) and in methanol solution (bottom trace). (b) High resolution study of one of the features of Fig. 4.1(a) (marked with an arrow) showing how some of the vibrational fine structure in the upper trace is lost in a polar solvent (lower trace).

In this chapter the emphasis will not lie with the detailed analysis of high resolution spectra. In principle this follows similar lines to those discussed in Chapter 2. The following pages will be concerned with seeing how the energy and intensity of a particular electronic transition are related to molecular constitution and examining some of the uses of low resolution spectra.

4.2 ABSORPTION OF LIGHT BY A MEDIUM

When a beam of light passes through a medium, generally the incident light intensity (I_0) will always be greater than the intensity of the light emerging from the other side of the sample (I). This is simply because the incident light beam is attenuated or reduced in intensity by reflections at the air–sample interface, scattering by any particulate matter suspended in the medium, and direct absorption of light by the material itself. In normal circumstances the absorption process is the major factor involved. It can be readily envisaged that the amount of light absorbed will depend upon (i) the concentration of absorber, (ii) the path length of the absorbing medium, and (iii) a factor dependent upon the physical nature of the material itself. Algebraically this dependence is written as:

$$A(\lambda) = \log(I_0/I) = \epsilon(\lambda)cl$$

where A is defined as the *absorbance* or *optical density* of the sample, c is the concentration of absorber in the medium (usually in units of mol m^{-3}), l is the path length, and $\epsilon(\lambda)$ is the molar absorption coefficient* at the wavelength λ. This expression is known as the Beer–Lambert law or more usually simply as *Beer's law*. The ratio I/I_0 is defined as the transmittance or transmission (T). Thus $A = -\log T$. The linearity of this function ($A \propto c$) is unfortunately not universally found to be the case. For some systems concentration may not be exactly linearly proportional to measured absorbance owing to a variety of factors.

Hydrogen bonding, ion pair formation, solvation, and chemical reactions can cause incorrect calculations of sample concentration in the solvent medium. For example, the measured weight of acid which is then dissolved in a solvent may be somewhat different from the amount of free non-hydrogen bonded acid in solution. The difference between the two values is generally concentration dependent. This gives rise to a quadratic dependence of A on c. The other two factors mentioned above, reflections and scattering, may also lead to non-linear behaviour in the A versus c plot. Instrumental factors such as stray light will also tend to reduce the measured absorbance as the concentration is increased. In addition, if the slit width of the spectrometer is too high compared with the spectral width of the absorption band, non-linear behaviour will result.

Beer's law may be used for example to determine the relative proportions of components in a multicomponent mixture. For a two-component mixture the total absorption at a given wavelength (λ_1) will be the sum of the absorbances of the two components A and B, namely:

$$A(\lambda_1) = A_A(\lambda_1) + A_B(\lambda_1) = \epsilon_A(\lambda_1)c_A l + \epsilon_B(\lambda_1)c_B l$$

and at another wavelength (λ_2):

$$A(\lambda_2) = \epsilon_A(\lambda_2)c_A l + \epsilon_B(\lambda_2)c_B l$$

* Also known as extinction coefficient

The molar absorption coefficients ϵ_A and ϵ_B at the two wavelengths concerned may be obtained by determining the absorbances of known solutions of the pure components. The concentrations c_A and c_B may then readily be obtained by solving these two simultaneous equations. For an *n*-component mixture the absorbance of the mixture needs to be measured at *n* distinct wavelengths, to provide enough data to determine all *n* concentrations.

4.3 INSTRUMENTATION

Instrumentation for u.v. and visible absorption spectroscopy has developed to a stage where the majority of work is now carried out on double beam ratio recording or direct reading grating spectrophotometers with readout directly in optical density or absorbance units (*A*).

A deuterium or hydrogen lamp is used for the region 190–340 nm whilst tungsten filament lamps are used for the region 320–800 nm. Source changeover is either manual or automatic. Various double beam systems are possible but commercial instruments are almost exclusively of the ratio type. Double beam systems provide the advantage of high stability with a consequence of better photometry with time and with the added benefit of being able to do differential spectroscopy directly. On the more routine instruments single grating monochromators are used either of the Littrow or Czerny–Turner type having a limiting resolution of about 0.2 nm over the whole of their range. Detection is by means of a single photomultiplier located close to the sample to receive as much scatterd light as possible. A photometric range of 0–2*A* (i.e. from an optical density of 0 to 2) is usual with integral scale expansion to give up to 0.1*A* full scale when an analogue recording device is used. Typical instruments of this type are the Beckman model 25, the Unicam model SP 1800, and the Varian model 635, all of which offer a wide range of accessories for multi-sampling, microsampling, and semi-automatic operation. For more stringent requirements in sampling, photometry, resolution, or stray light in the ultraviolet, visible, or near infrared regions, an instrument such as the Beckman ACTA M VII is necessary.

The optical arrangement is shown in Fig. 4.2. The u.v. source is a deuterium lamp while the visible and near infrared regions are covered by tungsten filament lamps. A dual grating monochromator is used which provides for a limiting resolution of better than 0.05 nm in the visible and u.v. regions (190–800 nm) and better than 0.3 nm in the near infrared (800–3000 nm). Stray light is less than 0.0001 per cent from 240 to 500 nm and less than 0.1 per cent at range limits, which allows a potential photometric range up to 6*A*. The flexibility of the modern monochromator is further enhanced by having an infinitely variable programme slit system giving a wide dynamic range to the energy throughput of the system to suit the nature of the sample under investigation. The double beam system provides wide separation of the two optical light paths in a large sample compartment which allows the introduction of helium cryostats, or other bulky equipment for special applications. Both light

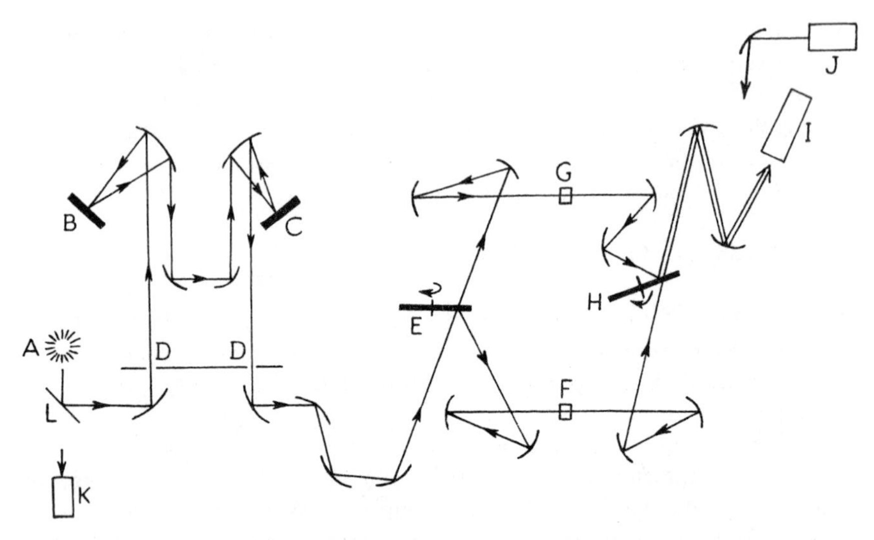

Fig. 4.2 Optical layout of Beckman Acta M VII u.v./visible/near i.r. spectrometer. (Diagram and technical details by courtesy of Mr. W.A. Scott and Beckman Instruments.)

Wavelength selection of u.v./visible light from the source (A) is made by a double monochromator consisting of two diffraction gratings (B, C) and adjustable slits (D). This beam is chopped by a sectored disc (E) and the beam sent either through the sample (F) or reference (G) cells. By means of a second chopper (H), located after the sample, either the sample or reference beam impinges on the detector (I). For near-infrared use the light path is reversed. Light from the source (J) travels through the system and is detected in a similar way on the detector (K) via rotation of the mirror (L). Thus for u.v./visible work the radiation is monochromated before passing through the sample, whereas for near-i.r. work the radiation is monochromated afterwards.

beams pass through a focus in the centre of the sample compartment which is a convenience for micro sampling or for using the instrument as a densitometer. The detector is a shielded end-on photomultiplier for the u.v.–visible region while for the near infrared region a lead sulphide cell is used. In the u.v.–visible region radiation travels from the source through the monochromator, through the double beam system to the detector, whereas in the near infrared region the radiation direction is reversed. Five photometric ranges from 0.1*A* full scale to 3*A* full scale and an additional five expanded absorbance ranges of 0.01*A* to 0.2*A* are provided. Readout of concentration or per cent transmission are also available. Wavelength scanning is by means of a stepper motor which provides the capability for electronic switching of scanning speeds over the range 1/64th of a nanometre per second to 4 nm per second. With a stepper motor also driving the chart recorder the chart presentation can be electrically switched over the range 1–100 nm per inch.

With increased use of digital data processing of spectral information modern

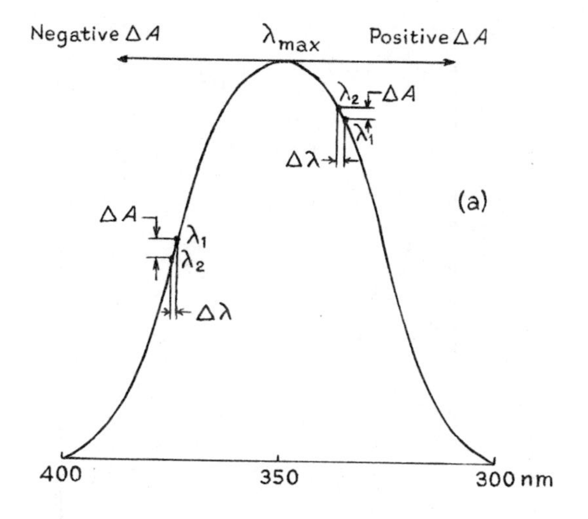

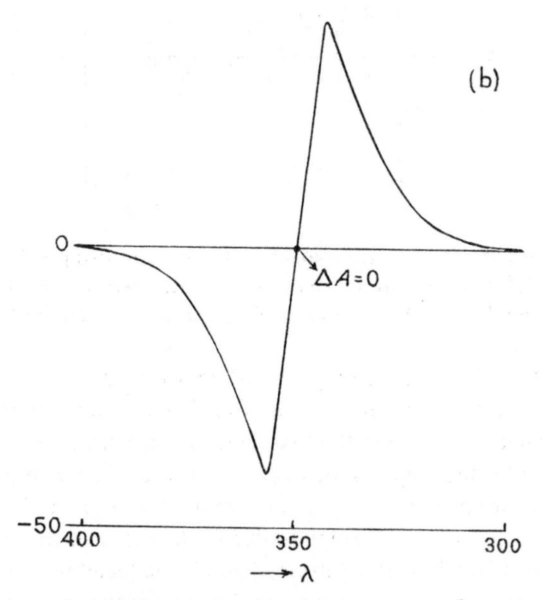

Fig. 4.3 Double wavelength spectroscopy. (a) Shape of a typical absorption band showing positive and negative values of ΔA. (b) Appearance of the first derivative spectrum.

instruments usually have provision for interfacing with computers, either on-line to small computers or off-line via punched tape.

Of particular interest is the ability to display a first derivative ($dA/d\lambda$) spectrum. The recorder electronically plots the difference ΔA between the

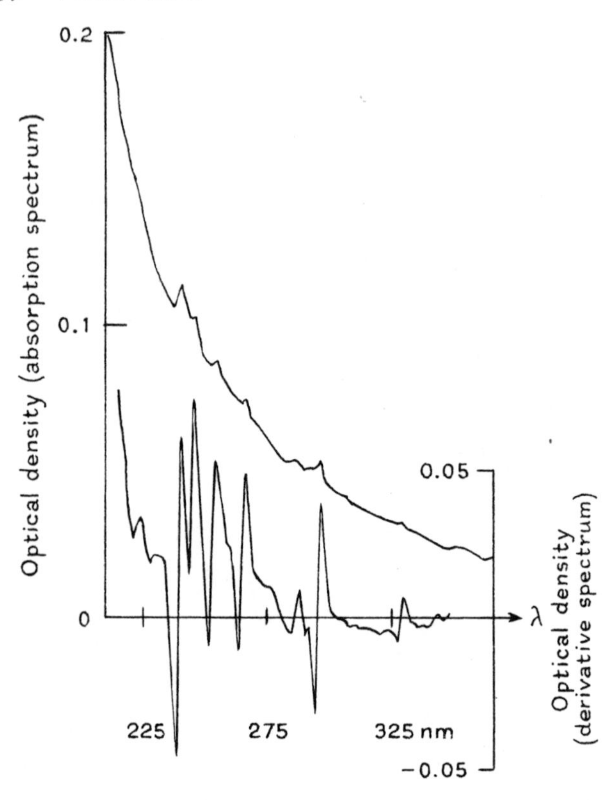

Fig. 4.4 Absorption and derivative spectra of iron atoms trapped in solid argon at 20 K. Note different optical density scales for the two spectra. [Adapted from Poliakoff, M., and Turner, J.J., *J. Chem. Soc., Faraday II*, **70**, 93 (1974).]

absorbance at λ_2 and that at λ_1 (slightly different from λ_1). As can be seen from Fig. 4.3 this may be either positive or negative. The maximum in the absorption band of course occurs at zero in the first derivative spectrum. An example of the use of the derivative technique is seen in Fig. 4.4, which shows the absorption spectrum of iron atoms trapped in a matrix of solid nitrogen at 20 K. The steep rise in absorption to lower wavelength is due to the fact that the matrix is highly scattering. Some of the iron atom absorptions almost blend into the background in the direct absorption case but can be readily seen in the first derivative. (For example the slight 'shoulder' on the upper spectrum at about 225 nm is fully visible in the lower trace.)

A machine which has been developed in recent years is the double beam, double wavelength spectrometer (for example the Perkin Elmer 356) which differs from the conventional type of spectrometer described above in that the diffraction grating is cut in half. Its two halves may be scanned independently. The sample beam is incident on one half of the grating and the reference beam on the other. When the two halves are locked together, the machine can be used as a

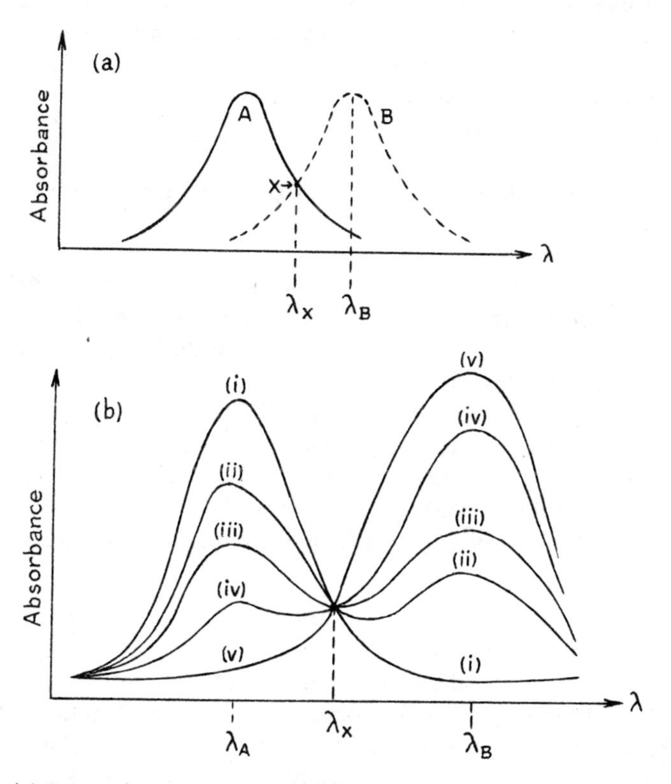

Fig. 4.5 (a) Schematic illustration of the absorption spectra of A and B in a transformation process A → B. Point X is the isosbestic point and remains at the same optical density throughout the reaction (b) as A is converted into B.

normal double beam spectrometer. If, however, the two halves are scanned with a small constant wavelength difference between them and both beams sent through the sample, the spectrum obtained is a close approximation to the first derivative of the absorption spectrum. Here with reference to Fig. 4.3, λ_2 is the sample beam wavelength and λ_1 is the reference beam wavelength.

The double wavelength technique is of particular importance when the sample is highly scattering. One often needs to measure the change in absorbance due to a reaction (brought about by a pH change or oxidation for example of the type A → B). For such a process the spectrum of the mixture at various stages of the reaction is shown in Fig. 4.5. Point X is called an *isosbestic point* (a point of equal absorbance during the transformation A → B). If now the reference wavelength is set at the wavelength of the isosbestic point (λ_X) the sample wavelength at λ_B and both beams are allowed to pass through the reaction cell, then the increase in absorbance at λ_B as the reaction proceeds may be readily recorded. If the reaction medium is highly scattering, any change in absorbance due to production of B may be easily seen since both the sample and reference beams are affected equally by scatter but unequally by absorption. In the conventional double beam arrangement it would be difficult to decide how much of the change in recorder signal was due to scatter and how much was due to absorption.

4.4 THE ELECTRONIC STATES OF POLYATOMIC MOLECULES

Electronic spectra arise form transitions between states, each of which describes a particular molecular electronic charge distribution. In order to be able to understand and interpret an electronic spectrum it is therefore necessary first to look at the electronic structure of the molecule to see how these states arise. This chapter is not the place to discuss at length the molecular orbital and other theoretical methods of quantitatively obtaining molecular orbital energy levels and state energies, but it will be found useful to look briefly at the molecular orbital structure of the simplest of all molecules, H_2.

By using the LCAO (linear combination of atomic orbitals) approach:

$$\psi_j = \sum_i c_{ij}\phi_i \tag{4.1}$$

where ψ_j describes a molecular orbital composed of contributions (weighted by the set of coefficients c_{ij}) from the atomic orbitals ϕ_i located on the atoms in the molecule. The ψ_j are normalized, i.e. $\int \psi_j^* \psi_j \, d\tau = 1$. For the simple diatomic H_2 molecule a bonding orbital (where the two hydrogen 1s atomic orbitals overlap in phase) is found:

$$\psi_b = \frac{1}{\sqrt{2(1+S^2)}}(\phi_{1s_1} + \phi_{1s_2}) \tag{4.2}$$

and an antibonding orbital:

$$\psi_a = \frac{1}{\sqrt{2(1-S^2)}}(\phi_{1s_1} - \phi_{1s_2}) \tag{4.3}$$

Here S is the overlap integral $\int \phi_{1s_1}^* \phi_{1s_2} \, d\tau$ between the two 1s orbitals on atoms 1 and 2. [In general, however, the c_{ij} of Equation (4.1) cannot be obtained this simply.] Making use of the quantum mechanical expression for the energy ϵ:

$$\epsilon = \int \psi_j^* \mathcal{H} \psi_j \, d\tau \tag{4.4}$$

where $\mathcal{H}$ is the Hamiltonian operator, the energy of the bonding orbital is given by substitution of (4.2) into (4.4):

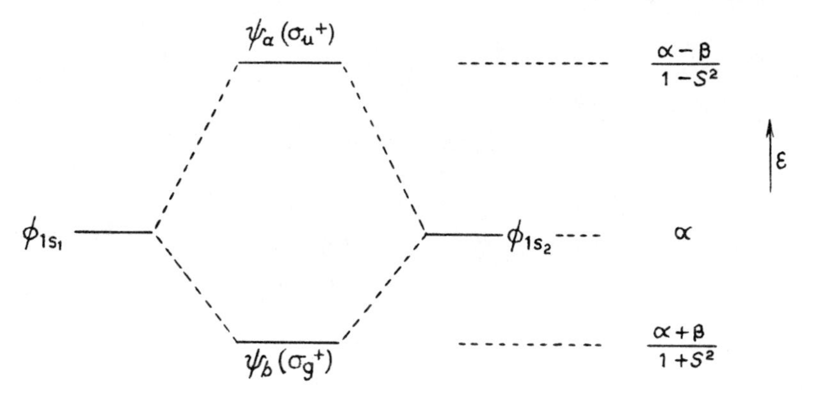

Fig. 4.6 Molecular orbital levels of the H_2 molecule.

$$\epsilon_b = \frac{1}{2(1+S^2)} \int \phi_{1s_1}^* \mathcal{H} \phi_{1s_1}\, d\tau + \int \phi_{1s_2}^* \mathcal{H}_{1s_2}\, d\tau + 2\int \phi_{1s_1}^* \mathcal{H} \phi_{1s_2}\, d\tau \quad (4.5)$$

The integral $\int \phi_{1s_1} \mathcal{H} \phi_{1s_1}$ is equal to the energy of an electron in a hydrogen atom 1s orbital and is conventionally given the symbol α. It is called the *Coulomb integral*. The integral $\int \phi_{1s_2}^* \mathcal{H} \phi_{1s_1}\, d\tau$ represents an interaction energy between the electrons in atomic orbitals ϕ_{1s_1} and ϕ_{1s_2} and is given the symbol β. It is called the *resonance integral*. ϵ_b is thus simply given by $(\alpha-\beta)/(1-S^2)$. These relative energies are shown in Fig. 4.6 on a molecular orbital diagram.

As well as writing the molecular orbitals as ψ_b or ψ_a, thus denoting whether they are bonding or antibonding between the two hydrogen nuclei, it is useful to use labels σ_g^+ and σ_u^+ which describe the symmetry properties of the resultant molecular orbital. These labels are generally of more use than ψ_a or ψ_b in polyatomic molecules since it is often difficult to decide between which pairs of atoms the molecular orbitals are bonding or antibonding. Also it has been demonstrated in Chapter 2 of Vol. 2 that the symmetry description of molecular properties is a very powerful tool. By filling the bonding orbital with two electrons the configuration $(\sigma_g^+)^2$ is achieved. The total orbital energy is given by $2(\alpha+\beta)/(1+S^2)$, and the total electronic energy will differ from this by the amount of the interelectronic repulsion energy. (This has been neglected in the very basic discussion here.) Since the Pauli principle restricts the two electrons to have opposed spins, the total electron spin of the system (S) is zero. The spin multiplicity $(2S+1)$ is equal to 1 and the state is a singlet. The electronic state is further described by the overall symmetry properties of the charge distribution of the two electrons. Since they both occupy the same molecular orbital the symmetry of the resultant state is given by the direct product $\sigma_g^+ \times \sigma_g^+ = \sigma_g^+$ (i.e. ${}^1\Sigma_g^+$). Promotion of an electron into the σ_u^+ antibonding orbital leads to a configuration $(\sigma_g^+)(\sigma_u^+)$. In this case the two electron spins may be parallel (triplet) or antiparallel (singlet). The spatial symmetry of the charge distribution is again given by the product of the species of the distributions of the two

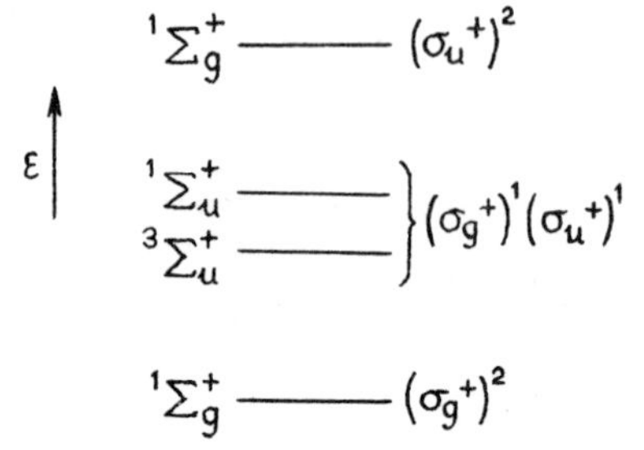

$^1\Sigma_g^+$ —— $(\sigma_u^+)^2$

ε

$^1\Sigma_u^+$, $^3\Sigma_u^+$ —— $(\sigma_g^+)^1(\sigma_u^+)^1$

$^1\Sigma_g^+$ —— $(\sigma_g^+)^2$

Fig. 4.7 Schematic arrangement of the four electronic states of the H_2 molecule arising from occupation of the orbitals of Fig. 4.6.

individual electrons, i.e. $\sigma_g^+ \times \sigma_u^+ = \sigma_u^+$. The states $^1\Sigma_u^+$ and $^3\Sigma_u^+$ thus arise. In general the triplet state always lies to lower energy than the singlet with the same electronic configuration. This can be understood if it is visualized that in the singlet state the electrons may occupy the same region of space whereas the Pauli principle requires that they remain further apart in the triplet, with therefore a smaller repulsion energy between them. (Things are in fact a bit more complicated than this; see [4.4].) If both electrons occupy the σ_u^+ orbital then a $^1\Sigma_g^+$ state results. These four states are represented schematically in Fig. 4.7.

The determination of the species of the electronic states of polyatomic molecules from a given electronic configuration is usually a much more difficult task. For example the configuration e^2 (two electrons occupying a doubly degenerate orbital of species e) gives rise to three electronic states $^3A_2$, $^1A_1$, and ^1E. The general method of derivation may be found in [4.5].

Electronic absorption spectra arise from transitions from the ground (lowest energy) state to those higher in energy (excited states). The energy separations, as have been seen, are dependent upon the energies of the orbitals containing the electrons and the electron–electron interactions for a given electron configuration. The group theoretical labels describe the symmetry properties of the electronic charge distribution, the exact energy of which may generally only be achieved after a considerable amount of time-consuming mathematics and computation. Depending upon the level of sophistication required, chemists use the observed transition energies (i) via rules of thumb to assign structures to unknown organic molecules [4.6, 4.7], (ii) to look in detail at the molecular orbital structure of a molecule, (iii) to obtain geometrical parameters such as bond lengths of angles from the observed fine structure as described in Chapter 2 (it was noted in the introduction to this chapter that this was a difficult task for most polyatomics), (iv) to obtain information about the electronically excited states of the molecule since these are useful in photochemistry, (v) in various applications such as kinetic studies where quantitative data concerning the concentration of various species can be obtained. Some of these applications will be discussed in the following sections.

4.5 INTERPRETATION OF THE ABSORPTION SPECTRA OF ORGANIC COMPOUNDS

In the previous section the electronic states corresponding to the various low energy orbital electron configurations for the hydrogen molecule were briefly derived. In general, the molecular orbital structure of an organic molecule will consist of a set of σ bonding orbitals at very low energy which are generally all occupied, some π bonding orbitals at higher energy followed by non-bonding orbitals (n) if any. To higher energy lie the π^* antibonding orbitals followed at higher energies still by the σ^* antibonding orbitals. The electronic transitions observed correspond to transitions between states of one electronic configuration to states of another [4.8].

Table 4.1

Chromophore	*Notation*	*Transition energy* (kJ mol^{-1})	*Absorption maximum* (nm)
σ Electrons			
>C—C< and >C—H	$\sigma \rightarrow \sigma^*$	800	~ 150
Lone pair electrons			
—Ö—	$n \rightarrow \sigma^*$	650	~ 185
—N̈<	$n \rightarrow \sigma^*$	600	~ 195
—S̈—	$n \rightarrow \pi^*$	600	~ 195
>C=O:	$n \rightarrow \sigma^*$	400	~ 300
>C=O:	$n \rightarrow \sigma^*$	630	~ 190
π Electrons			
>C=C<	$\pi \rightarrow \pi^*$	630	~ 190

Table 4.1 shows some typical transition energies and wavelengths for a series of *chromophores*. This term is used to describe the system containing the electrons responsible for the absorption. The highest energy transition is of course from the lowest bonding orbital (σ) to the highest antibonding orbital σ^*. As can be seen, most of the simple unconjugated systems give rise to absorptions of such short wavelength that they are not at all useful for routine detection. However,

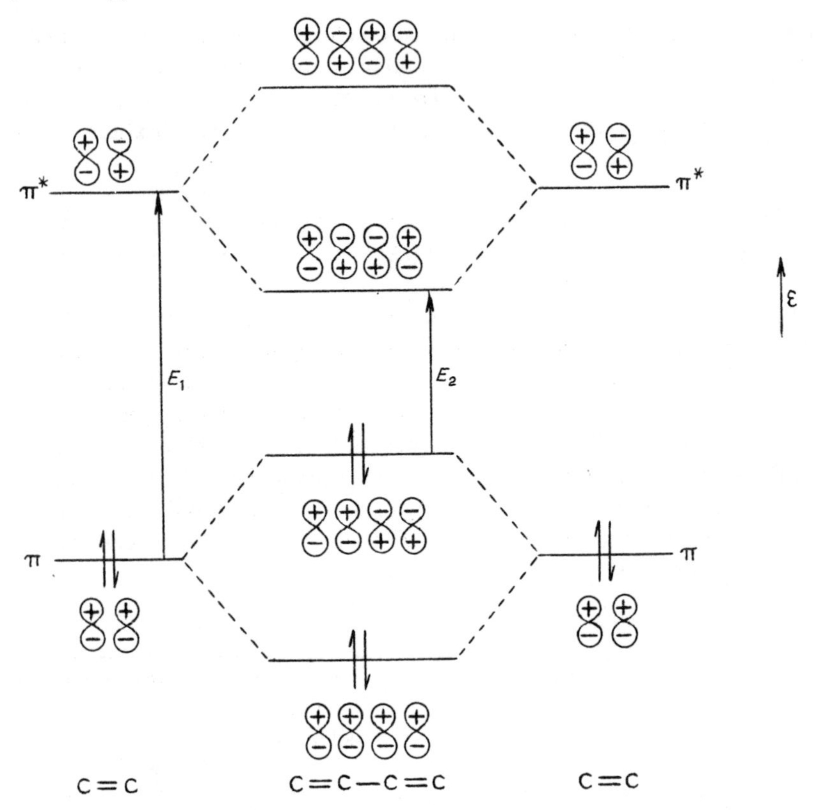

Fig. 4.8 Origin of the shift to longer wavelengths (lower energy) of the $\pi \rightarrow \pi^*$ transition on conjugation.

if the chromophore is conjugated then the wavelength of the $\pi \rightarrow \pi^*$ transition is in general longer than that of the isolated system. For example the $\pi \rightarrow \pi^*$ transition of an isolated double bond is typically found at ~ 190 nm (i.e. outside the range of most routine u.v. instruments) whereas in butadiene where the two double bonds are conjugated the absorption is easily observed at 217 nm. The mechanism for this is readily seen from Fig. 4.8. If two double bonds linking C_A and C_B, and C_C and C_D are brought together, the two π bonding orbitals, one bonding between C_A and C_B (C_A—C_B) and the other bonding between C_C and C_D (C_C—C_D), form bonding (C_A—C_B—C_C—C_D) and antibonding (C_A—C_B+ C_C—C_D) π orbitals between the two units. The former of course lies to lower energy since it is bonding between all carbon atoms. Similarly the two antibonding π^* orbitals (C_A+C_B) and (C_C+C_D) overlap to form two new orbitals, one bonding (C_A+C_B—C_C+C_D) and the other antibonding (C_A+C_B+C_C+C_D) between the two units as shown in Fig. 4.8. The lowest energy transition of this conjugated chromophore is smaller (i.e. the absorption occurs to longer wavelengths) than that of the isolated system. As more atoms are involved the absorption shifts to longer wavelength and eventually into the visible. This is the reason for the strong colours of many organic dyes which consist of long-chain conjugated chromophores. Generally as the absorption wavelength increases the extinction coefficient also increases. Thus $Ph(CH{=}CH)_3Ph$ has an absorption maximum at 358 nm with $\epsilon \approx 7.5 \times 10^4$ while $Ph(CH{=}CH)_7Ph$ has an absorption maximum at 435 nm with $\epsilon \approx 1.4 \times 10^5$. Usually a series of bands are seen close together in the spectrum as in Fig. 4.1; the figures quoted refer to the position of the band maximum of the most intense absorption. As the result of recording many thousands of u.v.–visible spectra of organic compounds, empirical rules [4.7] now exist for the calculation of band maxima for various types of transition as a function of the nature of the chromophore. For example, the value assigned to the parent open-chain diene is 214 nm for a spectrum recorded in an ethanol solution. To this must be added 5 nm for each alkyl substituent or part of a ring, 30 nm for another double bond in conjugation (i.e. a triene), 5 nm each for either a Cl or Br *auxochrome* (a substituent or a chromophore which shifts the absorption maximum to longer wavelength), etc. The comparison between the calculated band maximum and its observed value is generally very good (the error is between 1 and 3 nm).

For αβ-unsaturated ketone and aldehyde absorptions the rules are given in Table 4.2. Thus for the molecule (I) the longest wavelength absorption is at

(I)

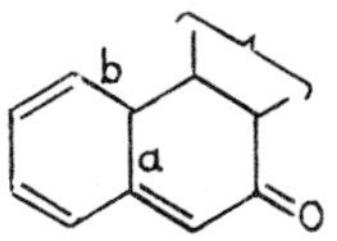

348 nm in excellent agreement with the value calculated [4.7] as follows:

Parent value	215 nm
β-Substituent (at a)	12
ω-Substituent (at b)	18
2 × Extended conjugation	60
Exocyclic double bond	5
Homoannular diene	39
	349 nm

Changing the solvent leads to another set of empirical corrections [4.7]. The origin of these solvent-induced shifts in the transition energies may sometimes be readily understood. In solution the molecule and chromophore are solvated by a sheath of solvent molecules. The excited electronic state is likely to be solvated differently from the ground state since now the electronic charge distribution of the molecule has been changed. Thus $\pi \rightarrow \pi^*$ transitions generally show a shift to longer wavelength (so called *red shift*) the more polar the solvent. This is of the order of 10 nm on going from a hexane to ethanol solution. Here the excited electronic state is likely to be more polar than the ground state and

Table 4.2 Rules for αβ-unsaturated Ketone and Aldehyde Absorption (in ethanol solution) [4.7]

$$\overset{\delta}{C}=\overset{\gamma}{C}-\overset{\beta}{C}=\overset{\alpha}{C}-C=O$$

Value assigned to parent αβ-unsaturated six-ring or acyclic ketone		215 nm
Value assigned to parent αβ-unsaturated five-ring ketone		202
Value assigned to parent αβ-unsaturated aldehyde		207
Increments for		
(a) a double bond extending the conjugation		30
(b) each alkyl group of ring residue	α	10
	β	12
	γ and higher	18
(c) auxochromes		
(i) —OH	α	35
	β	30
	δ	50
(ii) —OAc	α, β	6
(iii) —OMe	α	35
	β	30
	γ	17
	δ	31
(iv) —SAlk	β	85
(v) —Cl	α	15
	β	12
(vi) —Br	α	25
	β	30
(vii) —NR_2	β	95
(d) the exocyclic nature of any double bond		5
(e) homodiene component		39

thus stabilized (lowered in energy) more than the ground state by dipolar interactions with the solvent. On the other hand the $n \rightarrow \pi^*$ transition of the oxygen lone pair of the $>$ C=O group has a solvent shift in the opposite direction (shorter wavelength or *blue shift*) on going from a non-polar to a polar solvent. In this case the shift is due to the lesser extent to which the solvent may hydrogen bond to the oxygen atom of the carbonyl group in the excited state relative to the ground state. Other solvent shifts are not so readily explained but certainly arise through differential solvation of the ground and excited electronic states. The organic chemist generally uses u.v.–visible spectroscopy as an analytic tool in structure elucidation although he is often interested in the nature of the excited electronic states and their use in photochemistry.

4.6 SELECTION RULES FOR ELECTRONIC TRANSITIONS

Before moving on to polyatomic systems where a detailed analysis of the electronic spectrum may give information concerning the molecular orbital structure of the molecule, it is necessary to derive an expression for the intensity of an electronic transition. As seen in the chapter on atomic spectra in Vol. 1, selection rules will determine to a large extent whether a transition between two particular electronic states is observed or not in the absorption spectrum. It can be shown [4.2] that the intensity of an electronic transition [this is the integrated absorption intensity $\int_{\text{band}} \epsilon(\nu)\, d\nu$] is proportional to the square of the integral (see Appendix):

$$\int \Psi_{is}^{*} \mu \Psi_{fs}\, d\tau \tag{4.6}$$

where Ψ_{is} and Ψ_{fs} are the wavefunctions describing the electronic charge distribution in the initial and final electronic states respectively. Ψ is a multielectron function and is composed of contributions from the wavefunctions Ψ describing the molecular orbitals holding the electrons. It also contains a description of the total electron spin of the system. The symmetry description of the electronic charge distribution in the ground and excited states is also inherent in Ψ. μ is the dipole moment operator and transforms as the same irreducible representation as the cartesian coordinates x, y, and z. Equation (4.6) represents thus a transition dipole moment (between the two electronic states) and it is this that may interact with the electric vector of the incident light beam and result in the absorption of energy via an electric dipole process. For this integral to be non-zero it must be totally symmetric, i.e. it must belong to the highest symmetry irreducible representation of the molecular point group. (The evaluation of such triple products is discussed in Chapter 2 of Vol. 2.) By a knowledge of the symmetry species [4.9] of the electronic state involved it can be rapidly ascertained whether the integral is zero and thus whether the particular transition is forbidden. [Remember, however, even if group theory predicts a non-zero value, the integral of (4.6) may still be close to zero and the transition not observed.]

As an example consider the electronic transitions from the ground state of the H_2 molecule to the higher excited states derived in Section 4.4. For the $D_{\infty h}$ point group the dipole moment μ belongs to the representation $\Pi_u + \Sigma_u^+$. The electronic transition from the ground state $^1\Sigma_g^+$ to the state $^1\Sigma_u^+$ (arising from the configuration $\sigma_g^+\sigma_u^+$) gives rise to an integral of symmetry species $\Sigma_g^+ \times \Sigma_u^+ \times \Sigma_u^+ = \Sigma_g^+$, i.e. the totally symmetrical representation. Thus this transition is *symmetry* or *orbitally* allowed. On the other hand, the transition to the $^1\Sigma_g^+$ state (σ_g^{+2}) is symmetry forbidden since $\Sigma_g^+ \times \Sigma_u^+ \times \Sigma_g^+ = \Sigma_u^+$. (Another reason for the forbidden nature of this particular transition is that it corresponds to a two-electron jump. The theory will not be described here, but because of the nature of the wavefunctions Ψ, transitions involving the promotion of more than one electron are forbidden.) Since the molecule is constantly vibrating, this symmetry selection rule may be somewhat relaxed. During a non-totally symmetric vibration the molecular symmetry is lost and the electronic transition may thus become allowed during the time the molecule spends away from its equilibrium position. These *vibronic* bands are generally much weaker than the fully allowed electronic transitions.

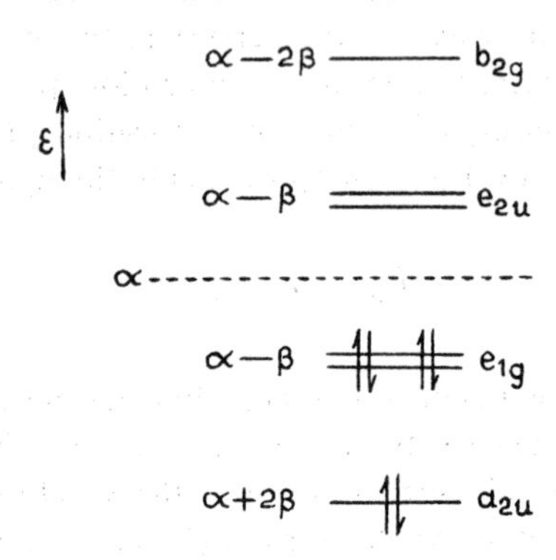

Fig. 4.9 The molecular orbitals of the π framework of benzene using Hückel molecular orbital theory.

The third possible transition for the hydrogen molecule example is the process ${}^1\Sigma_g^+ \rightarrow {}^3\Sigma_g^+$. This, however, is spin forbidden due to the fact that the integra (4.6) is zero unless Ψ_{fs} and Ψ_{is} describe systems of the same spin multiplicity. This leads to the selection rule $\Delta S = 0$ as in the atomic case. If *spin–orbit coupling* occurs, this selection rule is relaxed a little and transitions with $\Delta S = \pm 1$ may be observed, the intensity of such absorptions increasing with the size of the *spin–orbit coupling constant*. This increases with atomic number, 28 cm^{-} for C, 3000 cm^{-1} for Os for example. [The effect of spin–orbit coupling is to mix in a small amount of triplet character ($S = 1$) into a singlet ($S = 0$) ground state, for example such that Ψ_{is} of Equation (4.6) may be written $a\Psi_{sing} + b\Psi_t$ with $a \gg b$. Then the integral (4.6) with $\Psi_{fs} = \Psi_{trip}$ contains a non-zero component (on the basis of the $\Delta S = 0$ rule) of $b\int \Psi_{trip}\mu\Psi_{trip}\, d\tau$.]

The electronic spectrum of benzene illustrates these points. Only one strong band is seen in the spectrum, at 180 nm. Two weaker ones are observed at 200 and 260 nm, and a further band, much weaker than these two, at around 350 nm Transitions involving σ electrons occur at higher energies than the bands observe here, so these transitions must involve promotion of electrons held in π orbitals. By means of simple Hückel molecular orbital theory [4.10] a molecular orbital diagram can be obtained for the π network of orbitals. This is shown in Fig. 4.9. The molecular orbitals are labelled with their symmetry descriptions under the D_{6h} point group of the benzene molecule. The b_{2g} and e_{2u} orbitals are destabilized with respect to the free carbon p orbital, whereas the e_{1g} and e_{2u} orbitals are stabilized. (In this simple treatment, the overlap integral between adjacent carbon atoms has been ignored.) The electronic ground state arises from the configuration $(a_{2u})^2(e_{1g})^4$ and is ${}^1A_{1g}$. (For any configuration where all occupied orbitals are completely filled, a totally symmetric electronic state

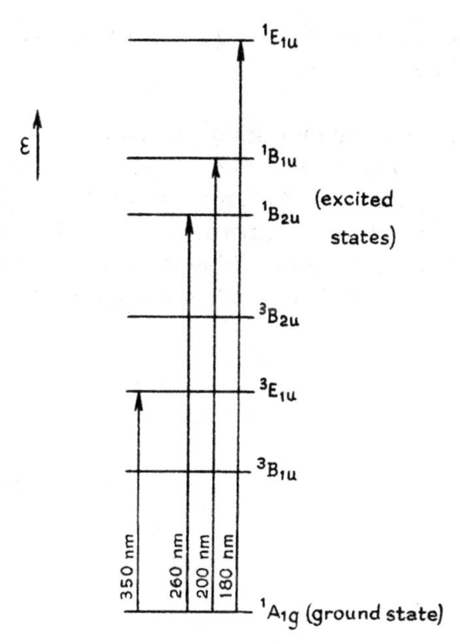

Fig. 4.10 The calculated order of the lowest excited states of benzene.

results). The lowest energy excitation will involve promotion of an electron from the e_{1g} to e_{2u} orbital. The promoted electron may have a spin parallel or antiparallel to the unpaired electron left behind in the e_{1g} orbital and so both singlet and triplet states are possible. In fact from this $(a_{2u})^2(e_{1g})^3(e_{2u})^1$ configuration, the states $^3B_{1u}$, $^3B_{2u}$, $^3E_{1u}$, $^1B_{1u}$, $^1B_{2u}$, and $^1E_{1u}$ can be derived. All that remains now is to make an estimate of the relative intensities of the transitions from the ground $^1A_{1g}$ state to these excited states by determining whether the integral (4.6) will be zero or non-zero. Firstly all transitions from the $^1A_{1g}$ ground state to the triplet states will be spin forbidden. Secondly in the D_{6h} point group μ transforms as $A_{2u} + E_u$. The reader may readily show by evaluating the necessary triple products that the transitions to E_{1u} states only are symmetry allowed. Thus the transition $^1E_{1u} \leftarrow {}^1A_{1g}$ is both symmetry and spin allowed and almost certainly gives rise to the strong band in the spectrum at 180 nm. The transitions to the states $^1B_{1u}$ and $^1B_{2u}$ are spin allowed but symmetry forbidden. As noted previously they may become vibronically allowed with much lower intensities. Transitions to these states can be assigned to the bands at 200 and 260 nm. The much weaker band at 350 nm is then due to the transition $^3E_{1u} \leftarrow {}^1A_{1g}$ which is symmetry allowed but spin forbidden. Since it occurs to lower energy than the other three transitions it must be assigned to a transition to a triplet state. The two transitions $^3B_{1u} \leftarrow {}^1A_{1g}$ and $^3B_{2u} \leftarrow {}^1A_{1g}$ are not seen in the spectrum since they are both symmetry *and* spin forbidden. A calculated energy level diagram for the electronic states of benzene is shown in Fig. 4.10. Note that the electronic spectrum tells nothing about the positions of the $^3B_{1u}$ and $^3B_{2u}$ states.

4.7 ELECTRONIC SPECTRA OF TRANSITION METAL COMPLEXES

This is a huge field in itself and there is not the space in this chapter to delve too deeply into the subject. The reader is referred to the books by Orgel [4.11] and Kettle and coauthors [4.12, 4.13] for a more detailed discussion of the theory and molecular approaches that have been used. As an indication of what can be observed under favorable conditions and how the spectrum can be interpreted, the spectra of the Cr^{3+} and Mn^{2+} ions will be considered. The Cr^{3+} ion in the

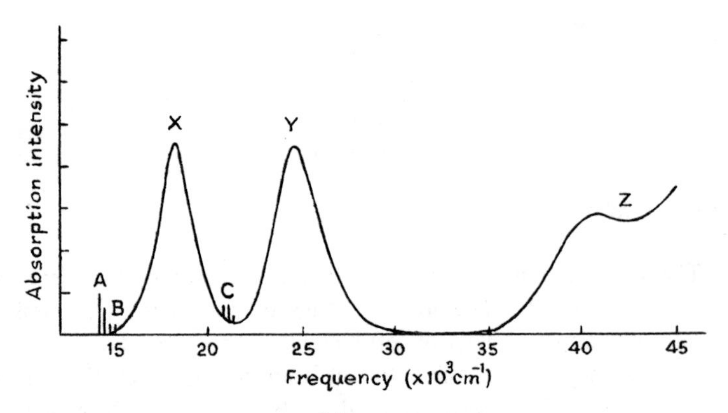

Fig. 4.11 Electronic spectrum of Cr^{3+} ions in alumina (ruby).

gas phase has the ground electronic configuration . . . $(3d)^3(4s)^0$, this being of lower energy than the electronic states derived from . . . $(4s)^2(3d)^1$. Using the Russell–Saunders coupling scheme described in Chapter 1 of Vol. 1, the electronic states derived from this d^3 configuration are ^4F, ^4P, ^2H, ^2G, ^2F, ^2D (twice), and ^2P. Experimentally the energies of these electronic states of the ion are found to be

^2F	36 700 cm^{-1}
^2H	21 200 cm^{-1}
^2D	20 400 cm^{-1}
^2G	15 200 cm^{-1}
^2P	14 200 cm^{-1}
^4P	14 200 cm^{-1}
^4F	0

where the ^4F state is (in agreement with Hund's rules) the electronic ground state. The absorption spectrum is found to consist of a series of sharp lines. Ruby results when the Cr^{3+} ion is present as an impurity in alumina crystals and the

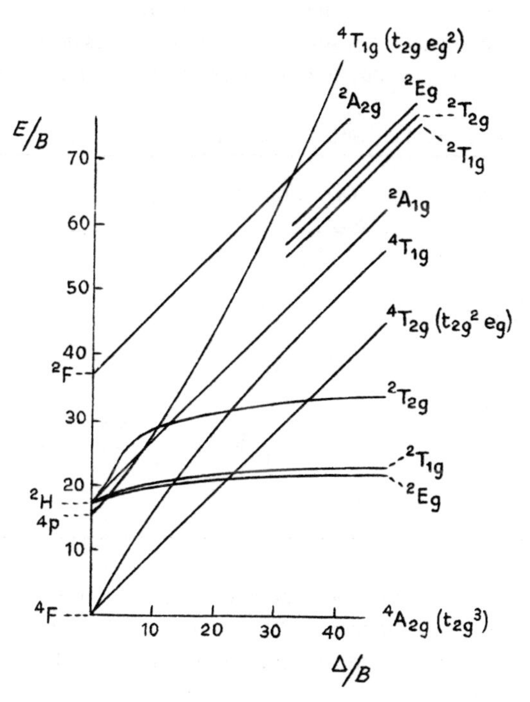

Fig. 4.12 Tanabe–Sugano diagram for a d^3 system.

spectrum (Fig. 4.11) changes dramatically; several absorptions are observed, some much broader than others. The general features of the spectrum are observed in many other systems such as $Cr(H_2O)_6^{3+}$ where the Cr^{3+} is octahedrally co-ordinated by other groups, the actual positions of the bands varying from compound to compound. In an octahedral environment the highest degeneracy allowed for an electronic state is three, so all the atomic terms with dimensions greater than three will split apart in energy and their degeneracy will thus be partially or completely removed. The ground state ^4F Cr^{3+} term for example splits apart into three electronic states $^4T_{1g}$, $^4T_{2g}$, and $^4A_{2g}$. The exact way these terms split apart can be found elsewhere [4.11, 4.12] but is generally represented on a *Tanabe–Sugano diagram.* Figure 4.12 shows such a diagram for the d^3 system. Δ is a parameter which gives an idea as to the 'strength' of the crystal field (more about this below) or the 'size' of the interaction between the transition metal atom or ion and the ligands, and B is a parameter which represents the magnitude of the interelectronic repulsions. On the left-hand side of Fig. 4.12 are the energy levels of the free ion and on the right-hand side is the region where the change in energy with Δ is linear. The electronic states of the coordinated ion are described by the symmetry species of the electronic charge distribution they represent. In the free gaseous ion the term classification described the angular momentum properties of the atomic charge distribution. It is noted that all the electronic states are labelled with a g (i.e. symmetric with respect to inversion). This arises simply because the states derive from electron configurations involving occupation of the centrosymmetric d orbitals.

Table 4.3 Intensities of electronic transitions

Transition type	*Oscillator strength, f*	*Molar absorption coefficient, ϵ*
Spin forbidden, Laporte forbidden	10^{-7}	10^{-1}
Spin allowed, Laporte forbidden	10^{-5}	10
Spin allowed, Laporte forbidden but with orbital mixing	10^{-3}	10^{2}
Spin allowed, Laporte allowed (e.g. charge transfer)	10^{-1}	10^{4}

In the O_h point group the dipole moment operator transforms as t_{1u}. It can be seen immediately that the integral (4.6) is thus equal to zero on group theoretical grounds for all transitions within the d manifold of orbitals. (In all cases ground and excited states are g, so the triple product is u.) All the d–d transitions of octahedrally coordinated systems are thus symmetry forbidden. This is called the *Laporte rule.* (They are also Laporte forbidden in geometries lackin a centre of symmetry because microscopically the d orbitals are still symmetric with respect to inversion through the centre of the metal atom.) As noted above these transitions may, however, become vibronically allowed (with an oscillator strength (see Appendix) of 10^{-4}–10^{-5}. See Table 4.3). As the molecule vibrates the strength of the crystal field (i.e. the value of Δ) changes and this gives a clue as to the relative widths of these vibronic transitions. If the transition energy changes markedly as Δ changes (i.e. the slope of the excited state on the Tanabe–Sugano diagram is different from that of the ground state) then a broad band is expected. The narrowest bands will derive from those transitions where the two slopes are equal. The electronic spectrum of ruby is thus readily explained. The most intense bands will be the spin allowed $\Delta S = 0$ transitions, $^4A_{2g} \rightarrow {}^4T_{2g}, {}^4T_{1g}, {}^4T_{1g}$. From the slopes of the Tanabe–Sugano diagram, these will be broad and are assigned to the bands X, Y, Z. The sharper but weaker transitions are assigned to the spin forbidden $\Delta S = -1$ transitions $^4A_{2g} \rightarrow {}^2T_{1g}$, ^2E_g (features A and B), and the transition $^4A_{2g} \rightarrow {}^2T_{2g}$ (feature C). These are weakly observed ($f \approx 10^{-7}$) because of spin–orbit coupling, and arise through transitions to states with slopes similar to that of the ground state. Other $\Delta S = \pm 1$ transitions include the higher energy transitions to the $^2A_{1g}$ etc. states

(Fig. 4.12). Since the slopes of these states are very different from that of the ground state they give rise to broad weak bands which readily merge with the background and are not seen. The transition energies may be fitted to the Tanabe–Sugano diagram to give values of Δ and B. Qualitatively for example the $^4T_{2g}$ state must lie lower than $^2T_{2g}$ since the broad band X lies between the sharp sets of features A, B, and C.

The electronic states of the coordinated Cr^{3+} ion must of course be related to the particular electronic configuration on a molecular orbital basis. In an octahedral environment, under the influence of the coordinated ligands, the five d orbitals, degenerate in the free atom or ion, split apart in energy into two sets, the triply degenerate t_{2g} set and the higher energy e_g set. The energy separation between them is dependent upon the strength of the crystal field and is equal to $\Delta\ cm^{-1}$ (sometimes called 10 Dq). The lowest energy electronic configuration for the Cr^{3+} system will be $(t_{2g})^3$. This gives rise to four electronic states, $^4A_{1g}$, $^2T_{1g}$, ^2E_g, and $^2T_{2g}$ as shown in Fig. 4.13. The electronic ground state is the $^4A_{1g}$ since

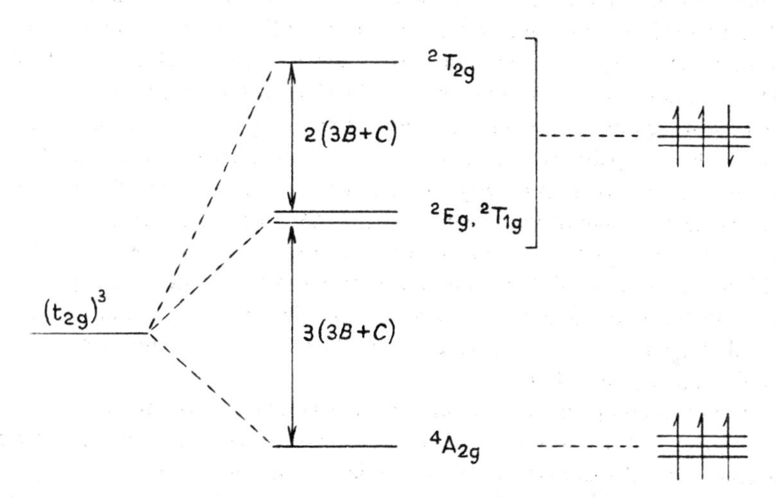

Fig. 4.13 Electronic states arising from the electronic configuration $(t_{2g})^3$.

here maximum spin multiplicity occurs and the electrons are kept apart by the Pauli principle. The energies of these four states differ only in electron–electron repulsion energies, their total orbital energies being equal. The functions B and C are called *Racah parameters* and measure the various types of electron–electron interactions which occur; they depend upon the d orbital wavefunction and are constants for a particular atom or ion. At high Δ on the Tanabe–Sugano diagram (i.e when the orbital energy is comparable to the electron–electron terms) these four states run parallel to one another. The $^4T_{1g}$ state however (arising from the configuration $t_{2g}^2e_g^1$), has a positive slope compared to the ground state, since the orbital contribution to the energy is Δ higher than that of

Table 4.4

Absorption	*Transition*
A lines	$t_{2g}^3\ {}^4A_{2g} \rightarrow t_{2g}^3\ {}^2E_g$
B lines	$\rightarrow t_{2g}^3\ {}^2T_{1g}$
X band	$\rightarrow t_{2g}^2 e_g\ {}^4T_{2g}$
C lines	$\rightarrow t_{2g}^2\ {}^2T_{2g}$
Y band	$\rightarrow t_{2g}^3 e_g\ {}^4T_{1g}$
Z band	$\rightarrow t_{2g} e_g^2\ {}^2T_{1g}$

these other states. The states arising from the configuration $t_{2g} e_g^2$ similarly have double this slope. Note that the states arising from the same electronic configuration on the right-hand side do not necessarily derive from the same atomic terms on the left-hand side. A summary of the ruby transitions is given in Table 4.4. To obtain a quantitative agreement, it is found that one has to chose the values of the parameters $\Delta = 17\,000\ \text{cm}^{-1}, B = 700\ \text{cm}^{-1}, C = 2800\ \text{cm}^{-1}$.

The octahedrally coordinated Mn^{2+} ion as in $Mn(H_2O)_6^{2+}$ for example in solution is a particularly interesting case. Here there are five d electrons and the d orbital configuration of maximum spin multiplicity is $(t_{2g})^3(e_g)^2$. This will be the electronic ground state as long as the orbital energy separation (Δ) is smaller than the electron pairing energy and gives rise to an electronic state ${}^6A_{1g}$. It is quite obvious that there is no way a d–d transition can occur which conserves the total electron spin of the system (i.e. with $\Delta S = 0$), since if an electron is promoted from the t_{2g} orbital it must enter the e_g orbital with opposite spin. The d–d transitions of the Mn^{2+} ion are thus both Laporte forbidden and spin forbidden. In consequence only very weak bands indeed ($\Delta S = -1$ transitions allowed by spin–orbit coupling) are seen in the spectrum, and manganous salts are only faintly pink coloured.

Tetrahedral complexes on the other hand generally give rise to much more intense spectra. The reasons for this will not be discussed here in detail but are concerned with the sort of mixing of atomic orbitals on the metal that may occur in this point group but not in the octahedral case. For example p and d orbitals both transform as t_2 in the tetrahedral point group (and may thus mix together) but in the octahedron p orbitals are ungerade, d orbitals are gerade, and no mixing is allowed. The p–d mixing in the tetrahedral case which gives rise to this increased intensity is also the factor which occurs on vibration of octahedral complexes allowing vibronic transitions. The d orbitals split apart in energy in the opposite way to that for the octahedron, i.e. the triply degenerate (t_2) set lie to higher energy that the doubly degenerate (e) set. The energy separation between the two is smaller than that for octahedral systems ($\Delta_{tet} \approx 4/9\ \Delta_{oct}$). Here lies a ready explanation of the experimental observation that many octahedrally coordinated Co^{II} ions are *pale red* whereas their tetrahedral counterparts are an *intense blue.*

The actual colours or, more strictly, the actual position of the d–d absorption bands in the spectrum are of considerable interest since the magnitude of interaction between the central metal atom and ligands will vary with the nature of the ligand. The larger the interaction, the larger Δ becomes and therefore the higher the transition energy. When the donor atoms of the ligands are arranged in the order of the magnitude of the values of Δ produced by them, the following *spectrochemical series* is obtained:

$$I < Br < Cl < S < F < O < N < C.$$

This order is one which is approximately obeyed in a whole range of different ligands where the ligand coordinates through these atoms. The nature of the central metal atom also influences the transition energy. Generally the order of Δ is M(II) < M(III) < M(IV) and for the oxidation number 2:

$$Mn < Ni < Co < Fe < V$$

for example.

These two series have led to an empirical set of rules which enable one to determine the value of Δ for any pair of ligand and metal. $\Delta = f \times g$ where f is ligand parameter and g a contribution from the central atom. A list of f and g values for some representative ions and ligands is shown in Table 4.5 for octahedral complexes. As can be seen the values bear out the two series noted above.

Table 4.5 Values of f and g for some selected ions and ligands

Ligand	f	*Ion*	g
F^-	0.9	V(II)	12.3
H_2O	1.00	Cr(III)	17.4
urea	0.9	Mn(II)	8.0
NH_3	1.25	Mn(IV)	23
en	1.28	Fe(III)	14.00
Cl^-	0.80	Co(III)	19.00
CN^-	1.7	Ni(II)	8.9
Br^-	0.76	Rh(III)	27.0
I^-	0.7	Re(IV)	35

$\Delta = f.g \times 10^3\ cm^{-1}$

Another series of interest to chemists is the *nephelauxetic series*, the variation of the Racah parameters as a function of metal and ligand. In general the value of the Racah parameter B decreases on coordination from the free ion value, i.e., $\beta = B_{free}/B_{complex} > 1$. This nephelauxetic series is in the order of the electronegativities of the donor atoms:

$$F < O < N < Cl < Br < S < I < Se$$

β can be written as $(1 - \beta) = h.k$ in a similar fashion to the value of Δ above, h and k being contributions to the effect from the ligand and central atom of the complex [4.14]. The sizes of Δ and β can often be correlated with the molecular orbital properties of the ligands.

The chemist may thus derive valuable information about the molecular orbital structure of the complex from the number, position (i.e. energy), width, and intensity of the electronic absorptions in the spectrum. When detailed quantum mechanical calculations of the molecular orbital structure are possible, correlation at the quantitative level is possible. Even when such calculations are lacking the qualitative analysis of the spectrum can allow conclusions to be drawn about the magnitude of the ligand field splitting (Δ), possible distortions from symmetric environments, bonding characteristics, and the relative ordering of the molecular orbital energy levels of qualitative theories.

4.8 CHARGE TRANSFER SPECTRA

In the examples discussed previously, the electron involved in the transition has been associated with the same atom or group of atoms in both the ground and excited states. For example, the butadiene transitions were from an orbital which described one arrangement of the four π orbitals on the carbon atoms, to another. The d–d transitions in the transition metal examples, as their name implies, occurred from a molecular orbital, largely metal d in character, to a similar metal-localized molecular orbital. The permanganate ion, however, has a well known deep purple colour; but since the electronic configuration of MnO_4^- is d^0, no d–d transitions can occur. The explanation of the permanganate colour (and indeed the intense colour of most inorganic indicators) is a different sort of electronic transition from those described above. They occur between electronic states where the electron involved in the transition moves from one group of atoms to another. These charge transfer transitions are usually very intense since the electron often has to jump a large distance in the transition thus giving rise to a large transition dipole moment in Equation (4.6). The transitions are not Laporte forbidden in transition metal complexes since *both* states are not concerned with the metal d orbitals. Figure 4.14 schematically indicates the process. If the interaction between the orbitals ϕ_A and ϕ_B (these may be orbitals located on a single atom or group of atoms) is small, the new orbital ϕ_C is composed mainly of ϕ_B, and ϕ_D is composed mainly of ϕ_A. In the MnO_4^- case, the lower energy orbital ϕ_B is an orbital located on the oxygen ligands whereas ϕ_A is a metal d orbital. The transition $\phi_C \rightarrow \phi_D$ is called a *ligand to metal charge transfer transition* and results in the oxidation state of the metal decreasing by 1. In several other systems (e.g. metal carbonyls) where there are high-lying empty ligand orbitals, ϕ_B may be a metal d orbital and ϕ_A a ligand orbital. In this case a metal to ligand charge transfer transition involves an increase of 1 in the metal oxidation state. In fact, the transfer of charge is never complete and all electronic transitions involve a certain amount of electron redistribution in the excited state compared with the ground state. The term charge transfer is reserved for those transitions where this process is very marked.

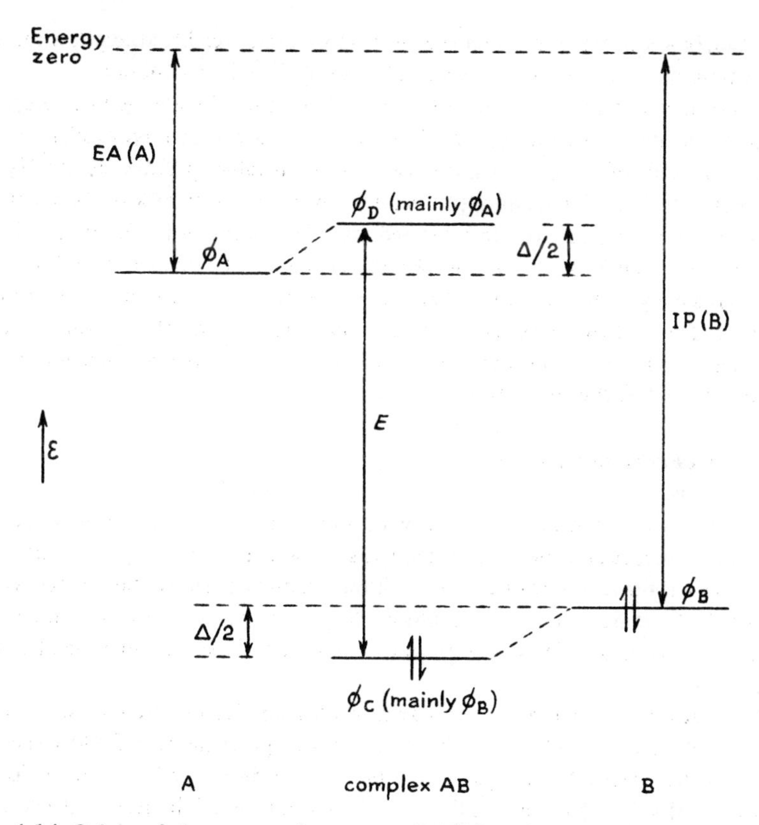

Fig. 4.14 Origin of charge transfer spectra. EA(A) = electron affinity of A; IP(B) = ionization potential of B; Δ = stabilization energy of AB.

The simple ideas of Fig. 4.14 can be used to explain the spectra of so called charge transfer complexes [4.15]. If I_2 is dissolved in benzene, a new electronic absorption is seen which is characteristic of neither iodine nor solvent. A 1 : 1 complex is formed through overlap of a low energy pair of electrons in a benzene 'donor' orbital (ϕ_B) and an empty higher energy 'acceptor' orbital (ϕ_A) on the I_2 molecule. The resultant complex gives rise to a new electronic absorption of energy IP(B) − EA(A) + Δ, where the symbols represent the ionization potential of a donor electron IP(B), the electron affinity of the acceptor molecule EA(A), and the total stabilization energy of the complex (Δ). If methyl groups are introduced into the benzene ring, the ionization potential of the donor is reduced. The effect of this change is readily observed in a shift of the absorption band to longer wavelength (lower energy) in accordance with these ideas.

4.9 ELECTRONIC SECTRA OF SHORT LIVED SPECIES

The electronic spectra of stable molecules are generally readily studied, whether in solution, in the gas phase, or as a single crystal. The determination of the spectra of unstable species is more difficult. Under this title are included electronically excited molecules in a triplet or excited state of spin multiplicity different from that of the ground state. Very often unstable species such as CH_3, CH_2, BH_3, etc. are of tremendous interest from the point of view of their structure and importance as reaction intermediates. The means of production and study of such transients, including charged species, can be divided into three types: (i) steady state generation, (ii) matrix isolation, and (iii) rapid detection methods. The first two approaches require standard u.v.–visible recording equipment; the third necessitates the rapid recording of the electronic spectrum during the lifetime of the short-lived species.

4.9.1 Steady state methods

A steady state concentration of short lived species may be obtained under certain circumstances and their electronic emission and absorption spectra obtained. If a furnace is made a part of the u.v. cell for example a steady state concentration of the C_3 radical may be detected in the gaseous phase above graphite at 3100 K. If silicon is added to the graphite then the SiC_2 species may be observed in addition.

The oldest known sources of emission spectra are electric discharges. These may be obtained by applying a suitable a.c. or d.c. potential (e.g. 2000 V) across the two metal electrodes in a gas at low pressure. A discharge functions by initial ionization of some of the molecules, then by excitation of the resulting electrons and ions, and finally by bombardment of other molecules with these excited particles. In this manner excited molecules are produced which may lose their excess energy by emission. If the energy of the discharge is sufficiently great, fission of the molecules into radicals or ions can result. The energy conditions required for a given type of discharge are greatly dependent on the ionization potential of the gas examined and the pressure of the gas within the discharge. Most types of discharges tend to split polyatomic molecules into diatomic species. One milder kind of discharge is that of the Tesla type which may include mild fission of the polyatomic molecule or even just electronic excitation of the polyatomic molecule itself.

In the Schüler tube the discharge is struck in a gaseous mixture of the parent molecule and a carrier gas such as helium (Fig. 4.15). The liquid nitrogen traps A and B are situated so that no gas other than helium can reach the electrodes. In addition the spent products of the discharge are removed here and fresh parent molecules enter the discharge at C. By adjusting the rate of addition of material at C the steady state concentration of fragments can be adjusted. The windows D allow observation of the absorption or emission spectra of the

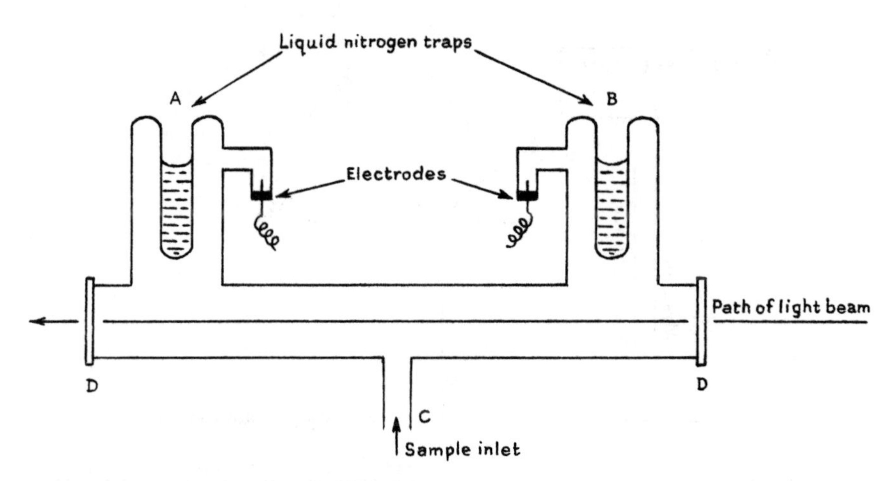

Fig. 4.5 The Schüler tube for the observation of the spectra of a steady-state concentration of short-lived species produced typically from an organic molecule through which a discharge has been passed.

products of the discharge. A similar method, somewhat limited by low intensity, is the use of high energy electrons to produce molecular fragments. This may be achieved using a low pressure, hot cathode discharge tube. CH is one of the best known radicals and may be obtained in almost any discharge, which is not too mild, through hydrocarbon compounds. Since the moment of inertia of the CH radical is small, the rotational structure is plainly resolvable [4.16].

Steady state concentrations of short-lived species may often be obtained in flow tubes [4.17]. For example, if a glass tube is connected to a pump, and nitrogen gas leaked through the tube, then a microwave powered discharge may be induced in the flowing nitrogen to produce about a 1 per cent concentration of nitrogen atoms. A yellow glow is observed around the discharge and downstream of it arising from the process:

$$N + N \rightarrow N_2^* \xrightarrow[\text{yellow}]{\text{emit}} N_2$$

If NO is now bled into the tube downstream of the discharge under the right conditions, a green glow is observed since:

$$N + NO \rightarrow N_2 + O \qquad \text{(1 collision in 4)}$$

$$O + NO \rightarrow NO_2^* \xrightarrow[\text{green}]{\text{emit}} NO_2$$

The electronic spectra of these short-lived species (excited electronic states) may then be recorded.

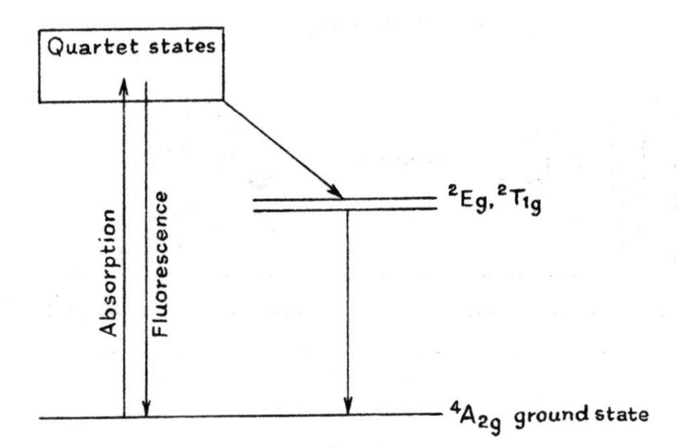

Fig. 4.16 Method of populating the low-lying doublet states of ruby for either (i) investigation of the electronic spectra arising from these states, or (ii) laser action.

It was noted in a previous section that the electronic ground state of the Cr^{3+} ion in ruby was a quartet whereas there were two low-lying doublet states ($^2T_2$, ^2E). Transitions to these states were forbidden but were weakly observed because of spin–orbit coupling. By Einstein's laws of absorption and emission of radiation (see Appendix), since the transition probability for absorption to this state from the ground state was low, so the probability of decay back to the ground state was also low, i.e. the 'lifetime' of these states could be quite long (it is this factor which allows the ruby laser to function). Thus by optically pumping the ruby crystal in one of the spin allowed bands so that a significant population exists in the $^2T_2$ and ^2E states as in Fig. 4.16, the electronic absorption spectrum of ruby in these excited states may be readily obtained.

4.9.2 Matrix isolation

The matrix isolation technique has been described in Chapter 4 of Vol. 2. In brief, if an unstable species is surrounded with an inert rigid material, e.g. Ar or Xe at cryogenic temperatures, it may be examined spectroscopically at leisure in this trapped state. An example will illustrate the method. Jacox and Milligan [4.18] have observed an electronic transition of the HNF radical trapped in solid Ar. Mixtures of Ar and F_2 (150:1) and Ar and H (or D)N_3(200:1) were cocondensed on a spectroscopic window at 14 K. The matrix was then photolysed using a medium pressure mercury lamp. Using a chlorine filter (such that the wavelength of the irradiating light was $280\text{ nm} < \lambda < 400\text{ nm}$) the HN_3

could be photolysed:

$$HN_3 \xrightarrow{h\nu} NH + N_2$$

and the production of the NH radical followed by recording its infrared spectrum. Photolysis without the chlorine filter allows photodissociation of F_2, and HNF could then be formed by reaction of the released F atoms and NH:

$$NH + F \xrightarrow{\text{diffusion}} HNF$$

The concentration of the reactants is such that the F atom only has to diffuse a short way to the adjacent matrix cage to react with the NH. Identification of the HNF species was achieved by looking at the isotopic infrared spectrum using H(D), ^{14}N (^{15}N) and deciding that HNF was the only molecule consistent with the observed data. Concurrent with the growth of infrared features assigned to HNF on photolysis was the growth of an absorption in the region 390–500 nm. A list of the observed bands is given in Table 4.6 for both HNF and DNF. A

Table 4.6

HNF		*DNF*		*Assignment*		
λ (nm)	Δ (cm^{-1})	λ (nm)	Δ (cm^{-1})	$v_1'\ v_2'\ v_3'$		$v_1''\ v_2''\ v_3''$
469.5		495.5		0, 0, 0,	←	0, 0, 0
	1081		764			
471.2		476.5		0, 1, 0	←	0, 0, 0
	1060		777			
448.8		459.5		0, 2, 0	←	0, 0, 0
	1006		836			
429.4		442.5		0, 3, 0	←	0, 0, 0
	1019		820			
411.4		427.0		0, 4, 0	←	0, 0, 0
	997		794			
395.2		413.0		0, 5, 0	←	0, 0, 0

vibrational progression in the bending mode of the excited state is observed. The average figure for this frequency is 1033 cm^{-1} (for HNF) and 798 cm^{-1} (for DNF). The corresponding frequencies in the electronic ground state (from the infrared spectrum) are 1432 and 1069 cm^{-1}. The assignment column of Table 4.6 gives the values of the three vibrational quantum numbers (v_1, v_2, v_3) for the three vibrational modes (ν_2 = bending frequency) in the ground and excited states. The fact that a progression is seen in the bending frequency indicates that the bond angle HNF is significantly different in ground and excited electronic states (compare the discussion in Chapter 2, but replace the diatomic bond length by HNF angle). This conclusion is supported by qualitative molecular orbital considerations.

In this particular case it was fortunate that bands did not occur containing progressions involving ν_1. If this had been the case there may have been so many bands present that resolution would have been impossible.

In situ photolysis is not the only means of obtaining the electronic spectra of unstable species. If an N_2F_4–Ar mixture is passed through a hot tube and the products are condensed at low temperature the spectrum of NF_2 may be observed in the matrix due to the gas phase equilibrium:

$$N_2F_4 \rightleftharpoons 2NF_2$$

which occurs to the extent of about 2 per cent at room temperature, but may be shifted to the right-hand side on raising the temperature. Similarly the gas phase sample may be photolysed during condensation or on its way to the cold window.

4.9.3 Rapid detection methods

Norrish and Porter have developed the flash photolysis technique for the detection of free radicals which have only a short life before decaying either by unimolecular or other routes [4.19]. The process involves generating an intense flash of light which produces extensive photolysis and liberates a high concentration of fragments. In the case of short-lived species there is no point in illuminating the molecules for any longer time than is necessary to achieve a stationary state concentration of the transients. An absorption spectrum is then very quickly taken using as the source of radiation an intense continuum from which a sufficient concentration of molecules may absorb in a very short time. The object is to record the absorption spectrum of the transient before it dimerizes or reacts. By this approach very simple species may be identified by means of their absorption spectrum. The method is known as flash photolysis and the experimental apparatus is shown in Fig. 4.17.

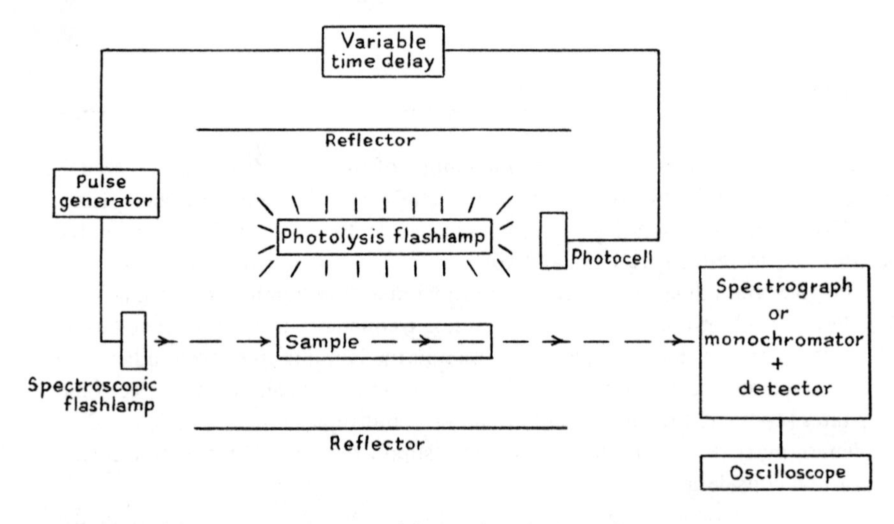

Fig. 4.17 Schematic arrangement of flash photolysis equipment.

A high concentration of fragments is generated by means of a photolysis flash which is produced in a discharge tube containing argon or xenon at low pressure. A high intensity flash is produced of duration 10^{-5} to 10^{-4} s. In general, the higher the flash energy, the longer its duration. The light is concentrated on to the sample cell by means of an elliptical reflector. The generated species may be rapidly detected by observation of its electronic absorption spectrum. This is done by means of another discharge tube containing argon which is placed along the line of the cell. The delay between the spectroscopic flash and photolysis flash is determined by means of the photocell and electronic delay circuit. Light from the photolysis flash falls on the photocell which initiates an electronic timer. This governs the operation of the pulse generator which fires the spectroscopic flash. By adjustment of the timer the spectroscopic flash may be produced between 10^{-5} and 1 s after the photolysis flash. Thus only species with a lifetime greater than 10^{-5} s can be detected. Using a laser flash, however, this limit may be dramatically reduced to a few nanoseconds. The species in the cell may be then detected on the spectrograph by means of its electronic absorption spectrum. Either photographic or photometric recording of the spectrum is possible. In the former the entire spectrum may be recorded after one flash under favourable conditions. In the latter a 'point by point' spectrum is plotted out by changing the frequency selected by the monochromator for each flash. Using this method, however, the sample absorption as a function of time may be displayed on an oscilliscope. Emission spectra of the transient can be recorded in favourable circumstances. Ramsay [4.20] has studied the CHO radical by flash photolysis of several aldehydes. The absorption bands occurred in the 450–750 nm region and two red degraded bands with heads at 613.82 and 562.41 nm were studied in detail. These bands had a simple structure and were composed of only single P, Q, and R branches. From the rotational analysis it was shown that the bands were due to a transition from a lower state where the valence angle was about 120° to an upper electronic state where the nuclei were linear. A rotational analysis was made on all the sharp bands of the CHO and CDO radicals and the internuclear distances and bond angles determined from the rotational constants are given in Fig. 4.18.

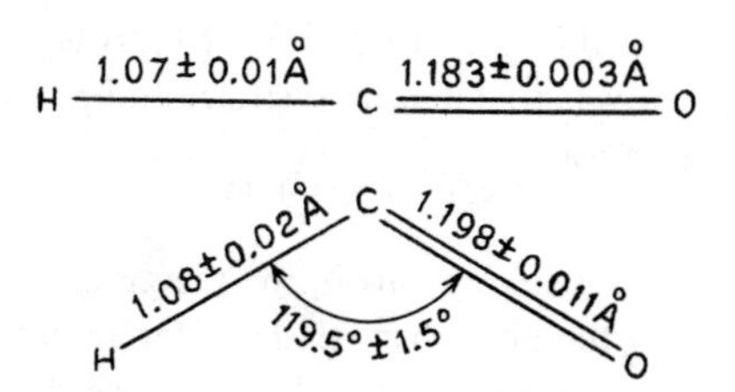

Fig. 4.18 Molecular parameters for CHO in its ground (top) and excited (bottom) states.

The electronic spectra of short-lived species in solution may be followed in precisely the same way. Here, a quartz cell containing the solution replaces the gas cell above.

Molecular fragments may also be rapidly produced in a shock tube. This consists of a long, narrow cylindrical tube divided into two sections of unequal length separated by a thin membrane of aluminium. The larger section is filled with H_2 or He at several atmospheres pressure and the smaller section held at a pressure of a few torr. When the membrane is suddenly punctured a shock wave travels forward, the temperature of which can be very high indeed. 10^4–10^5 K. This is sufficient to produce molecular fragments and ions. Short-lived species may also be produced in other high energy situations, e.g. pulsed radiolysis.

One ingenious method for obtaining the spectrum of transient species is the use of a pulsed dye laser [4.21]. The output intensity of the laser depends upon the optical gain associated with the stimulated emission process in the resonator containing the laser. If the reaction cell containing formaldehyde for example is placed inside the laser cavity and flashed, the gain of the dye laser will be severely reduced at those wavelengths were the short-lived species CHO, produced in the flash, absorbs. The output intensity of the laser as a function of wavelength thus has superimposed on it the absorption spectrum of CHO. The method is restricted to use with transient species whose lifetime is longer than the pulse length of the laser (0.3 μs in this case). The method can, of course, be used with stable molecules.

4.10 SOME APPLICATIONS TO KINETICS

Fast reactions may be followed using the flash photolysis technique either in the gas phase or in solution by varying the delay between the spectroscopic and photolysis flashes and observing the intensity of the spectrum as a function of time. Alternatively a high intensity spectroscopic source may be used and the time dependence continuously monitored at a given wavelength. Estimates of the lifetime of transient species and the possible mechanism of the processes occurring may often be made. For example, Norrish [4.22] has studied the mechanism of the reaction:

$$C_2H_2 + O_2 \rightarrow 2CO + H_2$$

One mixture employed contained 13 torr C_2H_2, 10 torr O_2, and 1.5 torr NO_2. The function of the latter was to act as a sensitizer and provide oxygen atoms as a result of the photolytic flash:

$$NO_2 \rightarrow NO + O$$

The concentration of radicals taking part in the reactions is shown in Fig. 4.19, as a function of time after the photolysis flash. After an induction period of about 400 μs the radical concentration rapidly increases. This represents an explosion; after this a relatively slow decay of radicals is observed.

Photolysis of $Cr(CO)_6$, in solution in the presence of ligand L, leads to

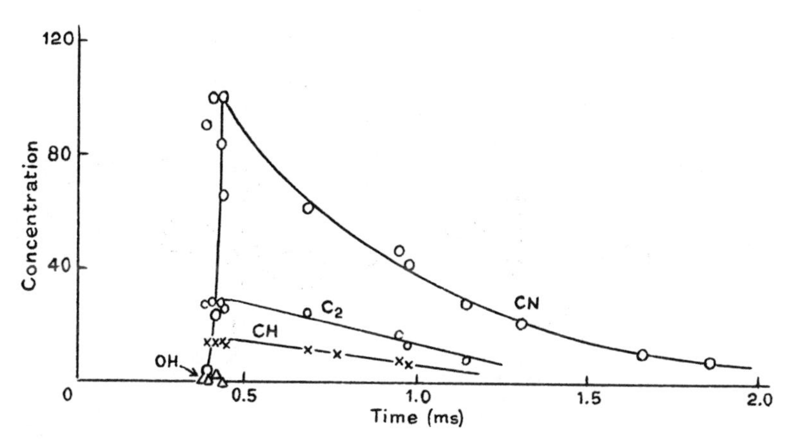

Fig. 4.19 Concentration of radicals as a function of time for an initial mixture of 13 torr C_2H_2, 10 torr O_2, and 1.5 torr NO_2. [Adapted from Norrish, R.G.W., Porter, G., and Thrush, B.A., *Proc. Roy. Soc., A,* **216**, 165 (1953).]

production of $Cr(CO)_5L$, probably via the intermediate $Cr(CO)_5$:

$$Cr(CO)_6 \xrightarrow{h\nu} Cr(CO)_5 \longrightarrow Cr(CO)_5L$$

$Cr(CO)_5$ gives rise to an absorption band in the visible part of the spectrum. By setting the monochromator such that the absorption intensity of this band is determined photometrically as a function of time on an oscilloscope with a time-base set to 100 μs, the dependence of the rate of decay of the signal on the nature of the solvent, ligand L, CO concentration, etc. gives information concerning the mechanism of reaction [4.23].

In general, electronic spectroscopy is also useful in following the rates of much slower reactions which may take place in normal solution cells where the sampling intervals may be minutes, hours, or even days. These less spectacular uses of the technique are quite important. Reasonably fast reactions may be followed by the use of flow techniques either in the gas phase or solution. Figure 4.20 shows spectra taken of a methanol solution of 2×10^{-3} molar *cis*-$Co(en)_2Cl_2$ as it isomerizes to the *trans*-isomer. This is a reaction readily followed using standard equipment. Two isosbestic points, at 446 and 477 nm, are apparent. Measurements of the optical density of the mixture at different wavelengths are sufficient to determine the concentrations of the two components as a function of time.

The *isosbestic point* (a point of equal absorbance) occurs when two comp pounds in solution are in chemical equilibrium and both molecules contribute to absorbance in a particular wavelength region. There will be at least one wavelength where the absorbance will be a function of the sum of the concentrations of the two species but will not depend upon their relative concentrations. On repeated scanning of a reaction mixture the absence of an isosbestic point shows

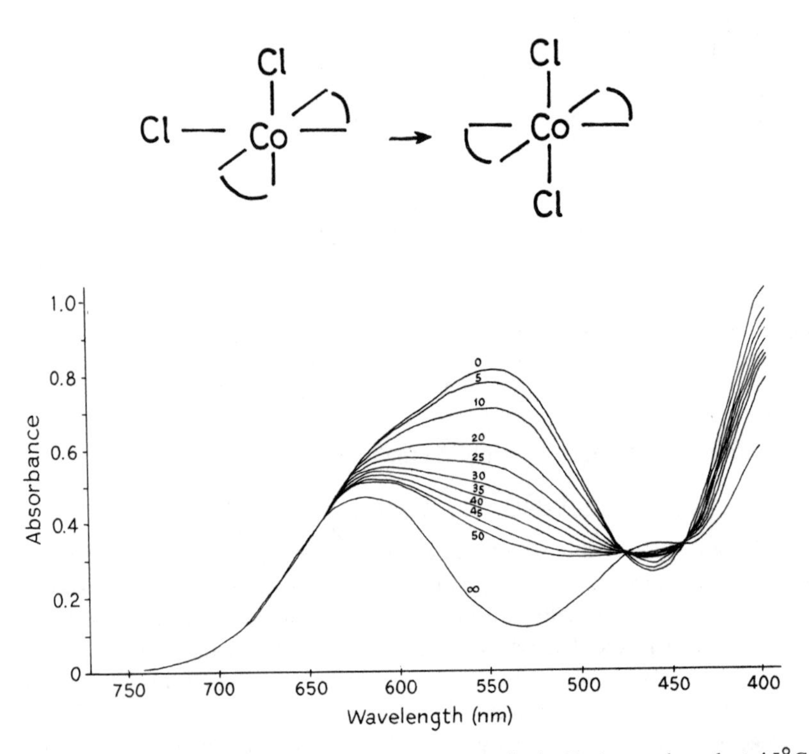

Fig. 4.20 Thermal *cis* → *trans* isomerization of $Co(en)_2Cl_2$ in methanol at 45°C. Reaction times in minutes.

that more than two compounds are contributing to the absorption spectrum. The positive observation of an isosbestic point is not evidence that only two compounds are present. A third (or more) compound may be present but not absorbing in the wavelength region concerned. For the isosbestic point:

$$A_{\text{ib}} = \epsilon(\lambda_{\text{ib}})(c_{\text{A}} + c_{\text{B}})l$$

where $\epsilon(\lambda_{\text{ib}})$ is the molar absorption coefficient of either A or B; l is the path length.

From the measurement of the rates of reaction at different temperatures, and different reagent concentrations, the order of the reaction may be obtained which may give pointers to possible reaction mechanisms. However, the examination of the kinetics of a reaction is usually only the first stage in the elucidation of a reaction mechanism.

4.11 OPTICAL ROTATORY DISPERSION AND CIRCULAR DICHROISM

Optical activity in a molecule can be expected if and only if the molecule cannot be superimposed upon its mirror image. In terms of the molecular symmetry properties which prevent such superimposition, the molecule must not possess an improper rotation axis $\mathbf{S}_n$ with $n \geqslant 1$. The most common axes of this type are for $n = 1$ (equivalent to a plane of symmetry, σ) and $n = 2$ (equivalent to a centre of inversion, i). Two molecules related in this way (the molecule and its mirror image) are called *enantiomorphs*, e.g. the molecules (II) and (III).

The simplest manner in which optical activity may be observed is well known. If a beam of monochromatic plane polarized light is incident on a solution containing an excess of one enantiomorph, the plane of polarization of the transmitted light is found to have been rotated by the solution. In order to discuss optical activity in molecules further it is necessary to describe the physics of light polarization in a little detail.

When viewed along the direction of propagation, a beam of plane polarized light has its electric vector E (which varies sinsusoidally with time) always confined to a single plane. There is also an oscillating magnetic vector H in a plane perpendicular to this which will not be considered further. One important way to regard plane polarized light is as the resultant of two beams of light, one right and one left circularly polarized. A circularly polarized beam of light is one where the electric vector rotates at a constant rate around the direction of propagation, completing one revolution during each sinusoidal cycle of E. (The vector may move to the right or to the left giving rise to right and left polarizations.(Figure 4.21 shows schematically how this occurs. The combination of Fig. 4.21(a) with Fig. 4.21(b) leads to the plane polarized case of Fig. 4.21(c) which, when viewed along the propagation direction, shows the E vector confined to a single plane. By way of contrast, unpolarized light is shown in Fig. 4.21(d) where the relationship between the directions which E points as the beam propagates is a random one. Thus, viewed along the propagation direction, there is no preferred direction in which E has pointed over the sinusoidal cycle. It is thus unpolarized. One property of these two circularly polarized beams is that like the enantiomorphs above they are non-superimposable mirror images of each other. The physical result of this is that the interaction of the two circularly polarized beams of light with an enantiomorph will depend upon the sense of the polarization. The two important differences that will be discussed here are

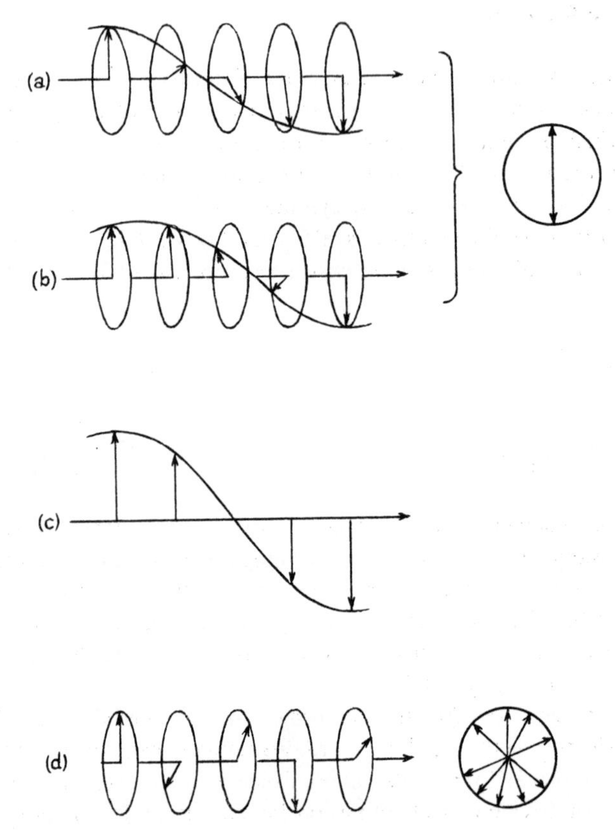

Fig. 4.21 Properties of polarized light, (a, b) left and right circularly polarized light, their combination giving plane polarized light (c); (d) unpolarized light showing the random directions taken by the E vector during the sinusoidal wave cycle.

the following. (i) The refractive indices for right and left circularly polarized light, n_r and n_l, will in general be different. (ii) The extinction coefficients at a given wavelength are different for the two beams, ϵ_r and ϵ_l.

Consider first the difference in refractive index. The time taken for a beam of light to pass through a medium of length l and refractive index n is simply given by nl/c where c is the velocity of light (in vacuo). If an enantiomorph has different values of n for left and right circularly polarized light, one circularly polarized component will travel through the medium faster than the other. After passage through the material there will then be a phase difference between the two circularly polarized components, i.e. the two beams will be out of step. This phase difference is $(n_l - n_r)\pi/\lambda$ per unit path length. As is shown in Fig. 4.22(a), (b) this gives rise to a tilting of the plane of polarization of the emergent light

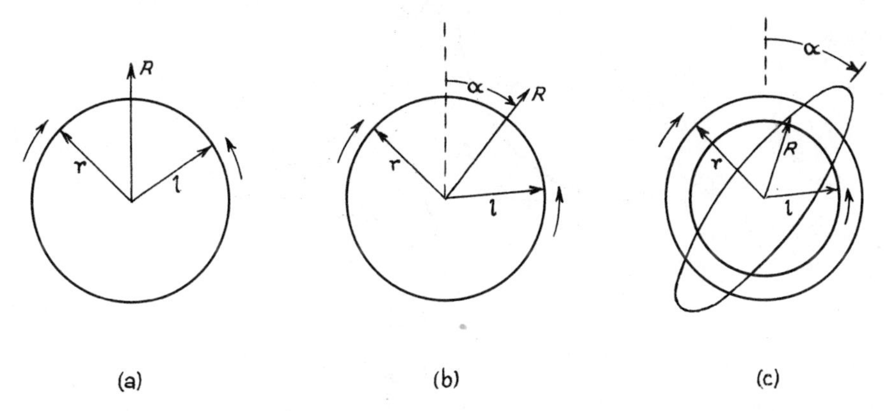

Fig. 4.22 (a) Right (r) and left (l) circularly polarized light combined in phase to give a resultant R which always points along the same axis to give light polarized in the vertical plane. (b) Retardation of the left circularly polarized light due to inequivalence of n_l and n_r. The resultant R now points in a plane tilted by an angle α with respect to (a), i.e. the plane of polarized light has been rotated. (c) Retardation of the left circularly polarized light and preferential absorption by the left circularly polarized light ($\epsilon_p > \epsilon_r$) leads to the resultant R describing an ellipse (elliptically polarized light) whose major axis is related to the resultant of (a) by the angle of rotation (α), and whose minor axis represents the difference between (ϵ_l and ϵ_r), i.e. the circular dichroism.

relative to that of the incident light. This therefore is the origin of the classical rotation of the plane of polarized light by optically active materials. The variation of the angle of rotation (α) with wavelength of light will depend on the change in the difference ($n_l - n_r$) with wavelength. In the region of an electronic absorption band (involving a transition associated with the atom located at the asymmetric centre), ($n_l - n_r$) changes as shown in Fig. 4.23(a) for one enantiomorph and in exactly the reverse way for the opposite enantiomorph [Fig. 4.23(b)]. This variation of α with wavelength is called *optical rotatory dispersion* (o.r.d).

The difference of refractive index is not quite the whole story. Since the extinction coefficients for the two beams are different, the relative intensities of the two emergent circularly polarized components are different compared with the incident beam of light. Thus one rotating electric vector is not exactly equal in magnitude to the other. The result of this is that the electric vector when viewed along the propagation direction describes an ellipse with the major axis describing effectively the "plane of polarization" and the minor axis representing the difference ($\epsilon_r - \epsilon_l$) [Fig. 4.22(c)]. This difference may be measured and is called *circular dichroism* (c.d.). (It is generally smaller than shown in the figure, so it is still approximately correct to refer to plane polarized light rather than the elliptic polarization shown.) Figure 4.23 shows what may be expected

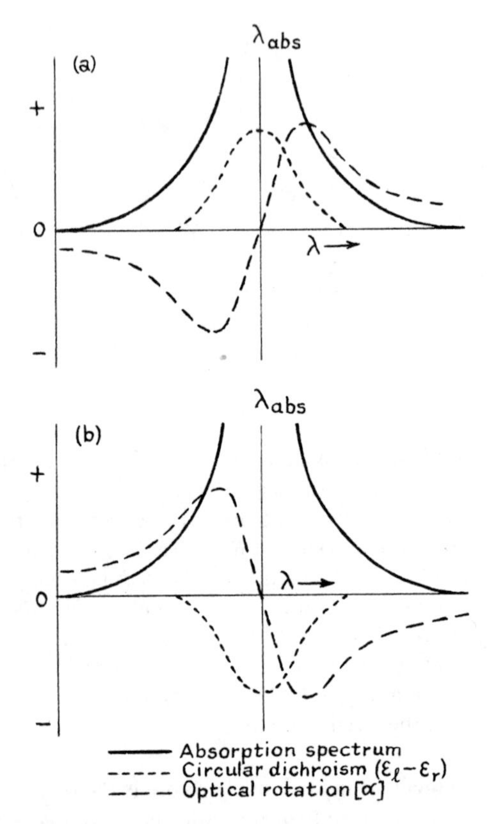

Fig. 4.23 Schematic description of (a) positive and (b) negative Cotton effects as might be expected for two dissymmetric enantiomorphs, showing the absorption spectrum as would be obtained with unpolarized light, the circular dichroism and the optical rotatory dispersion curve.

for a pair of enantiomorphs. O.r.d. and c.d. are often known jointly as the *Cotton effect* [4.24], and the size of the effect is equal but opposite for the two enantiomorphs. In a magnetic field, all materials become optically active and give rise to the phenomenon of *magnetic circular dichroism* (m.c.d.).

The major use of o.r.d. and c.d. is in determining the absolute configuration of molecules. By looking at the c.d. of a particular band due to a chromophore in a series of dissymmetric molecules, rules have been derived to predict the absolute configuration of a molecule given the sign of its Cotton effect. Owing to the prevalence of the carbonyl group in natural product chemistry the 300 nm absorption band of the C=O group has received a lot of study, and the so-called *octant rule* has been the result. A set of orthogonal axes may be set up relative to the C—O group such that the unit lies along the z axis with the O atom to positive z, and the space around the CO group split up into eight regions

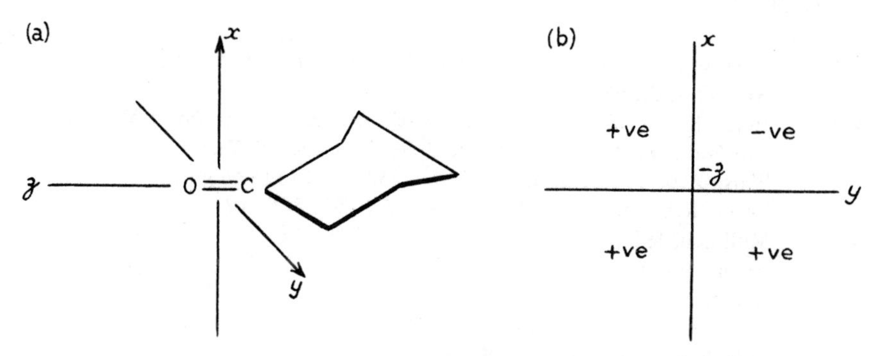

Fig. 4.24 Illustration of the octant rule for the 300 nm carbonyl absorption band. (a) Definition of the axis system for a CO group (attached to a cyclohexane ring). (b) Contribution to the Cotton effect for (non-fluorine containing) substituents lying in the four octants in the $-z$ hemisphere.

(octants) bounded by the xy, yz, and xz planes (Fig. 4.24). The octant rule says that the presence of substituents in the eight regions of space gives rise to defined positive or negative contributions to the Cotton effect. Since most of the molecule in the carbonyl case is likely to lie in the $-z$ hemisphere, these four octants are most important and their contributions for non-fluorine containing substituents to the Cotton effect are shown in Fig. 4.24(b). Fluorine-containing substituents give rise to effects of opposite sign to these shown here. Thus for example if the bulk of the molecule occupies the $-x, -y, -z$ octant in an enantiomorph a negative Cotton effect is expected. The absolute stereochemical configuration of the carbonyls or any other molecular type described by one of these rules is put on an absolute basis simply by the X-ray structure determination of one molecule with a known c.d. Similar ideas can be used to determine the absolute configurations of transition metal complexes, for example the molecules (IV) and (V).

REFERENCES

4.1 Herzberg, G., *Electronic Spectra of Polyatomic Molecules*, Van Nostrand, Princeton (1966).

4.2 Innes, K., in: *Molecular Spectroscopy; Modern Research* (ed. Rao, K.N. and Mathews, C.W.), Academic Press, London (1972).

4.3 Walsh A.D., *J. Chem. Soc.*, 2260 (1953), and following papers.

4.4 Snow, R.L. and Bills, J.L., *J. Chem. Ed.*, **52**, 585 (1974).

4.5 Mulliken, R.S., *Phys. Rev.*, **43**, 279 (1933).

4.6 Fleming, I. and Williams, D.H., *Spectroscopic Methods in Organic Chemistry*, McGraw-Hill, London (1966).

4.7 Scott, A.I., *Interpretation of the Ultraviolet Spectra of Natural Products*, Pergamon, Oxford (1964).

4.8 Murrell, J.N., *The Theory of the Electronic Spectra of Organic Molecules*, Methuen, London (1963).

4.9 Schonland, D., *Molecular Symmetry*, Van Nostrand, London (1965).

4.10 Streitwieser, A., *Molecular Orbital Theory for Organic Chemists*, Wiley, London (1961).

4.11 Orgel, L.E., *An Introduction to Transition Metal Chemistry; Ligand F Field Theory*, Methuen, London (1966).

4.12 Kettle, S.F.A., *Coordination Compounds*, Methuen, London (1969).

4.13 Murrell, J.N., Kettle, S.F.A. and Tedder, J.M., *Valence Theory*, Wiley, New York (1965).

4.14 Jørgensen, C.K., *Absorption Spectra and Chemical Bonding in Complexes*, Pergamon, Oxford (1962).

4.15 Mulliken, R.S., *J. Amer. Chem. Soc.*, **74**, 811 (1952).

4.16 Hoskins, W.D., *Trans. Faraday Soc.*, **30**, 221 (1934).

4.17 Thrush, B.A., *Chem. Brit.*, **2**, 287 (1966).

4.18 Jacox, E. and Milligan, D.E., *J. Chem. Phys.*, **46**, 184 (1967).

4.19 Norrish, R.G.W. and Porter, G., *Nature*, **164**, 685 (1950); Porter, G., *Proc. Roy. Soc., A*, **200**, 284 (1950).

4.20 Ramsay, D.A., *J. Chem. Phys.*, **21**, 960 (1953).

4.21 Atkinson, G.H., Laufer, A.H. and Kurylo, M.J., *J. Chem. Phys.*, **59**, 350 (1973).

4.22 Norrish, R.G.W., *Disc. Faraday Soc.*, **14**, 16 (1953).

4.23 Kelly, J.M., Hermann, H. and Koerner von Gustorf, E.A., *Chem. Comm.*, **105** (1973).

4.24 Mason, S.F., *Chem. Brit.*, **1**, 245 (1965).

5 Fluorescence and phosphorescence spectroscopy

5.1 INTRODUCTION

The transitions between the various electronic and vibrational states which give rise to the phenomena of fluorescence and phosphorescence are best illustrated in the form of a state diagram commonly referred to as a Jablonski diagram (Fig. 5.1). As may be recalled from the chapter of this book dealing with electronic abosrption spectroscopy in a simple polyatomic molecule, the electrons occupy molecular orbitals surrounding the nuclei which form the skeleton of the molecule. Each orbital is associated with a definite energy and angular momentum due to the motion of the electrons around the nuclei. In addition, electrons themselves possess the property of angular momentum (spin) due to rotation about their axes, and this spin angular momentum is conveniently measured in units such that each electron has associated with it $\frac{1}{2}$ unit of spin angular momentum. In a simple system one can consider the angular momentum due to motion of the electrons about the nuclei and that due to electron spin as independent. The Pauli exclusion principle dictates that, in any non-degenerate molecular orbital, only two electrons can be accommodated, and this only when the two electrons have values of spin angular momenta which have opposite sign. Thus one electron will have $+\frac{1}{2}$ (usually labelled α, or $\uparrow$) and the other $-\frac{1}{2}$ (usually labelled β or $\downarrow$) unit of angular momentum, and the net spin angular momentum is thus zero. The situation is illustrated in the box to the left of the state labelled S_0 in Fig. 5.1 with respect to the highest occupied molecular orbital in the lowest energy state of the molecule, i.e. that obtained by sequentially allotting electrons into the molecular orbitals in ascending order of energy until all available electrons are accommodated (Aufbau principle). States arising from such electronic configurations are conveniently labelled in terms of multiplicity which

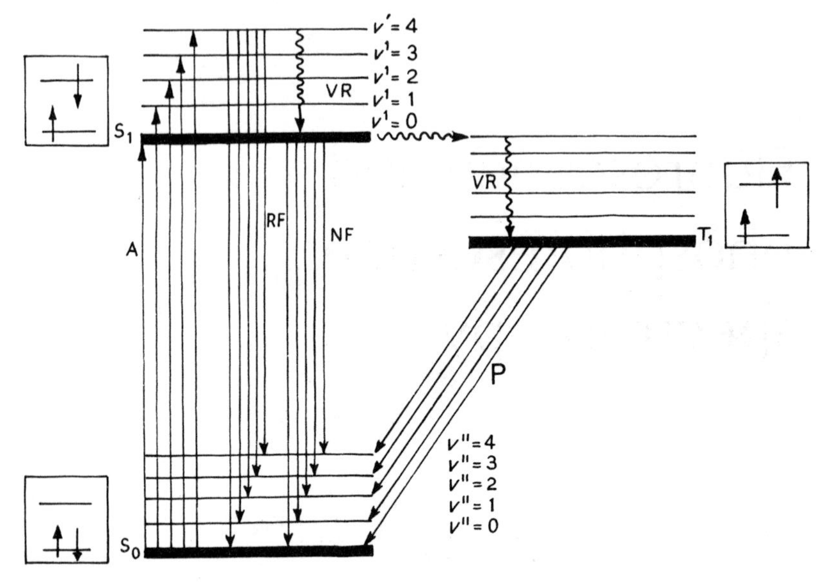

Fig. 5.1 Jablonski diagram depicting the important radiative processes in a typical organic molecule.

describes the total spin angular momentum in the system and which is given by $2S + 1$, where S is the total spin angular momentum. As we can see in the state labelled S_0, for all electrons which are paired in orbitals, including the highest occupied molecular orbital illustrated in Fig. 5.1, the net spin is zero, and the multiplicity is thus unity. Such states are referred to as singlet states, labelled S. The subscript zero in the simple nomenclature adopted here indicates that this is the lowest energy state (ground state) of the system.

As shown in Fig. 5.1, there exist higher energy molecular orbitals which are not occupied in the ground state of the molecule. It is possible to induce the promotion of an electron from an occupied to an unoccupied orbital by the absorption of electromagnetic radiation, typically in the visible and ultraviolet regions of the spectrum. The transitions normally observed occur through interaction of the electric vector of the electromagnetic radiation and the oscillating electrons in the molecules, and are referred to as electric dipole transitions. The state resulting from such an absorption is also illustrated in Fig. 5.1, and it can be seen from the orbital picture that the spin angular momenta of the electrons do not change during the electronic transition. The excited electronic state is thus labelled with the subscript 1 to indicate that it is the first excited singlet state.

Noting that we now have the two electrons we are considering in different molecular orbitals, it can be seen that the restrictions of the Pauli exclusion principle no longer apply, and it is thus possible for an electron to invert spin. This is illustrated in the orbital diagram on the right of Fig. 5.1, and results in an electronically excited state which has a total spin angular momentum of unity, thus has a multiplicity of 3, and is referred to as a triplet state labelled T_1. The majority of organic molecules are represented by the states shown in Fig. 5.1. The triplet state is always *lower* in energy than the corresponding singlet state having the same orbital distribution of electrons. This is due to the electron avoidance properties of the electrons in singlet and triplet states, [5.1]. Thus in singlet states the two electrons have a finite probability of occupying the same region of space, whereas in the triplet state this probability is zero, minimising coulombic electron repulsion energies.

The study of the physical and chemical fates of electronically excited atoms and molecules is the science of photochemistry. Here, although we are principally concerned with the radiative processes which arise from these excited states, it is necessary to have some understanding of the non-radiative decay processes of excited states in order to rationalize certain aspects of luminescence behaviour.

Having defined states in terms of electron spin, we can proceed to a definition of the terms fluorescence and phosphorescence. If during the emission process the electron undergoing the transition does not change its spin, the emission is referred to as *fluorescence* (labelled F in Fig. 5.1). If, however, spin inversion is necessary for the transition to occur, as is the case in a radiative transition between an excited triplet state and a singlet ground state, the emission is termed *phosphorescence* (labelled P in Fig. 5.1). This definition replaces earlier uses of the terms fluorescence and phosphorescence which were based upon the lifetime of the emission.

In Fig. 5.1, the thick solid lines represent the energies of the various electronic states which arise purely from the positions and motion of the electrons. Of course, in molecular species, vibrations of the nuclei making up the molecular skeleton also occur, and the energy levels associated with these quantized molecular vibrations are represented by the fine lines drawn in Fig. 5.1. It would be possible further to refine this picture by including all of the rotational levels, but these will be left for discussion until later. In general, electronic transitions such as absorption, fluorescence, and phosphorescence may occur from any vibrational level in the initial state to any set of levels in the final state. It can be seen clearly that, in general, analysis of the spectral distribution of phosphorescence and fluorescence will result in an emission envelope in which vibrational bands associated with ground-state frequencies will appear. The position of the highest-energy (zero–zero) transition in the fluorescence spectrum will always be higher in energy than that of the phosphorescence spectrum. Not all electronically excited states, however, give rise to observable emission and the reason for this lack of emission is because radiative relaxation (emission) back to the ground state is always in competition with non-radiative processes. These processes are illustrated in Fig. 5.2. Perhaps the most difficult non-radiative process to make general statements about is the process which results in a transfer from the potential energy surface of the excited state of the original molecule to the potential energy surface of some other species and which may eventually lead to product formation.

The other non-radiative processes do not result in the destruction of the

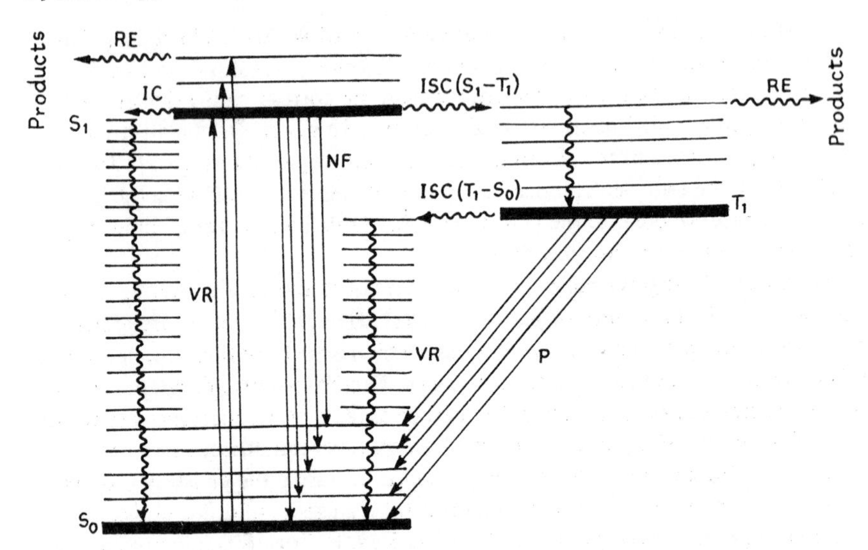

Fig. 5.2 A state diagram showing the important electronic and vibrational relaxation processes in competition with emission.

molecule but instead the electronic energy is converted into other forms of internal energy, mainly vibrational energy, and in the process a lower energy electronic state is populated. It should be noted that these non-radiative processes are in fact in each case the sum of two processes which occur in sequence. The initial step is a crossing to an isoenergetic level of the lower-lying electronic state followed by collisional vibrational relaxation (VR). As in the case of radiative processes, non-radiative transitions are defined in terms of whether or not change in spin angular momentum occurs during the transition. For a non-radiative transition between states of the same multiplicity the term 'internal conversion' (IC) is reserved. When spin inversion occurs, the somewhat cumbersome term 'intersystem crossing' (ISC) is universally employed. In Fig. 5.2 there are illustrated two frequently met intersystem crossings, and it is thus usual to indicate which is being referred to by denoting the states involved in the transition, i.e. $S_1 \rightarrow T_1$, $T_1 \rightarrow S_0$. On a Jablonski diagram, wavy lines always refer to non-radiative processes, and solid lines to radiative processes.

There are a number of other definitions which will be useful in the discussion below. The emission processes described above are *spontaneous* in that they occur subsequent to the act of absorption without the presence of a radiation field. Absorption is of course a *stimulated* process, since it cannot occur without the presence of a radiation field. Stimulated emission is, however, another possible fate of the excited electronic states of molecular and atomic species, and this arises when emission is induced by the presence of a radiation field. The probability of spontaneous emission is governed by the Einstein A coefficient, and that for stimulated emission, as for absorption, by the Einstein B coefficient, and the relationship between these parameters is developed in the Appendix as:

$$A_{\mathrm{if}} = \frac{8\pi h\nu^3}{C^3} \cdot B_{\mathrm{if}} \tag{5.1}$$

This relationship, which shows that the probability of spontaneous emission depends upon the cube of the frequency of the emitted photon, is of great importance since it dictates that spontaneous emission processes occur with high probability only for transitions between electronic states. Thus the photon energy which corresponds to the separation of vibrational and rotational levels is of such small magnitude that spontaneous emission of infrared or lower-energy photons can only be observed in special circumstances when competing processes are eliminated, usually at very low pressures in the vapour phase. For electronic transitions, however, the magnitude of ν^3 is usually such that spontaneous emission of luminescence can compete favourable with other decay processes even in condensed phases.

5.2 FLUORESCENCE

In the electronic absorption transition, the absorption occurs from a Boltzmann distribution of vibrational and rotational levels of the ground electronic state. For many simple molecules containing only first-row elements, the assumption is often made that only the zero-point level has significant population at room temperature. However inspection of Equation (5.2), which gives the population n_n of any level of degeneracy g_n, separated in energy by E_n from the lowest state of degeneracy g_m as:

$$\frac{n_n}{n_m} = \frac{g_n}{g_m} \mathrm{e}^{-E_n/kT} \tag{5.2}$$

reveals that, for polyatomic molecules with low-frequency vibrations, the assumption that only the zero-point level absorbs is invalid, although the assumption is nevertheless still often made. The states produced on absorption may have varying amounts of excess vibrational and rotational energy above the zero-point level of the upper state. We can distinguish two limiting cases of fluorescence, depending on the environment of the exciting molecule.

5.2.1 Resonance fluorescence

At very low pressures in the vapour phase the electronically excited molecules may not suffer any collisions during its lifetime, and is thus said to be 'isolated', Emission will thus occur from the initially populated level, and is termed *resonance fluorescence* (labelled RF in Fig. 5.1). Figure 5.3 illustrates a set of fluorescence transitions occuring from a specifically excited vibrational level of molecular iodine [5.2].

The pressure needed to achieve the condition of 'isolation' can be obtained from a consideration of the expression from the kinetic theory of gases for the

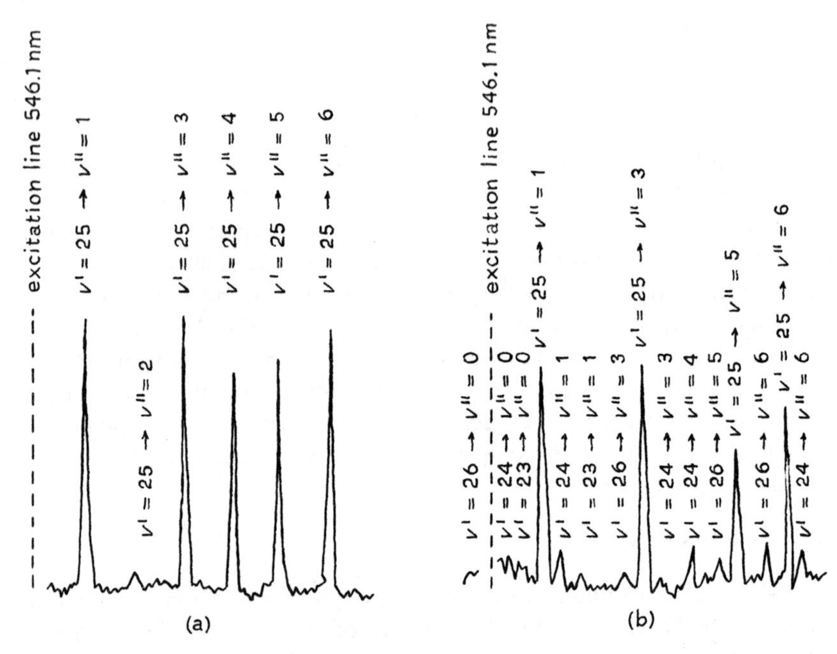

Fig. 5.3 Fluorescence from iodine vapour at 25C. (a) 0.02 torr* I_2, (b) 0.02 torr I_2, 0.05 torr neon. Excitation at 546.1 nm (Hg lamp) produces $v' = 25$ level, from which in (a) resonance fluorescence is observed. In (b) many other emitting vibrational levels of the upper state are produced by collision with neon. [Based on Fig. 3.11 in *Photochemistry*, Cundall, R.B. and Gilbert, G., Nelson (1970)].

collision frequency of two unlike gases:

$$Z_{AB} = n_A n_B \pi \sigma_{AB}^2 \left(\frac{8kT}{\pi\mu}\right)^{\frac{1}{2}} \tag{5.3}$$

where $\pi\sigma_{AB}^2$ is the collision cross-section, k Boltzmann's constant, μ the reduced mass $= m_A m_B/(m_A m_B)$, and T the absolute temperature, and n_A, n_B are the concentrations of molecules A and B. The expression can be used to estimate the collision rate per unit concentration of excited species A, and a concentration of B, calculated such that during the radiative lifetime of A the probability of collision is much less than unity. Typically radiative lifetimes are of the order of 10^{-7}s, and pressures required for isolation are of the order of 10^{-1}–10^{-2} torr. The radiative lifetimes of triplet states are exceedingly long, and it is thus not possible to achieve the corresponding isolated molecule condition for triplet emission studies. Resonance fluorescence may also be observed in polyatomic species, and that obtained under moderate resolution for excitation of a number of levels in a simply substituted benzene molecule under isolated molecule

* 1 torr = 133.2 N.m^{-2} at 273 K.

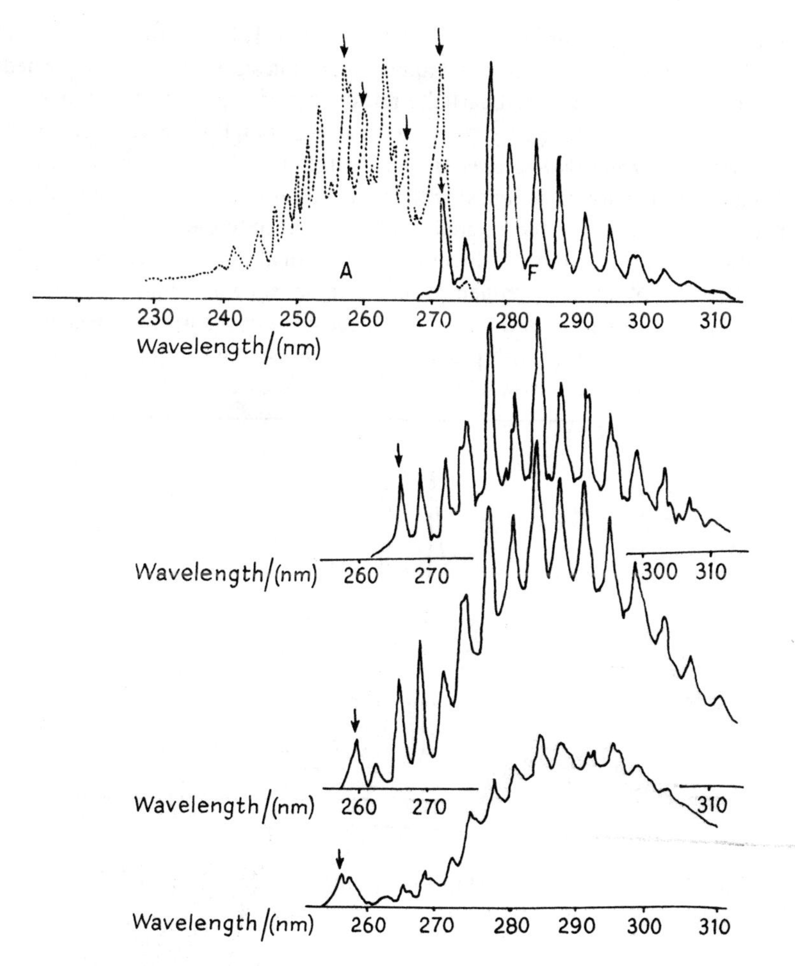

Fig. 5.4 Resonance fluorescence from 'isolated' *p*-difluorobenzene for excitation in different absorption bands shown in top figure. Arrows indicate exciting radiation in each case. [Based upon Fig. 3 in *J. Photochem.*, (1974/5) **3**, 365.]

conditions is shown in Fig. 5.4. Measurements of quantum yields of fluorescence and fluorescence decay times of 'isolated' molecules also yield information of much interest, but this will not be discussed here.

5.2.2 Normal fluorescence

In contrast to the isolated molecule situation, if a molecule is excited to some vibrational level of the S_1 state under high pressure conditions (typically > 1 torr), or in a condensed medium, the initially excited molecule will suffer many collisions before electronic relaxation occurs. Vibrational relaxation will therefore be complete prior to the emission process, and thus the emitting levels ara a Boltzmann distribution in the upper electronic state. Naturally intermediate situations exist in which a partially relaxed set of upper emitting levels is populated, and this necessarily gives rise to a very complex spectrum. It can be seen that the *normal fluorescence* (labelled NF in Fig. 5.1) will always be red-shifted (moved to lower energies) with respect to absorption because of the vibrational relaxation in the excited state, and that only the $v' = 0 \leftrightarrow v'' = 0$ transition is common to both absorption and normal fluorescence *spectra*. Provided that the vibrational frequencies in the upper state are similar to those in the lower state, the normal fluorescence spectrum and absorption spectrum will have a mirror-image relationship to another on an energy (frequency or wave-number) scale. This is illustrated in Fig. 5.5 for the case of anthracene.

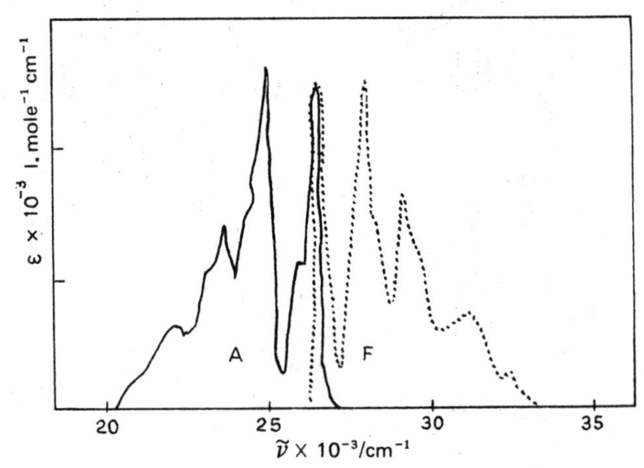

Fig. 5.5 Mirror image relationship for anthracene fluorescence (F) and absorption (A) spectra in solution.

5.2.3 Intensities of transitions

The intensity of each individual transition between vibrational levels for an allowed electronic transition (see below) is giverned by the overlap of vibrational probability factors in the upper and lower states termed *Franck–Condon factors* which depend upon geometry changes in the upper and lower electronic states. Maximum intensity in absorption and emission occurs when these overlap factors are maximized, and this is discussed below. The difference in position of maximum of emission (Fig. 5.6) and of zero–zero transition on an energy scale is frequently referred to as the *Stokes loss* [5.3] and is a measure of the change in geometry between the equilibrium configurations of the two states between which the transition occurs. This is illustrated for a hypothetical diatomic molecule in Fig. 5.7, in which vibrational probability distributions (proportional to the square of vibrational wavefunctions) are shown. The intensity of the transition in absorption is maximized when the transition is vertical, i.e. there is no

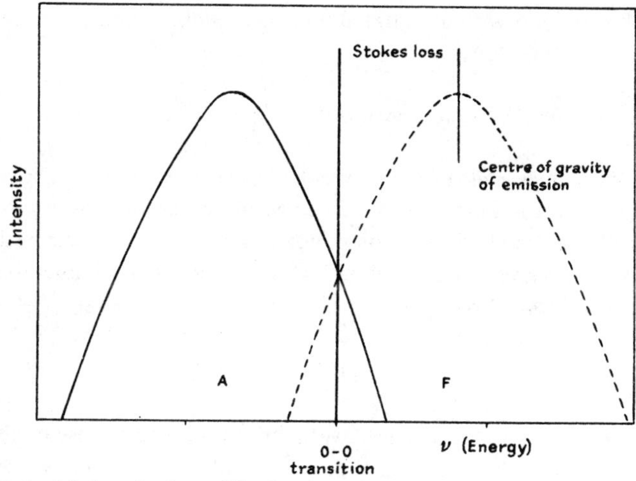

Fig. 5.6 Pictorial description of Stokes loss.

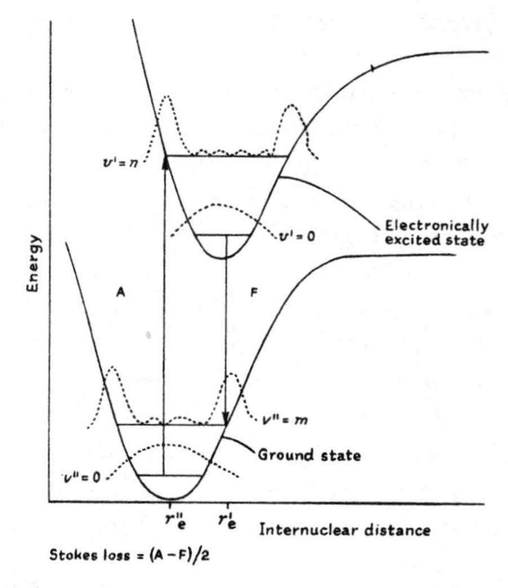

Fig. 5.7 The significance of the Stokes loss in terms of potential energy curves of upper and lower electronic states of a hypothetical diatomic molecule.

change in internuclear distance during the transition. This is marked A in the diagram, from $v'' = 0 \rightarrow v' = n$. If the excited electronic state has a different equilibrium internuclear distance r_e' from that of the ground state r_e'' (usually $r_e' > r_e''$), then, following vibrational relaxation in the upper state, the most probable emission occurs vertically from the zero-point level to some other level to the ground-state $v'' = m$, because this maximizes the transition probability. The Stokes loss is then related to the energies of the transition marked A and F in the diagram. Of course, transitions to other vibrational levels in the ground state also are observed but with diminished intensity owing to poorer overlap of vibrational wavefunctions.

Selection rules for electronic transitions

Quantum mechanics leads to the mathematical expression of the above qualitative explanation for the probability of an electronic transition as proportional to R_{if}^2, where R_{if}, termed the transition moment integral between the initial state i and final state f is given by Equation (5.4), as developed in Chapter 4 on electronic absorption spectroscopy.

$$R_{\text{if}} = \int_{-\infty}^{+\infty} \Psi_{\text{ef}} \hat{M} \Psi_{\text{ei}} \mathrm{d}\tau_{\text{e}} \int_{-\infty}^{+\infty} \Psi_{\text{nf}} \Psi_{\text{ni}} \mathrm{d}\tau_{\text{n}} \tag{5.4}$$

where the Ψ_{e} represent electronic wavefunctions, the Ψ_{n} represent vibrational wavefunctions, $\hat{M}$ is the electronic dipole moment operator, and where the Born–Oppenheimer principle of separability of electronic and vibrational wavefunctions has been invoked. The first term in Equation (5.4) is an integral involving only the electronic wavefunctions of the system, and the second terms, which are simply vibrational overlap integrals, when squared are the familiar Franck–Condon factors. We have already discussed briefly the fact that for any given electronic transition the intensities of the vibrational bands in absorption and emission will be governed by these Franck–Condon factors, but we might enquire what determines the magnitude of the electronic integral. The exact evaluation of the electronic transition moment integral is extremely difficult, but general statements can be made about the conditions under which the value is zero or finite, which corresponds to a *forbidden* and an *allowed* transition respectively; in other words selection rules are generated. Firstly we can say something about transitions between different spin states by separating electronic wavefunctions into space and spin parts. The electronic part of the transition moment integral $(R_{\text{if}})_{\text{e}}$ then is given by (5.5) in which the final term represents the overlap of the spin wavefunction of the initial and final states:

$$(R_{\text{if}})_{\text{e}} = \int_{-\infty}^{+\infty} \Psi_{\text{esf}} \hat{M} \Psi_{\text{esi}} \mathrm{d}\tau_{\text{es}} \int_{-\infty}^{+\infty} \Psi_{\text{sf}} \Psi_{\text{si}} \mathrm{d}\tau_{\text{s}} \tag{5.5}$$

where Ψ_{es} represents the space part of the electronic wavefunctions and Ψ_{s} the electronic spin wavefunctions.

Pure spin wavefunctions are orthonormal, i.e.

$$\int_{-\infty}^{+\infty} \Psi_{\text{s}}(\alpha)\Psi_{\text{s}}(\alpha) = \int_{-\infty}^{+\infty} \Psi_{\text{s}}(\beta)\Psi_{\text{s}}(\beta) = 1 \tag{5.6}$$

$$\int_{-\infty}^{+\infty} \Psi_{\text{s}}(\alpha)\Psi_{\text{s}}(\beta) = 0 \tag{5.7}$$

Thus an electric dipole transition is *spin-forbidden* if the initial and final states have different multiplicities. The singlet–singlet transitions shown in Fig. 5.1, viz. absorption and fluorescence, are spin-allowed, and thus have a high probability of occurring. The triplet–singlet transition shown (*Phosphorescence*) is spin-forbidden as is seen from (5.7) since the overlap integral of the spin wavefunctions of singlet and triplet states is zero. The transition is actually observed because the space and spin parts of the total electronic wavefunctions are not entirely separable, being mixed by spin–orbit coupling, which means that the rigorous selection rule implied by (5.7) is not in fact obeyed completely. Nevertheless, the rule accounts for the very long radiative lifetime of the triplet state (up to 100 s) compared with that of a typical singlet state (10^{-8} s).

Polarization

Next we can consider what factors affect the magnitude of the purely space part of the electronic transition moment integral, given by the first integral term in (5.5). $\hat{M}$ here is the dipole moment operator, which corresponds to the act of removing an electron from one orbital or region of space to another. The assumption is usually made that only the electron making the transition changes position, tion, and thus the integral to be evaluated is a one-electron term. The transitional dipole is a vector quantity, or in other words the change in position of the electron during the transition is in a fixed direction with respect to some system of coordinates in which the molecular frame is fixed. Thus the dipole moment operator is resolvable into three directions such that:

$$\hat{M}^2 = \hat{M}_x^2 + \hat{M}_y^2 + \hat{M}_z^2 \tag{5.8}$$

and in general only one of these components will be non-zero in value. Thus if a perfectly ordered system such as a fixed molecular single crystal is observed in absorption, using *plane-polarized* light, there will generally be one orientation of the crystal axes with respect to the plane of the polarization of the light which maximizes the absorption probability. Normally fluorescence involves the transition between the same two states as are observed in absorption, and thus the fluorescence is usually polarized *parallel* to the absorption (i.e. the light emitted has the same directional properties in terms of direction of the electrical vector as has the light absorbed). In many cases phosphorescence involves a transition in which the symmetry properties of the triplet state are different from those in the excited singlet state, and the transition dipole for phosphorescence is polarized in some direction other than that observed in absorption. Commonly this direction is *perpendicular* to the absorption. For a perfectly ordered rigid system, the intensity of emission will vary between the limits of 100 and 0 per cent as the plane of polarization of a detection system (see experimental section below) is varied from parallel to the plane of the exciting light through to perpendicular for a *parallel* transition, and vice versa for a *perpendicular* transition. Normally perfectly ordered rigid systems are not observed, however, and rather perfectly random distributions of the molecule to be studied are prepared in rigid media such as solvent glasses at low temperatures, or poly(methyl methacrylate) rigid media. In these cases a quantity termed the degree of polarization P defined as:

$$P = (I_{\parallel} - I_{\perp})/(I_{\parallel} + I_{\perp}) \tag{5.9}$$

where $I_{\parallel}$ is the intensity of emission observed polarized parallel to the absorbed radiation, and $I_{\perp}$ is that perpendicular to the absorbed radiation. The limits of value of P for a randomly distributed sample are $+ 0.5$ for a parallel transition and $- 0.33$ for a perpendicular transition [5.4]. Figure 5.8 shows the polarized phosphorescence spectra of phenanthrene in a poly(methyl methacrylate)

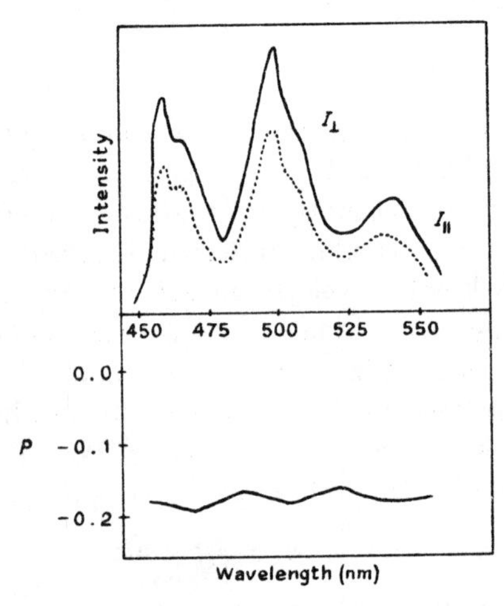

Fig. 5.8 Polarized phosphorescence spectra of phenanthrene in poly(methyl methacrylate) at 77 K. $I_{\parallel}$ is the intensity of component parallel to exciting plane; $I_{\perp}$ is the intensity of component perpendicular to exciting plane; P is the degree of polarization (see text). [Based upon Fig. 1 in *J. Chem. Soc., Faraday II,* **70**, 537 (1974).]

medium at low temperature from which it can be seen that the transition is strongly polarized perpendicular to the absorption. This information is of much importance in that it assists in the identification of the nature of the electronic states responsible for emission. In non-rigid matrices, the excited molecule may rotate before emission, leading to depolarization of the observed emission.

Symmetry-forbidden transitions

There are for many molecules electronic states of symmetry such that when the integration in the first term of Equation (5.5) is carried out, the value of the integral becomes zero. The transition is thus *symmetry-forbidden*. Despite this restriction, these transitions appear in absorption and emission due to the activity of one or more vibrational modes, termed *promoting modes*, which distort the symmetry of one or other of the states and thus render the integral of finite value, and the transition thus becomes observable. This phenomenon is known as *intensity borrowing*, and has been described by Herzberg and Teller [5.5]. Because of the necessity of population of a distorting (non-totally symmetric) vibrational mode either in the lower or upper state, the 0–0 transition is absent from both absorption and fluorescence in such cases. Benzene provides a classic example of a symmetry-forbidden transition which is induced by an asymmetric bending frequency mode $\tilde{\nu}_6$, of frequency 523 cm^{-1} in the upper S_1 state, and 605 cm^{-1} in the S_0 state. In absorption the main vibrational features observed are the progression $1 \times \tilde{\nu}_6$, $1 \times \tilde{\nu}_6 + 1 \times \tilde{\nu}_1$, $1 \times \tilde{\nu}_6 + 2 \times \tilde{\nu}_1$, etc., where $\tilde{\nu}_1$ is the totally symmetric carbon skeletal stretching frequency, referred to as the ring breathing mode (see Fig. 5.9), and has a frequency of 923 cm^{-1} in the upper state. It is possible to populate the zero-point level of the S_1 state by a 'hot-band' transition, shown dotted in Fig. 5.9, in which the required $\tilde{\nu}_6$ vibration is present in the ground-state. The emission observed from this level is shown in Fig. 5.9, and corresponds to the same progression as observed in absorption. Note, however, that the frequencies now observed are *ground-state* parameters which can be compared with values obtained from infrared and Raman spectroscopy. Electronic *absorption* spectroscopy thus gives information concerning electronically *excited* states, and *emission* spectroscopy yields corresponding information about *ground* electronic states. For small molecules in which rotational structure is within the resolving power of the instrument used to observe emission, information concerning the molecular rotational parameters in the ground state may also be obtained.

5.2.4 Non-radiative decay of fluorescent molecules

As indicated in Figs. 5.1 and 5.2, population of the excited singlet state of a molecule does not necessarily give rise to strong fluorescence, since there are other non-radiative decay channels open to the fluorescent state. It is usual to term the fraction of excited molecules which undergo any primary process, for example, fluorescence, as a quantum yield, Φ, which is defined as:

$$\Phi = \frac{\text{Number of molecules/unit volume/s undergoing process}}{\text{Number of photons/unit volume/s absorbed by the system}} \tag{5.10}$$

The meaning of this quantity in a kinetic sense is easily illustrated with the simple scheme shown, where A represents a molecule absorbing light, and superscripts refer to multiplicity.

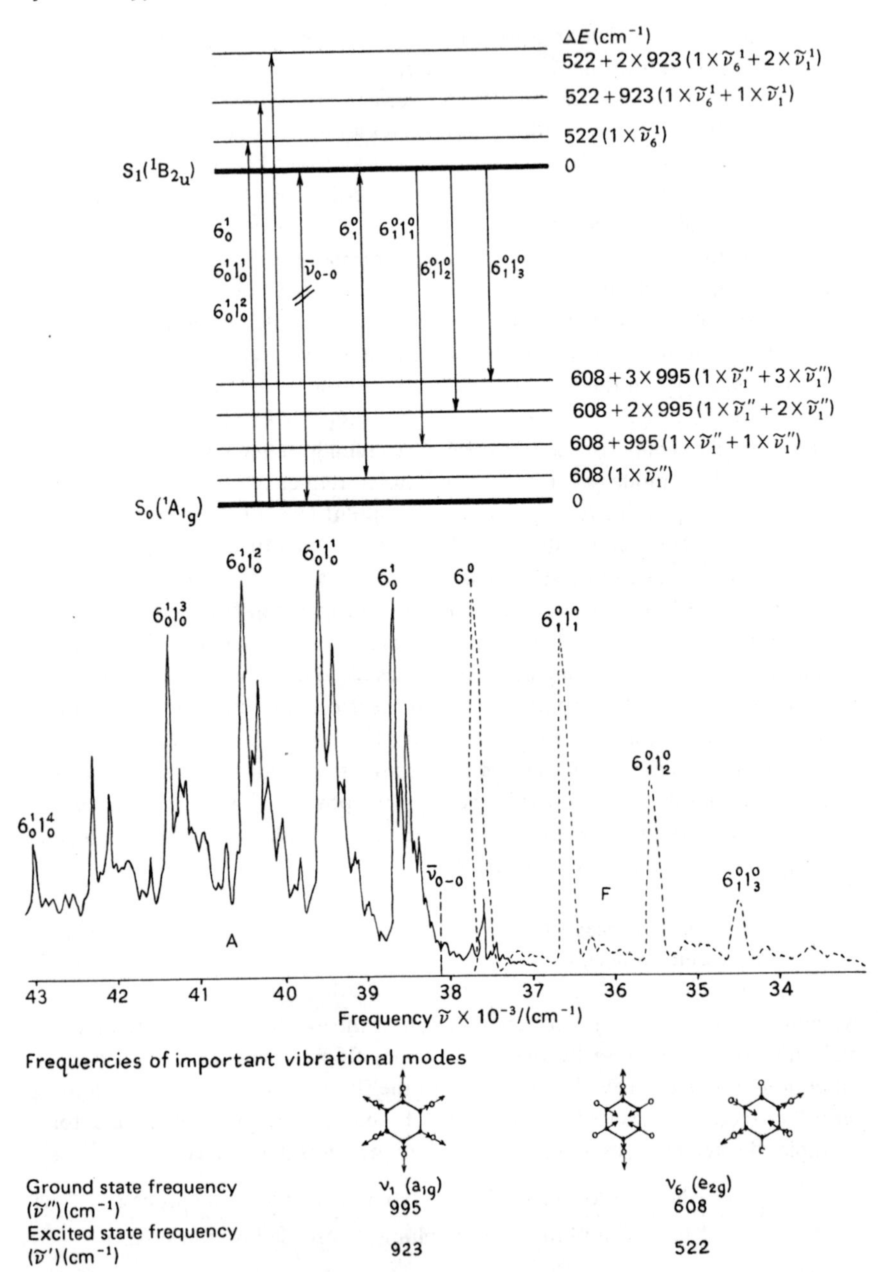

Fig. 5.9 Absorption and fluorescence of benzene vapour. (a) The main features in the vapour absorption spectrum and the fluorescence from the zero-point level of the upper state under isolated molecule conditions shown in (b). The zero–zero transition is absent because of symmetry restrictions. The spectral

Process	Rate	
$A + h\nu \rightarrow {}^1A$ Absorption	I_a	(5.11)
${}^1A \rightarrow A + h\nu$ Fluorescence	$k_F[{}^1A]$	(5.12)
${}^1A \rightarrow {}^3A$ Intersystem crossing	$k_{ISC}[{}^1A]$	(5.13)
${}^1A \rightarrow A$ Internal conversion	$k_{IC}[{}^1A]$	(5.14)
${}^1A \rightarrow$ Products	$k_{RE}[{}^1A]$	(5.15)

Applying the steady-state approximation to the concentrations of ${}^1A$ molecules we obtain the relationship:

$$d[{}^1A]/dt = I_a - (k_F + k_{ISC} + k_{IC} + k_{RE})[{}^1A] = 0 \qquad (5.16)$$

Thus

$$[{}^1A] = I_a/(k_F + k_{ISC} + k_{IC} + k_{RE}) \qquad (5.17)$$

and the quantum yield of fluorescence Φ_F is given by:

$$\Phi_F = k_F[{}^1A]/I_a = k_F/(k_F + k_{ISC} + k_{IC} + k_{RE}) \qquad (5.18)$$

Kinetically then, the quantum yield is a ratio of rate constants. We can also define the fluorescence lifetime or fluorescence decay time (τ_F) of a molecule as the time taken for the concentration of fluorescing molecules to fall to 1/e of any initial value, since all processes shown above actually contribute to the total decay. Some examples of the magnitude of Φ_F and k_F for a variety of organic molecules are given in Table 5.1.

$$\tau_F^{-1} = k_F + k_{ISC} + k_{IC} + k_{RE} \qquad (5.19)$$

Thus

$$k_F = \Phi_F/\tau_F \qquad (5.20)$$

The quantity τ_F, which is an experimentally determined parameter (see experimental section) is not to be confused with the natural or radiative lifetime, often given the symbol τ_0, which is kinetically equal to k_F^{-1}, and is the lifetime the molecule would have if fluorescence were the *only* decay path by which the concentration of the excited singlet molecule were depeleted. For most molecular systems there τ_0 is always greater than τ_F, except when there is no non-radiative decay, as in atoms and some small molecules at very low pressures, in which case $\Phi_F = 1$, and $\tau_0 \equiv \tau_F$. When experiments cannot be performed to evaluate k_F, use can be made of an approximate relationship developed by Strickler and Berg [5.8]:

features are built upon the $\tilde{\nu}_1$ (a_{1g}) totally symmetric ring breathing vibration and the doubly degenerate $\tilde{\nu}_6$ (e_{2g}) in-plane distortion mode. Thus in absorption one of the main progressions is the $6_0^1 1_0^n$ progression, and correspondingly emission from the zero-point level (populated by the hot-band 6_1^0 transition) gives rise to the $6_1^0 1_n^0$ progression. [Based upon Fig. 1 in *J. Chem. Phys.*, **52**, 5366 (1970), and *Proc. Roy. Soc. A.*, **259**, 499 (1966).

Table 5.1 Quantum yields and radiative rate constants in hydrocarbon solvents at room temperature

	Φ_F	$k_F \times 10^{-6}$ (s^{-1})
Benzene	0.06	0.81 [5.6]
Toluene	0.14	1.72 [5.6]
o-Xylene	0.16	1.67 [5.6]
α-Methylnaphthalene	0.21	1.0 [5.6]
Anthracene	0.36	80.0 [5.7]
9,10-Diphenylanthracene	1.0	120.0 [5.7]
Acetone	0.0009	0.56 [5.8]
Benzophenone	$< 10^{-4}$	–
Indole	0.39	167.0 [5.7]
2-Naphthol	0.32	2.44 [5.7]

$$k_F = 2.88 \times 10^{-9} n^2 \langle \nu^{-3} \rangle_{av}^{-1} \int \frac{\epsilon}{\nu} \, d\nu$$

where n is the refractive index. $\langle \nu^{-3} \rangle_{av}^{-1}$ the averaged frequency of the fluorescence, and $\int(\epsilon/\nu)d\nu$ the area under the *absorption* curve. The relationship stems from the relation between the Einstein coefficients for stimulated absorption and emission (B coefficient) and for spontaneous emission (A coefficient), given in the Appendix.

Table 5.1 gives an indication of the very great variation in k_F and Φ_F for a range of organic molecules. We should say something briefly concerning the magnitude of the rates of non-radiative decay processes, since, as shown in Equation (5.18), these rate constants may determine whether or not fluorescence will be observable from an excited electronic state. In quantum-mechanical terms the probability of a non-radiative process has been derived simply from first-order perturbation theory and is expressed in a form known as the Fermi–golden rule. [5.9]:

$$k_{NR} = \frac{2\pi}{\hbar} \beta^2 \rho F \tag{5.22}$$

where F is a Franck–Condon factor, i.e. the square of the overlap integral of the vibrational wavefunctions of the initial and final states, ρ is the density of vibrational states in the final electronic state which are coupled to the initial level, and β^2 is an electronic integral of the sort shown in Equation (5.5) for radiative transitions, but in which the operator has a different form. Very briefly, the factors influencing the rates of non-radiative decay are as follows.

(a) The closer in electronic energy the two coupled states are, the larger are the, non-radiative decay rate constant (energy gap law). Since for most molecules the gap between successive states in the singlet manifold, or triplet manifold (i.e. the S_1–S_2, S_2–S_3, T_1–T_2, T_2–T_3, etc. energy separations), gets progressively smaller, higher excited states than the S_1 or T_1 states do not usually exhibit luminescence, since non-radiative decay processes completely dominate. So far, azulene and its derivatives provide the only exception to this rule, this molecule exhibiting strong $S_2 \rightarrow S_0$ luminescence because the $S_2 \rightarrow S_1$ energy gap is unusually large, so that the non-radiative $S_2 \rightarrow S_1$ internal conversion is relatively slow, and the $S_1 \rightarrow S_0$ gap is unusually small, leading to efficient non-radiative deactivation of the S_1 state.

(b) If the density of states ρ is large enough, the non-radiative decay is *irreversible*, and will also occur in an *isolated* molecule (said to be in the *statistical* limit). For smaller molecules, *collisional* perturbation may be required to induce the non-radiative decay. This is illustrated in Fig. 5.10 for the case of glyoxal, in which at very low pressures only fluorescence is observed. As the pressure in the

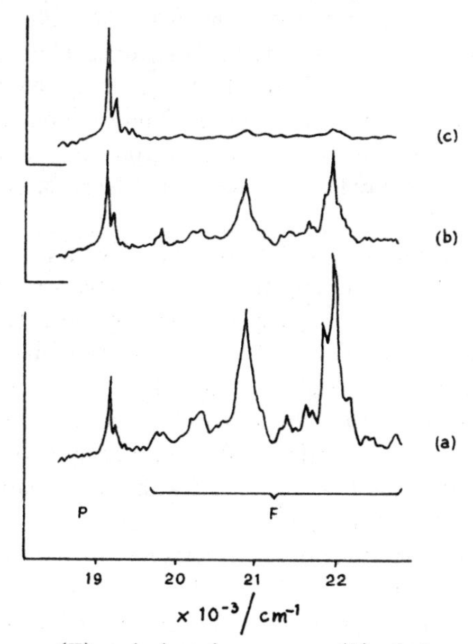

Fig. 5.10 Fluorescence (F) and phosphorescence (P) of glyoxal vapour. (a) 0.2 torr glyoxal, (b) 0.2 torr glyoxal plus 0.63 torr cyclohexane, (c) 0.2 torr glyoxal plus 29.2 torr cylohexane. [Based upon Fig. 1 in *Chem. Phys. Letters*, **8**, 232 (1971).].

system is raised, however, a collision-induced intersystem crossing is observed which leads to observable phosphorescence at the expense of the fluorescence. In small molecules (such as SO_2 and NO_2) in the isolated molecule limit, fluorescence is the only fate of the excited singlet state, but the lifetime of the state is often larger than expected on the basis of Equation (5.21) owing to the mixing in of adjacent triplet states and consequent 'dilution' of oscillator strength. Thus $\tau_F > \tau_0$ for such species.

For larger molecules non-radiative decay can occur in the isolated molecule, and thus for such cases $\tau_F < \tau_0$.

(c) The Franck–Condon factors play an important role in determining the rate constant of non-radiative decay. For undisplaced oscillators, these factors decrease very rapidly with energy gap between electronic levels, thus reinforcing the 'energy gap' law. The Franck–Condon factors can also be greatly reduced upon deuteriation of a hydrocarbon molecule. Deuteriation thus causes the rate of non-radiative decay to be reduced, and thus enhances the intensity of observed emission processes.

5.2.5 Medium effects on fluorescence spectra

The changes in fluorescence spectra on going from the vapour phase to a condensed phase can be quite dramatic, particularly for small molecules. For the reasons already discussed, well defined vibrational structure may be observed for fluorescence from low pressures of vapour. In fluid solution at room temperature both vibronic interactions and specific electronic interactions between solute and solvent will usually lead to a loss of resolvable structure in the emission spectrum profile.

As an example of the latter effect, solvent polarity may be cited. Many molecules have low-lying $^1n\pi^*$ and $^1\pi\pi^*$ states, and the relative energy levels of these different states can be altered greatly on changing the polarity of the solvent. Thus in increasingly polar solvent the $^1n\pi^*$ states increase in energy relative to the ground state, whereas $^1\pi\pi^*$ states become slightly lower in energy. The effect may be dramatic enough to invert the ordering of the states, and since in general $^1n\pi^*$ and $^1\pi\pi^*$ states have very different fluorescence properties, the observed fluorescence parameters, including spectral distribution, quantum yield, and lifetime will be greatly solvent-dependent.

As discussed above, the Stokes loss is a measure of the difference in equilibrium geometries between the relaxed excited and ground states of molecules. Since the electron distributions in the ground and excited states are different, inevitably the equilibrium configuration of the excited state will involve an arrangement of solvent molecules around the solute that is different from that in the ground state, and thus the Stokes loss is expected to be solvent-dependent also.

A wide range of specific sulute–solvent interactions has been documented, but perhaps the most important is hydrogen-bonding, which most commonly results in complete suppression (quenching) of fluorescence through enhancement of non-radiative decay routes. Thus benzo[*c*] cinnoline (I) fluorescence is quenched in solvents capable of hydrogen-bonding, e.g. ethanol, and it has been suggested that the quenching results from an increase in the intersystem crossing efficiency ($S_1 \rightarrow T_n$) of the hydrogen-bonded species relative to that of the non-hydrogen-bonded molecule [5.10].

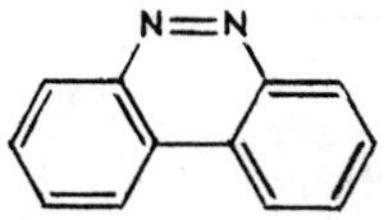

The acid strength of an electronically excited molecule may be different from that of the ground-state molecule and, like that of a ground-state molecule, solvent-dependent. Thus the relative pK_a values of the ground and S_1 states of 2-naphthol at neutral pH are such that, when the naphthol is excited, partial dissociation occurs in the S_1 state and fluorescence is observed both from naphthol and the naphtholate anion.

Concentration quenching; excimers and exciplexes

At high concentrations of many molecules, and particularly for aromatic hydrocarbons, the normal monomer emission is reduced in intensity (quenched) and new red-shifted emission maxima are observed. This emission is due to a fluorescence from an excited dimeric species which is termed an excimer, being a coalescence of the words 'excited' and 'dimer'. The excimer is formed, as shown in (5.23), by the interaction of an excited singlet state of the molecule with a ground-state partner of the same species.

$$\begin{array}{ccc} {}^1A_0 + A & {}^1(AA) & \\ \downarrow & \downarrow & \searrow A + A \\ A + h\nu_M & A + A + h\nu_D & \end{array} \qquad (5.23)$$

Thus, in a system capable of producing an excimer, it is expected that an increase in concentration will lead to an increase in excimer emission intensity at the expense of monomer emission. This situation is nicely demonstrated for pyrene in Fig. 5.11 in which the excimer emission on the right is seen to be favoured at high concentrations.

Excimer emission is red-shifted with respect to the normal monomer emission because of the binding energy of the excited and ground-state monomer molecules. No binding energy exists for two ground-state molecules, and thus emission from the excimer must produce a pair of ground-state molecules which are on a repulsive potential surface, dissociation of the pair occurring within the period of a vibration. In such a situation vibrational structure cannot be observed because the bandwidth is determined by the Heisenberg Uncertainty Principle as:

$$\Delta\nu = 1/2\pi\tau \qquad (5.24)$$

where τ is the lifetime of the final species, which in this case is extremely short. Excimer emission is recognizable therefore because it is red-shifted, broad, and structureless, and concentration-dependent.

The binding energies for excimers of aromatic hydrocarbons are usually of

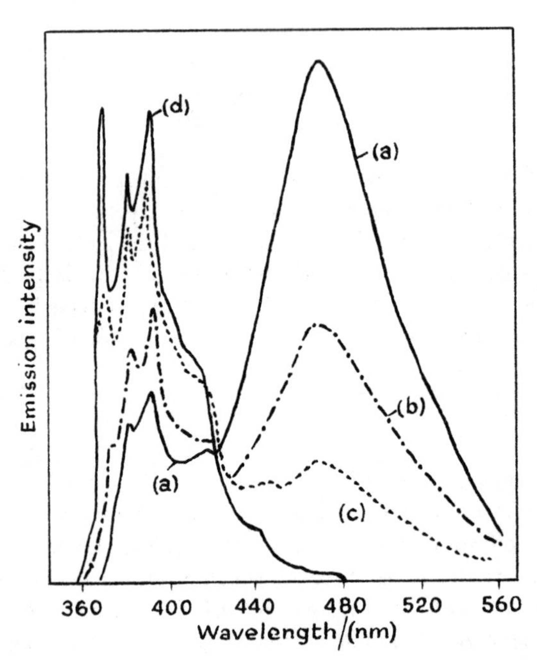

Fig. 5.11 Fluorescence of pyrene in ethanol, illustrating excimer emission: (a) 3×10^{-3} M, (b) 10^{-3} M, (c) 3×10^{-4} M, (d) 2×10^{-6} M. [Based on Fig. 131 Parker, C.A., *Photoluminescence in Solution*, Elsevier (1969).].

the order of a few tenths of an electron volt; e.g. naphthalene, pyrene, and acenaphthene have binding energies of 24.1, 38.6, and 19.3 kJ mol^{-1} respectively Close approach of monomer pairs is necessary to allow the charge-transfer and exciton resonance interaction which stabilize excimer formation. Thus steric factors or viscous solvents which preclude such interactions result in no, or inefficient, excimer formation.

Provided that the conformation of the molecule allows for this same close approach, *intramolecular* excimer formation is possible. Such is the case for 1,3-di-(1-naphthyl)propane (II) [5.12]. Both fluorescent intramolecular and

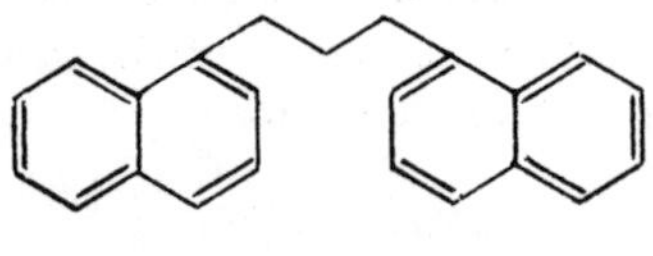

intermolecular *exciplexes* (excited complexes) are also known, the difference between an excimer and an exciplex being that an exciplex is formed by the interaction of an electronically excited monomeric species with a non-identical ground-state partner.

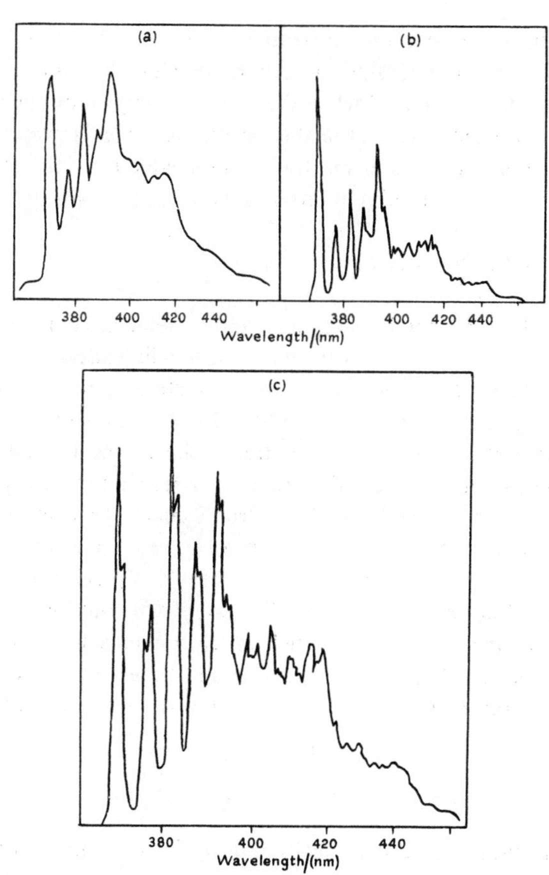

Fig. 5.12 Fluorescence spectrum from pyrene illustrating Shpol'skii effect: (a) 2×10^{-5} M in ethanol, 298 K; (b) 2×10^{-5} M in ether–pentane– ethanol, 77 K; (c) 10^{-4} M in n-hexane with 10 per cent cyclohexane, 77 K (Shpol'skii matrix). [Based on Figs. 137 and 139 in Parker, C.A., *Photoluminescence in Solution*, Elsevier (1969).]

The Shpol'skii effect [5.13]

The emission spectra of flexible molecules are usually broad and structureless because there are a large number of molecular conformations of the excited molecular conformations of the excited state in the medium which emit, and thus vibrational structure is only partially resolved. If a solvent is used, however, which at low temperatures has a crystalline structure in which the solute molecule adopts a unique configuration, vibrational structure becomes much more marked. This is demonstrated in Fig. 5.12 for pyrene in ethanol at room temperature, in ether–pentane and ethanol at 77 K, and in an n-hexane Shpolskii matrix (named after the poineer in this field) at 77 K, from which the improvement in structure in the spectrum can be seen. The optimum conditions for the appearance of the quasi-linear spectra, in which linewidths are as low as 2–3 cm^{-1}, are usually obtained when the volume occupied by the solute molecule corresponds exactly to that of the matrix molecule. However, the interpretation of Shpol'skii emission spectra is complicated by multiple emissions due to existence of solute molecules in different crystal environments.

Singlet and triplet energy transfer

Another type of interaction between an excited molecule (donor) and a different ground-state collision partner (acceptor) is one whcih leads to electronic energy transfer from the excited molecule to its ground-state partner. If, for example, both the donor and acceptor emit from their electronically excited states, then the result of energy transfer from the donor to the acceptor will be a quenching of the donor emission and a sensitization of acceptor luminescence. There is a variety of mechanisms by which such electronic energy transfer processes can occur, but two commonly met situations involve: (a) a long-range (up to 100 Å) interaction involving induced transition dipoles which practically is restricted to singlet states, and occurs often in the solid state; (b) a collisional interaction, termed the *exchange* interaction through which spin exchange can occur. Thus if the acceptor molecule A is a singlet in its ground state, (5.25) and (5.26) are both allowed by the exchange interaction, but only (5.25) can occur through

$$^{1}D^{*} + A \rightarrow {}^{1}A^{*} + D \tag{5.25}$$

$$^{3}D^{*} + A \rightarrow {}^{3}A^{*} + D \tag{5.26}$$

mechanism (a). Process (5.26) is usually referred to as *triplet–triplet* energy transfer. An example of phosphoresence quenching and sensitization is shown in Fig. 5.13.

5.3 PHOSPHORESCENCE

5.3.1 Phosphorescence and the nature of the triplet state

The pioneering work of Lewis, Calvin, and Kasha [5.14] established that the triplet state was paramagnetic, and produced long-lived luminescence of the same duration as the e.s.r signal from the state. It is now evident that phosphorescence is an electric-dipole transition analagous to that producing fluorescence, but one in which spin inversion occurs during the transition. It is pertinent to discuss briefly the ways in which triplet states may be populated.

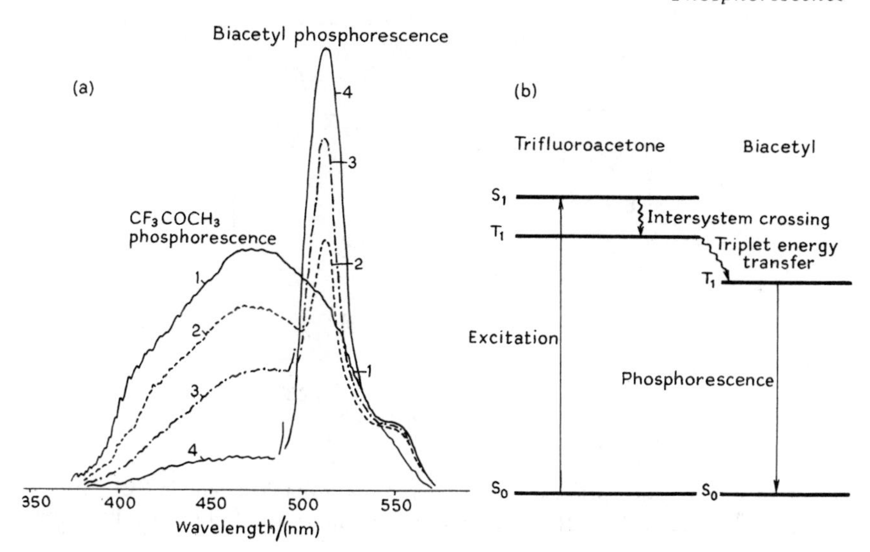

Fig. 5.13 Triplet–triplet energy transfer between trifluoroacetone vapour and biacetyl. (a) Phosphorescence spectra of (1) 120 torr CF_3COCH_3, (2) 120 torr CF_3COCH_3 plus 0.05 torr biacetyl, (3) 120 torr CF_3COCH_3 plus 0.1 torr biacetyl, (4) 120 torr CF_3COCH_3 plus 0.25 torr biacetyl. Sensitivity for biacetyl spectrum on (3) and (4) is 0.1 of that for (1) and (2). (b) Energy level diagram showing processes involved.

5.3.2 Population of the triplet state

The methods available are: (a) direct $S_0 \rightarrow T_1$ absorption; (b) through intersystem crossing from the optically pumped singlet S_1 state; (c) through triplet–triplet electronic energy transfer.

$S_0 \rightarrow T_1$ absorption

As stated previously, the restriction imposed by Equation (5.7) implies that electric dipole transitions between states of different multiplicity have zero probability. However, spin–orbit coupling induces allowedness into formally spin-forbidden transitions. Since spin–orbit coupling increases with the nuclear charge of an atom and is especially important for atoms with filled or nearly filled outer shells, in certain circumstances significant $S_0 \rightarrow T_1$ absorption intensity may be obtained. Thus, although the oscillator strength of the $S_0 \rightarrow T_1$ transition for compounds containing only carbon, hydrogen, and nitrogen is very small, the substitution of a bromine or iodine atom at a site where the electronic transition is localized results in considerably increased $S_0 \rightarrow T_1$ oscillator strength (Fig. 5.14). Paramagnetic species such as oxygen also induce spin–orbit coupling, and indeed the measurement of $S_0 \rightarrow T_n$ absorption spectra of molecules

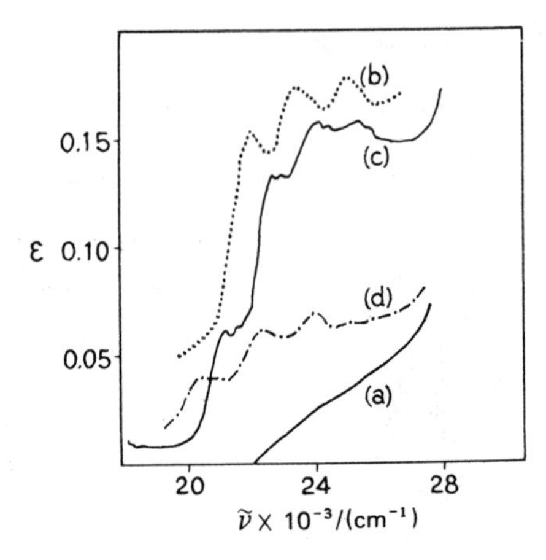

Fig. 5.14 S_0–T_1 absorption spectra of: (a) 1-chloronaphthalene in CCl_4; (b) 1-chloronaphthalene in C_2H_5I; (c) 1-iodonaphthalene; (d) 1-chloronaphthalene in CCl_4/O_2.

taken under a high pressure of oxygen is one of the few ways of obtaining information about non-phosphorescent triplet states. The mechanism by which oxygen induces the $S_0 \rightarrow T_n$ transition is one in which allowedness is introduced into the transition by mixing some charge-transfer character into the $S_0 \rightarrow T_1$ as a result of collisions with oxygen.

Intersystem crossing

For systems in which the excited singlet state is pumped optically, the observation of phosphorescence is dependent upon the magnitude of the $S_1 \rightarrow T_1$ intersystem crossing quantum yield, given kinetically using reactions (5.11)–(5.15) as:

$$\Phi_{ISC} = k_{ISC}/(k_F + k_{ISC} + k_{IC} + k_{RE}) \qquad (5.27)$$

If the possible fates of the triplet state can be represented as (5.28)–(5.30), the phosphorescence quantum yield is given by expression (5.31), where Φ'_P is the *quantum efficiency* of phosphorescence of the triplet state, given by (5.32).

		Rate	
$^3A \rightarrow A + h\nu_P$	Phosphorescence	$k'_P[^3A]$	(5.28)
$^3A \rightarrow A +$	T_1–S_0 ISC	$k'_{ISC}[^3A]$	(5.29)
$^3A \rightarrow$ products	Reaction	$k'_{RE}[^3A]$	(5.30)

$$\Phi_P = \Phi_{ISC}\Phi'_P \qquad (5.31)$$

$$\Phi'_P = k'_P/(k'_{ISC} + k'_{RE} + k'_P) \qquad (5.32)$$

Φ_{ISC} may vary from unity, as in, for example, benzophenone, to zero as in 9,10-diphenylanthracene, and we should thus consider what factors influence the magnitude of Φ_{ISC}, which are contained within the 'golden rule' expression Equation (5.22). Firstly the electronic term β^2 is the square of the electronic integral which mixes the two states S_1 and T_1 through in this case a spin–orbit coupling operator H_{SO}. The magnitude of this purely electronic interaction is shown by first-order perturbation theory to depend inversely upon the energy separation of the two states, and one would thus expect S_1–T_1 intersystem crossing to be a dominant process when the two states are close in energy, but to be inefficient or absent for molecules in which the S_1–T_1 separation is large. Table 5.2 shows that this is indeed the case. Direct spin–orbit coupling between

Table 5.2 Singlet–triplet splittings and intersystem-crossing efficiencies for organic molecules in solution.

	$\Delta E(S_1-T_1)$ (cm^{-1})	Φ_{ISC}
Benzene	8 800	0.25 [5.15]
Naphthalene	10 800	0.7 [5.16]
Anthracene	11 500	0.6 [5.17]
Ethylene	~ 24 000	0.0
1,3-Butadiene	~ 21 000	0.0
Acetone	3 000	1.0 [5.18]
Acetophenone	1 820	1.0 [5.19]
Benzophenone	1 525	1.0 [5.16]
Pyrazine	4 000	~ 0.9 [5.20]

the S_1 and T_1 states may be forbidden by symmetry restrictions, and in such cases second-order perturbations involving either vibronic coupling to a higher singlet state which is directly spin–orbit coupled to the T_1 state [Fig. 5.15(a)] or direct spin–orbit coupling to a T_2 state which is vibronically coupled to the T_1 state [Fig. 5.15(b)] may provide mechanisms for observed intersystem crossing [5.21]. Spin–orbit coupling between states of different electronic configuration, for example $n\pi^*$ and $\pi\pi^*$ states, is usually much stronger than that betwe between states of the same configuration [5.22], and this, coupled with the energy separation of the states, is sufficient to account for the fact that the rate constant for intersystem crossing, between the $^1\pi\pi^*$ and $^3\pi\pi^*$ states of naphthalene is only $10^5\,s^{-1}$, whereas that for the $^1\pi\pi^*$ and $^3\pi\pi^*$ states of benzophenone is $3.3 \times 10^{10}\,s^{-1}$.

Franck–Condon factors clearly influence the rates of intersystem crossing, and this point has been discussed earlier. It should be noted that, althought the discussion above has centred on intersystem crossing between singlet and triplet states which are most commonly met with in organic molecules, the term is also used for other transitions involving change in multiplicity, such as between quartet and doublet states frequently encountered in inorganic molecules.

Intersystem crossing may be enhanced by the same perturbations that give

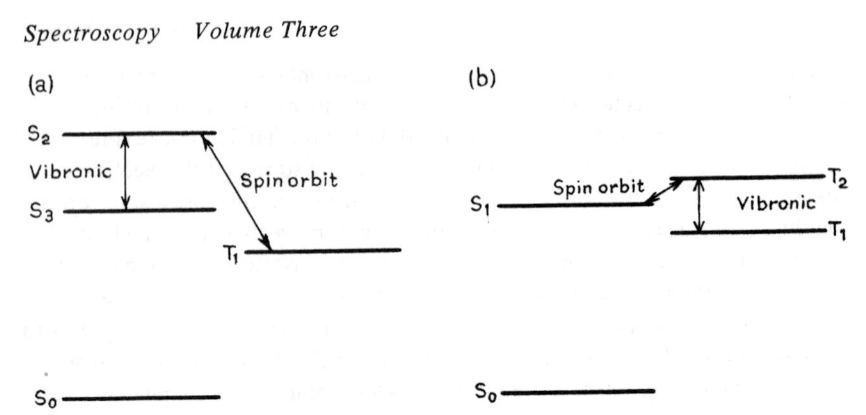

Fig. 5.15 Spin–orbit–vibronic coupling schemes.

rise to oscillator strength in the $S_0 \rightarrow T_1$ absorption, namely proximity of atoms of high nuclear charge (heavy atoms) and paramagnetic species. In the presence of such species, clearly phosphorescence is likely to be enhanced at the expense of fluorescence, as shown in Table 5.3 [5.23].

Table 5.3 Fluorescence Φ_F and phosphorescence Φ_P quantum yields of halogenated naphthalenes [5.22, 5.23]; all data refer to glassy solutions.

Compound	Φ_F	Φ_P	k_{ISC}
Naphthalene	0.55	0.06	10^5
1-Chloronaphthalene	0.29	0.30	–
1-Bromonaphthalene	0.002	0.27	5×10^8
1-Iodonaphthalene	0.0005	0.38	$< 3 \times 10^9$

5.3.3 Phosphorescence intensity

As Equation (5.31) implies, a large value of Φ_{ISC} from S_1 to T_1, while it necessarily leads to a higher probability of observing phosphorescence from a molecule, is not by itself a sufficient condition, since there are non-radiative decay paths open to the triplet state once formed.

Since both the radiative and non-radiative $T_1 \rightarrow S_0$ relaxation processes are spin-forbidden they are in general slow processes, and thus phosphorescence lifetimes, τ_P, are usually considerably longer than fluorescence lifetimes. For this reason phosphorescence is very susceptible to the presence of impurities such as molecular oxygen which lead to bimolecular quenching, and therefore phosphorescence is observed for relatively few molecules in fluid solution. However, when an appropriate (non-reactive) solvent is carefully purified phosphorescence may be observed in fluid media. However, by far the most widespread media for observation of phosphorescence spectra are mixtures of solvents which form clear glasses at low temperatures, usually 77 K. Under these conditions diffusion

Table 5.4 Phosphorescence spectra, lifetimes, and quantum yields, at 77 K

	Phosphorescence (0, 0 band)/ (cm^{-1})	*Phosphorescence lifetime, τ_P/(s)*	*Phosphorescence quantum yields*
Benzene	29 510	6.3 [5.23]	0.23 [5.24]
p-Dibromobenzene	–	3×10^{-4} [5.25]	–
Aniline	26 800	4.7 [5.25]	–
Naphthalene	21 250	2.3 [5.26]	0.05 [5.24]
1-Nitronaphthalene	19 200	0.05 [5.25]	–
2-Iodonaphthalene	20 500	2×10^{-3} [5.22]	0.20 [5.22]
Acetone	21 980 (max)	6×10^{-4} [5.25]	0.04 [5.25]
Acetophenone	26 000	8×10^{-3} [5.27]	0.6 [5.24]
Benzophenone	24 650	6×10^{-3} [5.18]	0.74 [5.24]

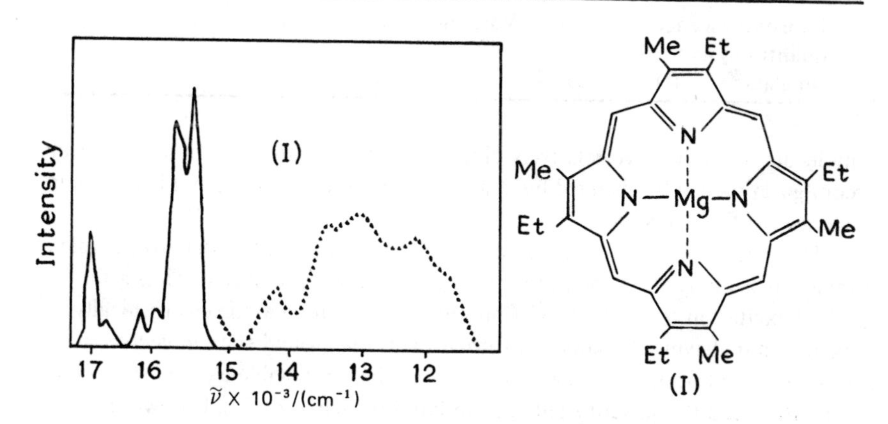

Fig. 5.16 Fluorescence (solid line) and phosphorescence (dotted line) spectra of magnesium etiopophyrin (I) in EPA glass. [Based upon Fig. 2 in *J. Chem. Phys.*, 32, 1410 (1960).].

is supressed, and many molecules exhibit phosphorescence with lifetimes in the 10^{-6} to 10 s range (Table 5.4). An example of the phosphorescence spectrum of an inorganic compound is given in Fig. 5.16, from which this emission can be seen to have vibrational structure like the fluorescence, but to be red-shifted with respect to it.

In organic molecules which have both $^3\pi\pi^*$ and $^3n\pi^*$ states, such as aromatic carbonyl compounds and aza-aromatics, the characteristics of the phosphorescence may be used to determine the electronic nature of the lowest-energy (and therefore emitting) state, using characteristics outlined in Table 5.5.

5.3.4 Solvent effects on

The effects of solvent polarity and viscosity on phosphorescence can be similar to those observed for fluorescence. Thus the phosphorescence properties of a

Table 5.5 Characteristics of $^3n\pi^*$ and $^3\pi\pi^*$ phosphorescence

Emitting state	$^3\pi\pi^*$	$^3n\pi^*$
Polarization of phosphorescence	Predominantly out-of plane	Predominantly in-plane
Vibrational structure in spectra	Variable, but aromatic ring frequencies often seen (in aromatic compounds)	In carbonyl compounds C = 0 stretching frequencies observable
Lifetime of phosphorescence (in glass at 77 K)	>1 s	10^{-2}–10^{-3} s
Phosphorescence quantum yield (in glass at 77 K)	Variable	Usually large

molecule which has two relatively close-lying triplet states of different electron configuration can be changed by change in solvent and consequent inversion in energies of the two states.

The most stable conformation of an electronically excited state of flexible molecule is rarely the same as that of the ground state molecule. Thus after initial excitation to the Franck–Condon excited state, i.e. the electronically excited state having the same conformation as the ground state molecule, a relaxation will occur to give a species having the most stable excited state conformation and the solvent shell appropriate for such a species. If however a viscous medium is employed, $S_1 \rightarrow T_1$ intersystem crossing and phosphorescence may occur before conformational relaxation. Thus, by varying the viscosity, phosphorescence may be observed both from relaxed and unrelaxed triplet states. This phenomenon is demonstrated for butyrophenone in Fig. 5.17. In a very viscous 1 : 1 isopentane–methylcyclohexane mixture at 77 K a high energy phosphorescence is observed (a) which may be attributed to phosphorescence from the unrelaxed $n\pi^*$ triplet state. In a less viscous 9 : 1 mixture of the same solvents at 77 K a lower energy phosphorescence is observed (b) corresponding to the phosphorescence of the relaxed $^3n\pi^*$ state.

In phosphorescence spectral studies, in contrast to fluorescence, perhaps the most important solvent effect and certainly the one most frequently encountered is that due to the presence of impurities. Two effects may result from this: (a) quenching of the phosphorescence of the emitting molecule by energy transfer or some other mechanism; (b) contributions to the total phosphorescence spectral profile from emission from the impurity. There is little to be said about these effects except to emphasize that extremely careful solvent and solute purification is necessary to avoid their occurrence.

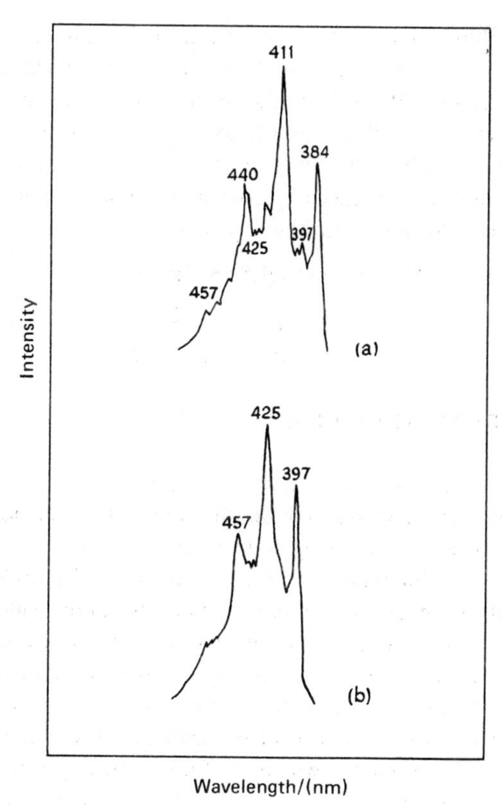

Fig. 5.17 Phosphorescence of butyrophenone at 77 K in (a) 1 : 1 and (b) 9 : 1 isopentane–methylcyclohexane. Wavelengths of bands in nm identified by numbers. [Based upon Fig. 1 in *Chem. Phys. Letters*, **13**, 545 (1972).].

5.3.5 Delayed fluorescence

Delayed emission is that which persists long after the excitation process, and clearly includes phosphorescence. The techniques for recording delayed emission spectra are outlined in the experimental section. It is possible however to observe delayed emission which by definition is long-lived, but which has the same spectral distribution as normal (very short-lived) fluorescence. Such emission is termed delayed fluorescence, and is of two types, E-type, first seen in eosin, and P-type, first seen in pyrene.

E-type delayed fluorescence arises from thermal reactivation of a triplet state molecule to the singlet state from which fluorescence occurs. Thus the spectral distribution is that of normal fluorescence, but the lifetime is associated with the triplet state. Since the reactivation process is thermal this phenomenon is confined to species in which the singlet–triplet energy separation is small.

P-type delayed fluorescence arises from the mutual annihilation of *two* triplet states to produce one excited singlet and one ground-state singlet molecule. This energy pooling process results in the initial formation of an excited dimer (eximer) which, if emittive, will give fise to excimer emission as well as normal monomer emission which have the same spectral characteristics as the excimer and monomer fluorescence produced by continuous illumination, but again, which have lifetimes associated with the triplet state.

$$\begin{array}{ccccc} {}^3P + {}^3P & \longrightarrow & {}^1(PP)^* & \longrightarrow & {}^1P^* + P \\ & & \downarrow & & \downarrow \\ & & P + P + h\nu_D & & P + h\nu_M \end{array} \qquad (5.33)$$

5.4 EXCITATION SPECTRA

Fluorescence and phosphorescence spectra are recorded using a fixed excitation frequency and scanning the wavelength distribution of the emission with an analysing monochromator or spectrograph. If the emission frequency is kept constant, and the excitation wavelengths are scanned, the fluorescence or phosphorescence *excitation* spectrum is obtained, and this can be made absolute if the spectral intensity of the excitation source and spectral transmittance of the excitation monochromator are known. Excitation spectra can provide valuable information since use is made of an emission measurement which is capable of very great amplification to record a spectrum which relates to an absorption process. Thus, for a compound for which the fluorescence quantum yield is independent of excitation wavelength, the corrected fluorescence excitation spectrum will be identical to the absorption spectrum, and thus the absorption spectra of extremely low concentrations of material can be obtained using the excitation technique which are impossible to obtain by direct absorption methods. Phosphorescence excitation spectroscopy is a very sensitive means of recording $S_0 \rightarrow T_1$ absorption spectra which in unperturbed systems have extremely small transition probabilities, and so are difficult to observe directly.

5.5 EXPERIMENTAL METHODS

5.5.1 Spectrofluorimeter design

Fluorescence and phosphorescence spectra are recorded conveniently on a spectrofluorimeter which will generally be fitted with a device for distinguishing between long-lived ($>$ 1 ms) and short-lived emission. A block diagram of a spectrofluorimeter is shown in FIg. 5.18, and consists basically of the following components.

(a) *Light source* to give excitation wavelengths typically between 200 and 800 nm. A xenon arc lamp of 150–500 W power run from a d.c. stabilized power supply is a favourite choice of source.

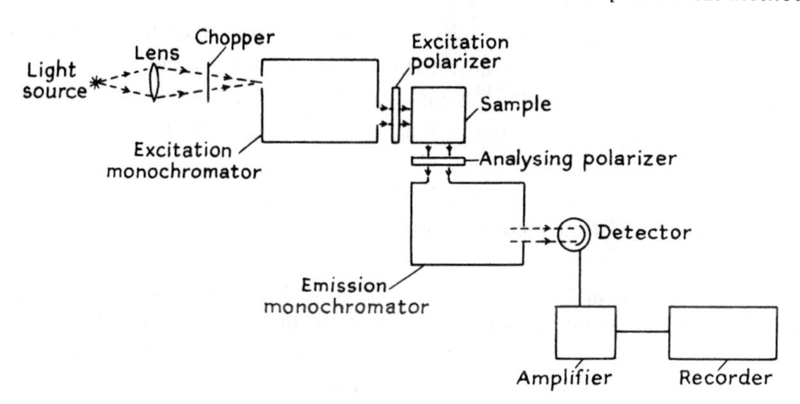

Fig. 5.18 Block diagram of spectrofluorimeter

(b) An *excitation monochromator* for the selection of excitation wavelengths, usually of high resolution grating type fitted with motor drive.

(c) An *emission monochromator* for the spectral analysis of the luminescence of similar or better characteristics to those of the excitation monochromator.

(d) A *flexible cell-housing* in which a variety of cells, often of fused silica, may be accommodated, together with low and high temperature control systems.

(e) *Detection system*, usually in the form of a sensitive photomultiplier, the electrical output from which is amplified and displayed on a strip-chart recorder. For greater sensitivity single photon-counting detection may be employed.

(f) *Correction system*. Emission spectral profiles obtained from a simple instrument of the above type will be distorted from their true shapes owing to the fact that the sensitivity of the photodetector and the transmission characteristics of the optical system are non-linear functions of wavelength of the light emitted. The simplest method of correcting spectra is to compare the instrumental response for a standard compound of known corrected spectral characteristics with that of the sample of interest, although spectrofluorimeters are now available which compensate electronically for the wavelength response of the system.

(g) *Phosphorimeter attachment*. This is used for the measurement of delayed emission characteristics, and consists of a mechanical or electronic device which interrupts the excitation beam periodically. If the detection system is gated so that emission is viewed only after a time has elapsed after the cut-off of the excitation, then short-lived emission such as prompt fluorescence will have decayed to zero intensity, and thus only long-lived emission will be observed. The decay time of the luminescence can also be measured using this technique, and for a mechanical device such as a rotating toothed chopper, the limit of measurable lifetime is of the order of 1 ms. Electronic devices can be used to measure decay times much shorter than this (see below).

(h) *Polarizers*. If polarized spectra are required, a prism polarizer can be inserted at the exit slit of the excitation monochromator which transmits plane polarized light. An analysing polarizer is then inserted on the emission side of the optical system, and usually this can be easily set to measure light which has the electric vector parallel to that of the excitation radiation, and perpendicular to it.

5.5.2 Emission lifetime measurements

The use of mechanical devices such as that just described to obtain emission lifetimes limits lifetime measurements to those emissions having lifetimes longer than a few ms. However, techniques are now available which allow the measurements of much shorter lifetimes. Two methods have been found to be particularly useful: phase modulation of the excitation source coupled with phase sensitive detection and single photon counting techniques.

(a) *Phase modulation*. Phase modulation involves the modulation of the exciting beam at a high frequency. The phase of the luminescence is then compared with that of the exciting light. For an emission decaying exponentially with a lifetime τ, the phase of p relative to that of the exciting beam is given by:

$$\omega\tau = \tan p$$

where ω is the modulation frequency. Thus by feeding the excitation and emission into an electronic circuit capable of determining the out-of-phase angle of the two signals, τ may be obtained.

(b) *Single photon counting*. Figure 5.19 is a block diagram of a typical single

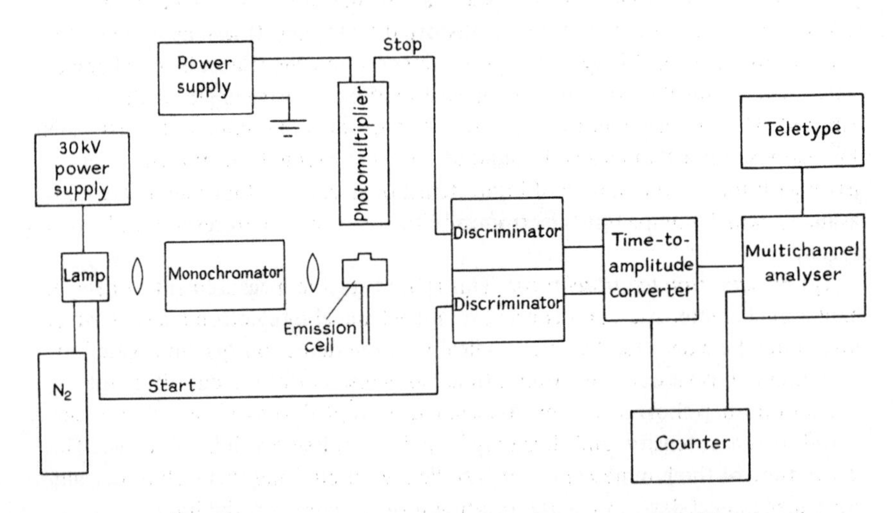

Fig. 5.19 Block diagram of emission lifetimes using system for obtaining time-correlated single-photon counting.

photon counting apparatus. Light from the nitrogen flash-lamp excites the sample, after passing through a monochromator. This flash is also used to start a counting sequence in a time-to-amplitude converter (TAC). In simple terms the start pulse triggers a voltage ramp where the voltage is proportional to time. The voltage ramp is stopped by a signal from the photomultiplier on observation of an emitted photon. The voltage attained by the ramp is stored in a multichannel pulse height analyser and the TAC reset for the reception of the next start pulse. In this way the multichannel analyser accumulates a record of the time-correlated emission of the sample. For simple systems this will be an exponential decay curve from which the emission life-time may be easily obtained. The system is useful in the $\mu s \rightarrow ns$ time-scale.

5.5.3 Time-resolved emission spectroscopy

If in the lifetime apparatus above a second monochromator is interposed between between the sample cell and photomultiplier, time-dependent emission spectroscopy may be carried out. Thus by incorporating a particular time delay into the detection system and using a stepping motor on the emission monochromator, the emission spectra may be measured at any time after excitation and, as pointed out earlier, such time-dependent spectra give valuable information about the relaxation processes in excited electronic states.

5.6 APPLICATIONS OF FLUORESCENCE AND PHOSPHORESCENCE

5.6.1 Analytical

Emission spectroscopy provides an extremely powerful tool for the quantitative and qualitative analysis of very small quantities of luminescent materials. The advantages of sensitivity of analytical methods based upon emission over those based upon absorption techniques is due to the fact that the emission measurement is absolute whereas the absorption measurement is relative. Thus in absorption one is measuring the difference in light transmitted through a sample and a reference blank. As the concentration of absorbing species becomes very small, there is considerable instrumental difficulty in measuring the difference in transmittance, and the method becomes inaccurate. In the case of emission methods, however, the light emitted is monitored by a photo-device which is capable of extreme sensitivity. The total intensity, say, of molecular fluorescence from a dilute solution in photons per second is given by:

$$Q = I_0(2.3\epsilon cl)\Phi_F$$

where I_0 is the incident intensity, ϵ the molar decadic extinction coefficient, c the concentration of the molecule to be estimated in moles/l, l is the absorption path length, and Φ_F the fluorescence quantum yield. This expression can be used to deduce that concentrations of emitter as low as 10^{-12} M can easily be estimated by these methods. In the case of atomic fluorescence, the limits of detectability of the elements Ag, Au, Cu, Mg, and Ni are of the order of 10^{-12} g, and those of Cd and Zn even smaller at 10^{-15} and 10^{-14} g respectively. These sensitive fluorescent and phosphorescent spectral measurements are used routinely in such diverse applications as monitoring trace elements in oils, soil extracts, blood, and urine; drugs in the blood stream;pollutants in the atmosphere, and indeed in any application where high sensitivity is an advantage [5.29, 5.30].

5.6.2 Scintillators

Any solution which emits fluorescence when excited by ionizing radiation may be used as a scintillation solution. A high-energy particle impinging on a dilute solution of a scintillator produces ionized and electronically excited solvent molecules. Most of the excitation energy may be transferred to the solute and thus produce solute fluorescence, which can be monitored. In this way the intensity of a flux of high-energy particles produced in any experiment may be monitored [5.31].

5.6.3 Probes

Since the emission efficiency and spectra of a molecule are sensitive to environmental effects, suitable fluorescent or phosphorescent molecules can be used to probe molecular environments. A rather interesting example of this application is the use of naphthalene fluorescence to monitor the kinetics of transfer of molecules from an aqueous medium into micelle systems in detergents.

5.6.4 Lasers

The principle of stimulation of emission from metastable states has led to the rapid development of powerful light sources which get their name from the process of Light Amplification by the Stimulated Emission of Radiation (LASER). As can be seen from the Appendix, for a two-level system the application of a high radiation density results in the population of initial and final levels becoming equal, since under these conditions spontaneous emission processes will be of negligible importance. In such circumstances there is no net absorption or emission of radiation, and the transition is said to be saturated. With a three-level system, however, it is possible to arrange for a population inversion of a metastable state with respect to a lower state to occur by optical pumping through an upper state, or by an electrical discharge, and under these circumstances stimulated emission from the metastable state can be induced. The discussion of the technology which has permitted this phenomenon to be exploited as practical powerful light sources is outside the scope of this book, but it should be noted that the principle on which the laser operates is at its simplest a three-level system such as is outlined in Fig. 5.1. The levels involved in the ruby laser are illustrated in Fig. 5.20, in which it can be seen that the optically pumped

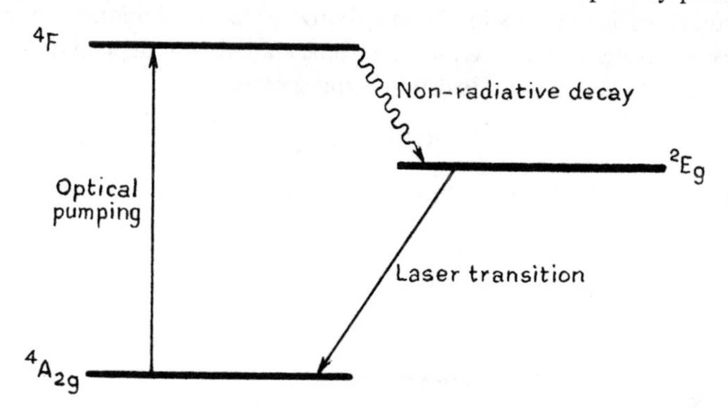

Fig. 5.20 Transitions involved in ruby laser. Optical pumping occurs through the ground ($^4A_{2g}$) state of the Cr^{3+} ion to the $^4F_{2g}$ and $^4F_{1g}$ states (shown as a single state for simplicity). Non-radiative decay to the ^2E_g state occurs, and the stimulated emission corresponds to the $^2E_g \rightarrow {}^4A_{2g}$ transition.

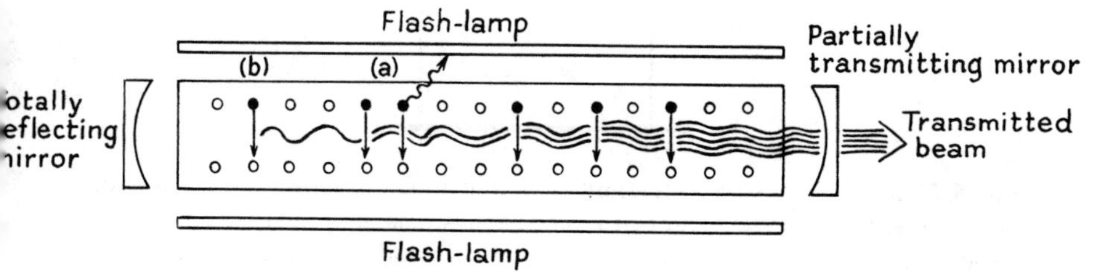

Fig. 5.21 Operation of ruby laser. Filled circles represent chromium ions excited by the flash-lamp pulse. The chromium ion emitting *spontaneously* in a direction other than along the laser cavity is lost to the system (a). Ion (b) emitting *spontaneously* produces a wave which passes along the laser cavity and encounters another excited ion which then gives out *stimulated* emission and the waves is thus reinforced. This process is repeated along the cavity, the wave eventually passing through the partially transmitting mirror to give the laser pulse.

transitions are the $^4A \rightarrow {}^4F$ transitions of the Cr^{3+} ion in the crystal, and the transition giving rise to the stimulated emission is the $^2E \rightarrow {}^4A$ transition. The optical principle of the laser is summarized in Fig. 5.21.

5.7 APPENDIX

Absorption and stimulated emission are thus processes which are induced by a radiation field, and if this has a density ρ, the net rate of a stimulated process between states labelled i and f is then:

$$\text{Rate} = n_i B_{if} \rho \tag{5.34}$$

where n_i is the concentration of molecules in the initial level, and B_{if} is termed the Einstein coefficient for stimulated radiative processes. Spontaneous emission processes occurring from an excited electronic state do not depend upon the radiation density ρ, and the rate can be expressed as:

$$\text{Rate} = n_i A_{if} \tag{5.35}$$

where n_i is as before, and A_{if} is now the Einstein coefficient and for spontaneous emission.

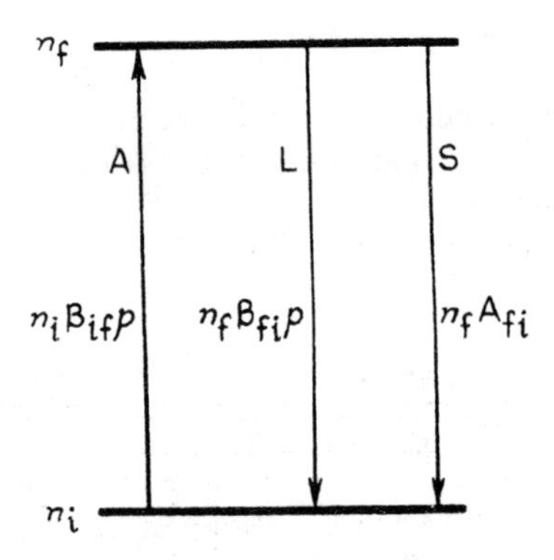

Fig. 5.22 Rates of absorption, stimulated emission, and spontaneous emission processes between levels i and f of molecular systems of populations n_i and n_f respectively. A is the absorption process, L is stimulated emission, and S is spontaneous emission. A_{fi} is the Einstein A coefficient, B is the Einstein B coefficient, and ρ is the radiation density.

In a two-level system shown in Fig. 5.22, given that $B_{if} = B_{fi}$, the net rate of change in the system undergoing both stimulated and spontaneous emission process is given by the left-hand side of Equation (5.36) and is:

$$n_i B_{if} \rho - n_f B_{if} \rho - n_f A_{if} = 0 \tag{5.36}$$

set equal to zero in an equilibrium situation. Initially the population in the upper state is given by the Boltzmann distribution:

$$\frac{n_f}{n_i} = \frac{g_f}{g_i} e^{-(\epsilon_f - \epsilon_i)/kT} = e^{-h\tilde{\nu}/kT} \tag{5.37}$$

for singlet states. For electronic transitions, this means that the initial population in the upper state n_f is negligible, and there is in the presence of a radiation field a net stimulated *absorption* of radiation followed by *spontaneous* emission. The relationship between the Einstein A and B coefficients is:

$$A_{if} = \frac{8\pi h\nu^3}{C^3} \cdot B_{if} \tag{5.38}$$

from which it can be seen that the probability of spontaneous emission depends upon $\bar{\nu}^3$. This spontaneous emission is likely for electronically excited states, but for vibrational transitions, or rotational transitions, where $\bar{\nu}^3$ becomes very small, spontaneous emission processes have very low probability.

It can be seen from Equation (5.36) that the application of a high radiation density ρ can only lead to the populations of upper and lower levels n_f and n_i becoming equal in the two-level situation, in which case the transition is saturated. It is only possible therefore to observe net stimulated emission from such a system if a pumping mechanism other than optical excitation is used to create a population inversion with respect to the lower state (i.e. $n_f > n_i$), or if a third pumping level is utilized so that the optical pumping frequency and stimulated emission frequencies are different.

REFERENCES

5.1 McGlynn, S.P., Azumi, T. and Kinoshita, M., *Molecular Spectroscopy of the Triplet State*, Prentice-Hall, New Jersey, (1969), p. 69.

5.2 Pringsheim, P., *Fluorescence and Phosphorescence,* Wiley-Interscience, New York, (1949).

5.3 Berlman, I.B., *Handbook of Fluorescence Spectra of Aromatic Molecules*, Academic Press, New York, (1969).

5.4 Albrecht, A.C., *J. Molecular Spectroscopy*, **6**, 84, (1961).

5.5 Herzberg, G. and Teller, E., *Z. Physik. Chem.,* **B21**, 410 (1933).

5.6 Cundall, R.B. and Pereira, L.C., *J. Chem. Soc. Faraday II,* **68**, 1152 (1972); Cundall, R.B. and Robinson, D.A., *ibid*., 1133.

5.7 Berlman, I.B., *Fluorescence Spectra of Aromatic Molecules*, Academic Press, New York, (1971).

5.8 Strickler, S.J. and Berg, R.A., *J. Chem. Phys*., **37**, 874 (1972).

5.9 Freed, K.F., *Fortsch. Chem. Forsch*., 105 (1972).

5.10 Rau, Von H. and Bisle, H., *Ber. Bunsengesell. Phys. Chem.,* **77**, 281 (1973).

5.11 Parker, C.A. and Hatchard, C.G., *Trans. Faraday Soc*., **59**, 284 (1963).

5.12 Chandross, E.A. and Dempster, C.G., *J. Amer. Chem. Soc*., **92**, 704 (1970).

5.13 Shpol'skii, E.V., *Sovt. Phys. Uspekhi*, **6**, 411 (1963), and references therein.

5.14 Lewis, G.N., Calvin, M. and Kasha, M., *J. Chem. Phys.,* **17**, 804 (1949).

5.15 Cundall, R.B and Pereira, L.C., *Chem. Phys. Letters*, **18**, 371 (1973).

5.16 Sandros, K., *Acta Chem. Scand*., **23**, 2815 (1969).

5.17 Bowers, P.G. and Porter, G., *Proc. Roy. Soc., A,* **299**, 348 (1967).

5.18 Murov, S.L., *Handbook of Photochemistry*, Dekker, New York, (1973).
5.19 Lamola, A.A. and Hammond, G.S., *J. Chem. Phys.*, **43**, 2129 (1965).
5.20 Jones, S.H. and Brewer, T.L., *J. Amer. Chem. Soc.* **94**, 6310 (1972).
5.21 Albrecht, A.C., *J. Chem. Phys.*, **38**, 354 (1963); van Egmond, J. and van der Waals, J.H., *Mol. Phys.*, **26**, 1147 (1973).
5.22 McGlynn, S.P., Azumi, T. and Kinoshita, M., *Molecular Spectroscopy of the Triplet State*, Prentice-Hall (1969), p.212.
5.23 Lim, R.C. and Lim, E.C., *J. Chem. Phys.*, **57**, 605 (1972).
5.24 Gilmore, E.H., Gibson, G.E. and McClure, D.S., *J. Chem. Phys.*, **23**, 399 (1955).
5.25 McClure, D.S., *J. Chem. Phys.*, **17**, 903 (1949).
5.26 Lim, E.C., Laposa, J.D. and Yu, J.M.H., *J. Mol. Spectroscopy*, **19**, 412 (1966).
5.27 Becker, R.S., *Theory and Interpretation of Fluorescence and Phosphorescence*, Wiley-Interscience (1969).
5.28 Wagner, P.J., May, M. and Haug, A., *Chem. Phys. Letters*, **13**, 545 (1972).
5.29 See, for instance, Parker, C.A., *Photoluminescence in Solution*, Elsevier, London (1969).
5.30 Winefordner, J.D., Schulman, S.G. and O'Haver, T.C., *Luminescence Spectrometry in Analytical Chemistry*, Vol. 38 of *Chemical Analysis*, ed. Elving, P.J. and Kolthoff, I.M., Wiley-Interscience, New York, (1972).
5.31 See, for instance, Birks, J.B., *Photophysics of Aromatic Molecules*, Wiley-Interscience, London (1970).

6 Astrochemistry

6.1 INTRODUCTION AND INSTRUMENTATION

Astrochemistry is concerned with the study of the matter which exists in the universe in, for example, planets, stars, and the interstellar medium. The absorption and emission spectra of these enable some of the atoms and molecules present in these systems to be identified. This leads to various theories, in particular, how the universe came to be and how life evolved. Often much speculation is involved; a knowledge of photochemistry and various types of chemical reactions is required besides an appreciation of the processes which may take place in conditions much more extreme than can be experienced on the earth; for example, the temperatures of stars range from 50 000 to 3000 K.

The following spectroscopic regions are yielding key information in astrochemical work: (i) microwave; (ii) far-infrared and infrared; (iii) ultraviolet and visible regions.

The principles and instrumentation of (i) and (ii) have been described in the pertinent chapters. In this account we shall be concerned only with the specialized instrumentation relevant to type (iii) in astrochemical studies. Nevertheless, we shal be concerned with some of the information resulting from (i), (ii), and (iii). The study of stellar spectra in the ultraviolet and visible regions may be carried out using either one of the spectrographs described in Chapter 2 or by a slitless type. In the slitless type of spectrograph, instead of a series of monochromatic images of the slit leading to bands and lines on the photographic plate, a series of monochromatic images of the source is produced on the plate. A simple form of slitless spectrograph is the objective prism type where light collected by a telescope is dispersed by a large prism and focused on to a photographic plate situated in the focal plane of the telescope.

Slit spectrographs of both the prism and grating types are employed in conjunction with large light-gathering telescopes of both the refracting and reflecting types to study stellar spectra. The refracting telescopes have lenses which focus the image of the source on to the slit of the spectrograph, but these suffer from the disadvantage that different wavelengths are refracted to differing extents. Reflecting telescopes, however, which have large concave mirrors, do not have this drawback.

Reflecting telescopes are constructed so that secondary mirrors can be brought into the light path, and this gives a variety of positions at which the spectra may be recorded. Four such arrangements of a reflecting telescope are illustrated in Fig. 6.1. One of these positions is termed the *prime focus arrangement* and is illustrated in Fig. 6.1(a). Light from the source enters the telescope

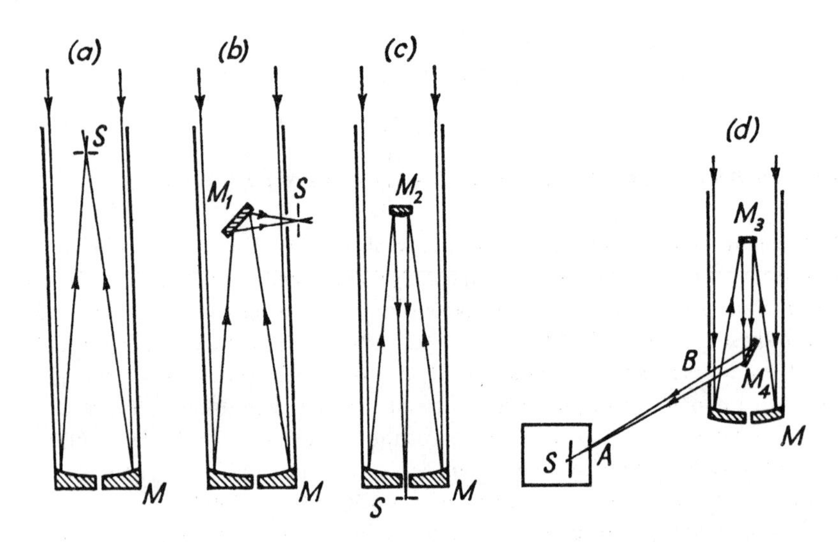

Fig. 6.1 (a) The prime focus, (b) Newtonian, (c) Cassegrain, and (d) coudé arrangements of a reflecting telescope.

and is reflected by a large parabolic mirror M on to the slit of a spectrograph S which is situated at the focus of the parabolic reflecting mirror. If the reflecting telescope is not one of the largest, however, this prime focus arrangement is restricted in its use, since any spectrograph placed at the focus of the parabolic reflector would seriously reduce the amount of light from the source entering the telescope. Another arrangement of a reflecting telescope is shown in Fig. Fig. 6.1(b) and is known as the *Newtonian mounting*. A plane mirror M_1 placed in the light path just below the focus of the parabolic mirror, M, causes the light to pass through a small hole provided in the wall of the telescope and on to the spectrograph slit at S.

The *Cassegrain arrangement* illustrated in Fig. 6.1(c) involves a convex hyperbolic mirror M_2 set in the convergent beam of light below the focus of the parabolic reflecting mirror M. The mirror M_2 reverses the beam of light along a path through a hole in the centre of the main parabolic reflector, the slit of the spectrograph being located at S.

Another commonly used position is the *coudé focus* illustrated in Fig. 6.1(d), in which the light beam from the parabolic mirror M is reversed by a convex mirror M_3 (similar to that used in the Cassegrain arrangement but of different curvature) to a further mirror M_4 which deflects the converging beam along the hollow axis AB of the telescope. The spectrograph S is situated in an air-conditioned, temperature-controlled laboratory below the main reflecting mirror M.

While all of the described foci are employed, the Cassegrain and coudé positions are favoured, since larger spectrographs may be used in these arrangements. It is easier to mount the spectrograph at the Cassegrain focus without causing strain to be placed on the reflecting mirror M which must be very accurately supported. In the coudé arrangement the spectrograph is completely separated from the telescope. However, on account of the shorter focal length, the prime focus and the Newtonian positions are best for investigating electromagnetic radiation of low intensity.

A powerful modern type of reflecting telescope is the *Schmidt type* the optical arrangement of which is illustrated in Fig. 6.2. A spherical mirror M is

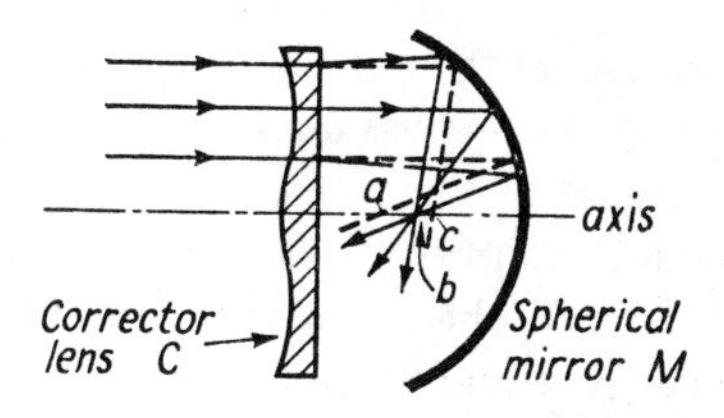

Fig. 6.2 The optical arrangement of a Schmidt telescope.

employed in conjunction with a correcting lens C. The lens C is plane on one side whilst the other side is convex in the central portion and concave in the outer sections. The convex section acts as a converging lens and the concave portion as a diverging lens. The correcting lens C forms no image and does not necessarily alter the focal length or the magnification of the system. Its object is to correct for spherical aberration of the main spherical mirror M.

With no correcting lens (broken lines) the rays cross the axis of the system at different points a, b, and c, while with the lens C in position the rays (full lines) cross the axis at the same point b, thus eliminating spherical abberration. The focal surface of the Schmidt telescope is curved so that the photographic plate or film must always be curved correspondingly to conform with the focal surface of the mirror M. The Schmidt telescope may be constructed to give a very short focal length, and in consequence a high photographic speed is possible. It also has good definition over a very wide field of view which is not possible with a parabolic mirror type reflector. In fact, Schmidt telescopes are often used for searching out objects to be viewed by the more restricted reflecting telescopes.

The optical arrangement of the Schmidt telescope makes possible high photographic speed and excellent definition of the image which are desirable features to have in the spectrograph camera itself, since a short-focus camera permits the entrance slit to be widened sufficiently to take in most of the star's image without losing resolution on the plate. The conventional arrangement of camera lens and plate holder may, therefore, be replaced in the spectrograph by a spherical mirror M and a correcting lens C of the same design as that employed in the Schmidt telescope. The plate holder P is mounted at the focus of the spherical mirror and curved so as to conform with the optical surface of this mirror. This arrangement is known as the *Schmidt camera*.

To illustrate the use of a Schmidt camera the type of spectrograph used to investigate the spectrum of the night sky or of nebulae (clouds of gas or of fine solid particles) may be considered. Before it enters the spectrograph, light from the source is reflected by a plane mirror M_1 [see Fig. 6.3(a)] provided with a shutter to a further mirror M_2. The shutter over the mirror M_1 limits its useful width, this mirror acting as the slit of the spectrograph. The distance between the mirrors M_1 and M_2 is 23 m, and that between the mirror M_2 and the spectrograph is the same distance. The distance of the slit from the first prism makes a collimating lens unnecessary. The spectrograph illustrated in Fig. 6.3(b) consists of two 60° quartz prisms and an *f*/1 Schmidt camera with a focal length of approximately 10 cm. Light from the mirror M_2 [see Fig. 6.3(a)] enters the spectrograph, is dispersed by the prisms, and passes through a corrector lens C. The spectrum is focused on the photographic film P by a spherical mirror M. The linear dispersion of such an instrument varies from 115 Å/mm at a wavelength of 3200 Å to 500 Å/mm at 5000 Å. The instrument described is based on a spectrograph employed by Struve [6.1] and his collaborators to study nebulae and is particularly suitable for the study of extended low-intensity sources.

The high intensity of electromagnetic radiation emitted by the sun makes feasible spectroscopic studies of a kind which are impossible for any other star. Spectrographs of high resolving power can be used to examine parts of the sun's surface. For example, one experimental arrangement makes use of a moving mirror mounted at the top of a tower; the mirror reflects the sunlight through an objective lens and focuses the light on to the slit of a grating spectrograph which is mounted in a laboratory below ground level.

For observations on the solar chromosphere, that is the thin envelope of relatively transparent gases which lies above the photosphere of the sun, a slitless spectrograph may be used enabling the whole of the solar image to be recorded simultaneously, which is not possible using a conventional slit instrument. These observations are made at the time of total solar eclipse (i.e. when the moon obscures the sun from view).

One of the advantages of a slitless spectrograph is the large field of view which can be recorded. Hitherto, the slitless arrangement has been used

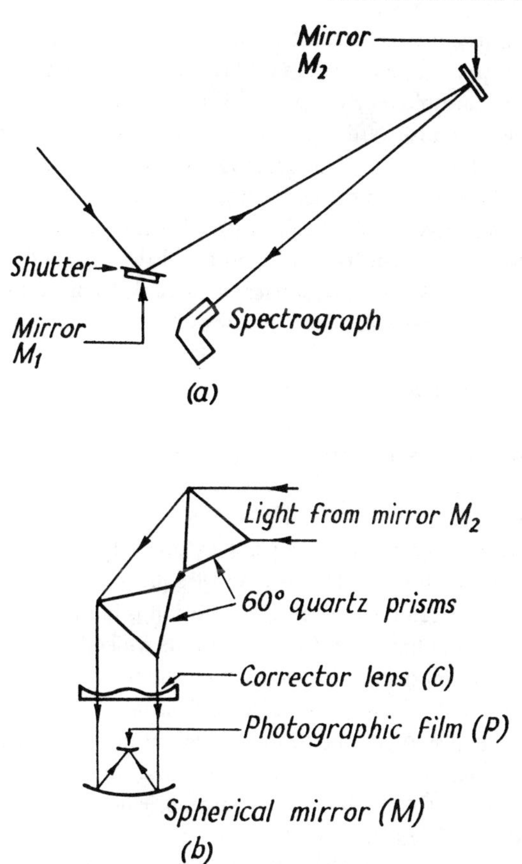

Fig. 6.3 (a) Mounting of spectrograph. (b) Optical arrangement of a spectrograph employing the Schmidt camera.

extensively for recording simultaneously the spectra of dozens of stars for the purpose of classification (see p. 221). In addition, slitless instruments are less wasteful of light than those employing slits and may therefore be used to investigate distant stars with low intensity in conjunction with a telescope of small aperture. Since the resolving power of a slitless spectrograph will be poor compared with that of the slit type, the former instrument is very inferior to the latter when fine structure studies are required. At one time it was impracticable to record a reference spectrum (e.g. iron arc lines) alongside the unknown spectrum using the objective prism arrangement. Without these reference lines exact wavelength determinations are impossible. However, Fehrenbach [6.2] has developed a technique which overcomes this limitation and enables accurate wavelength determination.

One of the big problems in the study of radiation absorbed or emitted by distant bodies is that the earth's atmosphere absorbs such a large part of the spectral range. In recent years this has been overcome to some extent by employing various types of space vehicles (see later) which receive the radiation above the earth's atmosphere. Even high altitude aircraft flights have proved useful, and one recent step forward has been to mount a Fourier spectrometer. The aircraft flies above the cold trap for water vapour in the atmosphere (above about 12 km) and this reduces the concentration of water vapour detected. Water is a strong absorber across the far-infrared region. This procedure is described by Fink and Larson [6.3] who obtained the spectra of mars, venus, and the moon. In the case of venus it became possible to place a lower or upper limit on water vapour or ice clouds in its atmosphere.

6.2 THE DOPPLER EFFECT

6.2.1 Measurement of very high temperatures

The frequency of emitted radiation depends on the velocity of the source relative to that of the observer. The observed frequency increases if the motion is towards the observer and decreases if the motion is in the opposite direction. In the case of a spectral line, only for those atoms which have no component of velocity in the direction of the observer will the frequency of the emitted light be equal to the natural frequency. If each of the atoms in a given gaseous system had roughly the same velocity due to the overall motion of the gas, the width of the spectral line would be unaffected, but the line would be displaced one way or the other according to the direction of motion. When, however, the centre of mass of the system is fixed, and the atoms are in random motion and have a Maxwellian distribution of velocities, the line is broadened but not displaced, that is the spectral line emitted by this gas consists of a range of frequencies symmetrically disposed about the natural frequency.

From the Doppler principle it follows that if a source of light of wavelength λ_0 is moving in the line-of-sight with velocity v relative to the observer, the apparent wavelength, λ, measured by the observer will be:

$$\lambda = \lambda_0(1 + v/c) \tag{6.1}$$

where c is the velocity of light. Motion away from the observer produces a shift to longer wavelengths, and such motion is taken as being positive.

Since the range of velocities increases with temperature, so must the range of frequencies comprising the spectral line increase. The temperature of the atoms may be calculated in terms of the broadening of the spectral line and the mass, m, of the atoms. For a Maxwellian distribution of velocities it may be shown that when the extinction coefficient, ϵ, is plotted against the wavelength, λ, the total width of the line at half maximum (i.e. 0.5ϵ) is given by:

$$\delta\lambda = 1.67\frac{\lambda_0}{c}\sqrt{\frac{2kT}{m}} \tag{6.2}$$

where $\delta\lambda$ is the width of the spectral line at half maximum in Ångström units, λ_0 is the wavelength of the centre of the line also in Ångström units, c is the velocity of light, k is Boltzmann's constant, and T is the absolute temperature of the atoms of mass m. From a knowledge, then, of the values of the constants c, m, and k the temperature T may be calculated from the measurements of $\delta\lambda$ and λ_0. By this method the temperature of the solar corona† was estimated at over a million degrees. Lengthy analysis is usually needed to separate thermal broadening from other line broadening phenomena such as:

(1) Self absorption. When radiation from a hot gas passes through a layer of the same material, the central portion of the emission line is preferentially absorbed.

(2) Turbulence, that is the chaotic mass motion of stellar material, which introduces a Doppler broadening not due to thermal effects.

(3) Collisional or pressure broadening. A radiating or absorbing atom may be perturbed by forces due to any close neighbouring atom, ion, or electron. The broadening produced depends on the pressure and on the nature of the perturbing species.

(4) Magnetic and electric fields cause broadening and splitting of spectral lines.

6.2.2 Determination of radial velocities [6.4]

By the application of the Doppler principle relative velocities of, for example, stars and nebulae can be determined. Thus, from Equation (6.1) it follows that:

$$\frac{\lambda - \lambda_0}{\lambda_0} = \frac{v}{c} \qquad (6.3)$$

that is

$$\frac{\Delta\lambda}{\lambda_0} = \frac{v}{c}$$

and

$$v = c\Delta\lambda/\lambda_0 \qquad (6.4)$$

Hence v may be determined from the experimental data.

Confirmation of the Doppler principle is obtained from numerous sources, but the shifts in planetary spectra are particularly suitable for this purpose. The relative motions of the planets are known very precisely, and in fact, are used for detecting systematic errors in line-of-sight velocity determinations given by a particular spectrograph.

The line-of-sight velocities of stellar bodies are determined by means of a spectrograph attached to a telescope at one of the positions already described (see p. 200). The spectrum is obtained, and on both sides of it is recorded the

† This is a tenuous gas cloud of pearly white colour enveloping the sun and extending for at least a million miles.

iron arc or other reference spectrum which serves to define a system of wavelengths of a source at rest relative to the observer. By the use of the Hartmann formula the wavelengths of the stellar lines are determined. $\Delta\lambda$ is then merely the difference between this wavelength value and the wavelength of the same line measured from a laboratory source.

With a spectrograph of dispersion 15 Å/mm at a wavelength of 4340.48 Å (i.e. the Balmer H_γ line of hydrogen) a velocity of 1 km/s corresponds to a shift on the photographic plate of 1 μ. Stellar velocities are usually quoted to an accuracy of 0.1 km/s.

6.2.3 Spectroscopic binaries

Another interesting application of the Doppler principle has been the discovery of pairs of stars so close together that no telescope has been able to resolve them. Such stars have been proved to be double by the behaviour of the lines in their spectra.

For two stars of approximately equal brightness revolving around each other, which possess spectral lines in common, when one star is receding and the other approaching the observer, each spectral line common to both stars is a doublet. This doubling of lines results from the frequency shift according to the Doppler principle. When, however, one star is in front of the other in the line of observation, the spectral lines are single.

Such double stars are called *spectroscopic binaries*, many hundreds of which are known. In Fig. 6.4(a) the spectrogram obtained from β-Aurigae may be observed. This spectrogram shows the doubling of the 3933.7 Å line of singly-ionized Ca, whereas in Fig. 6.4(b), which was taken when one star was in front of the other, a single line may be seen (Fig. 6.4).

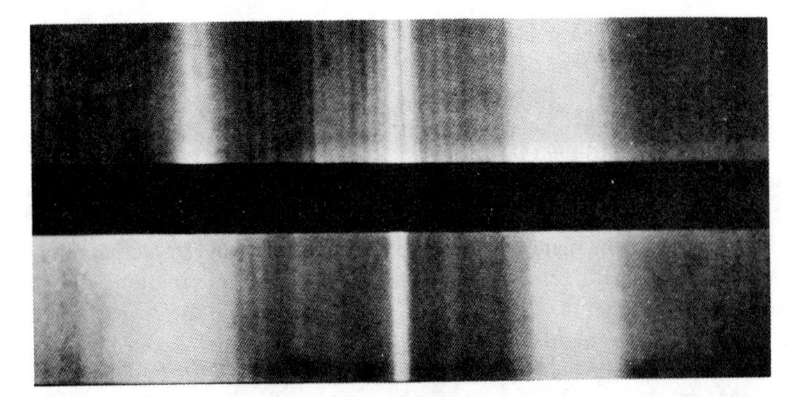

Fig. 6.4 Spectrogram of the binary star β-Aurigae. In the centre of the upper spectrogram (a) the doubling of the 3933.7 Å line of singly-ionized Ca may be observed while in the lower spectrogram (b) a single line may be seen. (*Courtesy of the Harvard Observatory.*)

6.3 PLANETARY ATMOSPHERES

6.3.1 Introduction

A planet is a solid body having no light or heat of its own. In the solar system nine planets are recognized: Mercury, Venus, Earth, Mars, Jupiter, Saturn, Uranus, Neptune, and Pluto.

The main aims of planetary science within our solar system are to classify materials on the planet in relation to their physical state and evolution and to compare the similarities and differences amongst the planets with a view to gaining better insight into the entire group. Studies have been made on all of these nine planets.

The solid material on these planets is composed mainly of Al, Ca, Fe, Mg, Si, and their oxides, while their atmospheres are composed mainly of compounds formed from the elements C, H, N, and O – in particular CO_2, H_20, N_2 and 0_2.

The studies of the atmospheres and planets may be conveniently divided into examination of their (a) lower atmospheres and (b) their upper atmospheres. In their lower atmospheres the temperature is well defined and there is a Boltzmann distribution of population for particular molecular vibrational and rotational states. The gases are well mixed together, and the effects of photodissociation and photoionization can generally be neglected since after the molecules have dissociated they swiftly recombine. However, in the upper atmosphere local thermodynamic equilibrium does not hold while the infrared energy cannot be efficiently radiated away.

Attempts are made to formulate theories of thermal structures of their atmospheres, to account for any cloud structure and to explain their chemical composition. The composition of a planetary atmosphere is, amongst other factors, dependent on: (a) the amount of sunlight it receives; (b) the composition and mass of the solid planet itself; (c) the conditions apertaining when the planet was first formed.

Before the spectra of planetary atmospheres are dealt with more fully it will be convenient to consider the absorption and emission spectrum of the Earth's atmosphere. The reason for this is that radiation from all extraterrestrial sources will suffer considerable, if not complete, absorption in the spectral regions that are rendered either semi- or completely opaque by the absorbing species in our own atmosphere. In fact, at wavelengths less than 3000 Å the earth's atmosphere is completely opaque.

One method of reducing undesirable absorption in the atmosphere when making spectroscopic studies is to mount the spectroscopic equipment on an aircraft which makes the study at high altitudes.

The infrared properties of the atmosphere can be studied by making spectroscopic observations from aircraft at about 12 km. Optical measurements have been made from aircraft and more recently Fourier spectroscopy has become feasible. Stair [6.3] has divided such atmospheric studies into four categories:

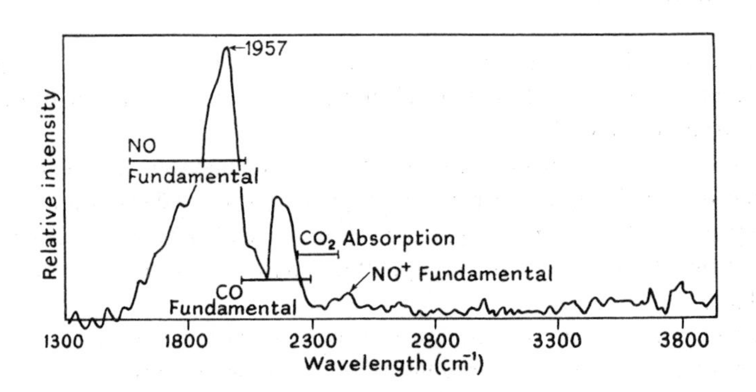

Fig. 6.5 Spectrum of hot air in the 2.5–8 μ region from a high altitude nuclear fireball. These data were obtained from 12 km altitude with an interferometer spectrometer using a bolometer detector (after Stair [6.3]).

(1) thermal equilibrium properties which can be studied both in transmission and emission; (2) non-thermal properties at high altitudes, e.g. infrared chemiluminescence; (3) auroral emissions; (4) artificial perturbations such as high altitude nuclear detonation. One such study was the spectrum of hot air (2.5–8 μ region) from a high altitude nuclear fireball. The spectra are given in Fig. 6.5 and were taken at 12 km altitude with an interferometer spectrometer using a bolometer as detector. CO, NO, and NO^+ were the main features observed.

6.3.2 Absorption spectrum of the Earth's atmosphere

The opaqueness of the Earth's atmosphere at wavelengths less than 3000 Å is partly due to ozone, which has a maximum concentration at a height of 20–30 km above the Earth's surface. The ozone formation is dependent on the photodissociation of the oxygen molecule by the absorption of wavelengths of less than 2400 Å. The O_3 absorption extends to approximately 2200 Å but at about 2400 Å the $A^3\Sigma_u^+ \leftarrow X^3\Sigma_g^-$ bands of oxygen are found. Below 1950 Å intense O_2 absorption (the Schumann–Runge system $B^3\Sigma_u^- \leftarrow X^3\Sigma_g^-$) produces opaqueness down to about 1300 Å. Below this wavelength other systems of O_2 and N_2 cause complete absorption.

At the infrared end of the electromagnetic spectrum intense rotation–vibration transitions of water vapour occur in the regions 0.94–1.85 μ, 2.5–3.0 μ, and 5–7.5 μ, and, in addition, rotational transitions of water vapour seriously deplete the solar spectrum at wavelengths greater than 16 μ and completely between 24 μ and ~1 mm. Other localized absorptions also occurring in the visible and infrared regions are due to CO_2, N_2O, CH_4, and O_2.

On moonless nights the Earth's atmosphere is still found to give a feeble illumination, and efforts have been made to account for this emitted electromagnetic radiation. Owing to the weakness of the source the type of instrument employed must have a high photographic speed, and consequently the dispersion and resolution are poor. The spectrograph described on p. 203 has been used for this type of investigation. The following are some of the difficulties encountered. tered. (i) It is not easy to determine by inspection of such spectrograms whether lines or bands are being observed. (ii) The extent of any individual band may be small, possibly owing to the low temperature of the source. (iii) Forbidden lines are observed which are not easily reproduced in the laboratory.

From the work done so far, the greater part of the spectra of the Earth's atmosphere is thought to be molecular in origin, though part of it is attributed to atoms.

The atomic lines which have been definitely identified are the O I $^1D_2$–$^1S_0$ transition at 5577 Å, the O I $^3P_{1,2}$–$^1D_2$ transition at 6300 Å and 6364 Å, and the sodium D-lines.

Oxygen above 100 km is in the atomic state. At night recombination of the O atoms occurs mainly between 90 and 120 km. Oxygen is more easily photo-dissociated than nitrogen, and this together with the absence of nitrogen lines in the spectrum is taken to indicate that nitrogen in the upper atmosphere is mainly in the molecular state.

From the twilight glow the violet N_2^+ bands may be detected in emission. The sodium D-lines and the red O I lines are also present, their intensity being greater than that observed in the night glow.

The spectra from auroral displays differ considerably from those of the twilight and night glows. Forbidden lines of O I and N I and molecular bands of N_2 and N_2^+ have been identified in auroral spectra. In addition the Balmer lines of hydrogen, the D-lines of sodium, and some He lines are occasionally found.

For the OH bands at 10 440 Å the height at which the radiation is emitted has been estimated at about 70 km. The rotational structure of these bands has been resolved; and Meinel [6.5, 6.6] determined the temperature of the OH stratum from the rotational intensity distribution and obtained a value of 260 ± 5 K.

6.3.3 Other planetary atmospheres [6.7, 6.8]

A number of criteria are available which enable the astronomer to decide whether a given planet is likely to have an atmosphere. The criteria depend on the observations of: (i) clouds, seasonal polar caps, twilight arcs, and atmospheric refractions, (ii) polarization and reflection phenomena, (iii) whether sufficient gravitational attraction exists on the planet for it to retain an atmosphere. This can be computed from the mass, radius, and temperature of the planet. These criteria are useful since they enable the astronomer to fix his attention on those planets where a definite possibility of an atmosphere exists.

The presence of absorption or emission bands in the spectrum of a planet is especially valuable in a number of ways. It gives definite proof of the existence of particular species (i.e. an atmosphere of some kind), identifies the substance(s), and enables estimates of the relative abundance of the various species to be made. On the other hand, spectroscopic observations may be limited by: (i) The transmission of the Earth's atmosphere. (ii) Sources of low spectral intensity. This leads to low dispersion spectra. (iii) H_2, N_2, and the inert gases are not revealed and CO_2, CO, and N_2O only if the source is bright enough to make infrared observations possible.

Gases which are common to the atmosphere of both the planet and the Earth can sometimes be detected if the source is sufficiently intense for high dispersion studies. This is achieved by virtue of the Doppler shift which results in a lack of coincidence of the two spectra.

The high percentage of water vapour and oxygen in the earth's atmosphere makes difficult the determination of these gases in extraterrestrial atmospheres. However, steps can be taken to avoid the terrestrial water vapour by, for example, working at high altitudes or on days of low humidity.

Table 6.1 summarizes the major constituents in the atmospheres of some of the planets in the solar system while Table 6.2 gives a fuller account for Mars, Venus, and the Earth.

Table 6.1 The main constituents in the atmospheres in the solar system

Venus	CO_2, H_2O
Mars	CO_2, H_2O, O_2
Jupiter	CH_4, NH_3
Uranus	CH_4, H_2
Neptune	CH_4, H_2
Saturn	CH_4, NH_3
Titan (largest satellite of Saturn)	CH_4

Big developments in astrochemistry have been in the use of techniques which overcome the absorption of radiation by the Earth's atmosphere. One way in which this may be achieved is to place the spectrograph in a suitable housing which is mounted on a rocket. The spectrograph could be, for example, an ultraviolet one covering the range 500–2500 Å. After passing through the Earth's atmosphere and taking the necessary spectra the rocket is recovered.

Barth et al. [6.10] used a spectrometer in Mariner 6 and 7 and obtained the spectrum of the Mars upper atmosphere in the 1100–4300 Å region, and they obtained as well the emission spectra of CO, O, and CO_2^+. Their spectra are given in Fig. 6.6 where may be observed, below 2600 Å the Cameron bands of CO, the Fox–Duffenback–Barker bands of CO_2^+ above 3000 Å, the CO_2^+ ultraviolet doublet band system at 2890 Å, and the 2972 Å line of atomic oxygen. The CO_2^+ has been produced through photoionization as a result of solar ultraviolet radiation. In the ionospheric layers of the Earth molecular ions have a prominent role, e.g. N_2^+, O_2^+, NO^+, H_2O^+, and NO_2^-, and as Herzberg [6.11] has indicated, the upper atmospheres of other planets would be expected to contain layers in which molecular ions are an important feature.

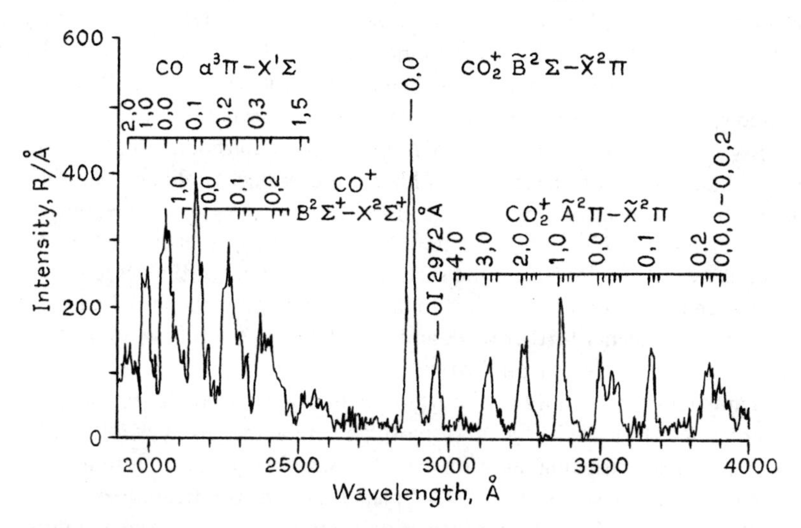

Fig. 6.6 Ultraviolet spectrum of the upper atmosphere of Mars [6.10]. The Cameron bands of CO lie below 2600 Å, the Fox–Duffenback–Barker bands of CO_2^+ above 3000 Å, the CO_2^+ doublet band at 2890 Å, and an atomic absorption oxygen line at 2972 Å. (After Barth *et al.* [6.10]).

The Soviet entry probes, Venera 4, 5 and 6, have been used to study the atmosphere of Venus. In this case a direct sampling of the planetary atmosphere showed that the mole fraction of CO_2 on Venus is in the range 93–97 per cent in the total atmospheric pressure range of 1.6–2.0 atm.

A considerable amount of data is now available on the gaseous matter in the

Table 6.2 Abundance of gaseous matter in kg/cm^2 [6.12].

Volatile	*Earth atmosphere*	*Venus atmosphere*	*Mars atmosphere*
CO_2	0.5×10^{-3}	95 ± 25	$(1.4 \pm 0.2) \times 10^{-2}$
H_2O	$(1-10) \times 10^{-3}$	10^{-2}	$(0.5-2.5) \times 10^{-6}$
O_2	0.23	$<6 \times 10^{-3}$	$<2.5 \times 10^{-5}$
N_2	0.75	<5	$<0.8 \times 10^{-3}$
Ar	1.3×10^{-2}	<5	$<0.3 \times 10^{-2}$
CO	$(1-10) \times 10^{-7}$	3×10^{-3}	0.7×10^{-5}

atmospheres of Venus and Mars, and a comparison is made with that of the Earth in Table 6.2.

It is apparent from Table 6.2 that CO_2 is an abundant compound in the atmospheres of all three planets. At present, though, it is not possible to make a comparison of the total amounts of CO_2 at or near the surfaces of these planets. It is possible, for example, that the Mars polar caps may contain large amounts of solid CO_2 while the surface of Venus may well have large amounts of carbonate deposits. If Venus has all its water in the atmosphere, as some think feasible, then the water on Venus is no more than 1/1000th of that on the Earth.

Particularly interesting studies have been made of water on Mars. It was detected spectroscopically in 1964 and later work indicated that the relative humidity might be as high as 50 per cent. One worker considered that the presence of liquid water on the surface of Mars was extremely unlikely. Another worker obtained spectroscopic evidence that the polar caps contain water. It is considered possible that large amounts of water may be present as permafrost or as hydrated minerals on the surface. This example of the detection of the presence of water on Mars serves to illustrate that after the initial detection of a species on a planet further work and speculation often follow. Murray [6.13] has described Mars from Mariner 9.

It is apparent that a study of the atmosphere of terrestrial planets leads us to a better appreciation of the Earth's atmosphere and its stages of evolution. Perhaps the most intriguing question is that of whether the Earth is unique or just a member of the series of terrestrial planets and in, say, the sequence between Venus and Mars. More information is required on the degassing process of the surface of each of these planets, and how they differ. Further, it is desirable to determine to what extent their atmospheres have been modified after outgassing. Theories are formulated and evidence produced for reactions between the constituent gases of the atmosphere and the planet's crust. It has been speculated that a large number of compounds may be present in the atmosphere of Venus, and some compounds which have not yet been detected but have been proposed are COS, HBr, H_2S, and $HgBr_2$; such proposals take into account the high surface temperature.

Thus it is evident that the information gained by spectroscopic means on a few species in the atmospheres of planets leads to various types of speculation.

6.4 SPECTRA OF NEBULAE AND FORBIDDEN TRANSITIONS

A nebula is defined as a cloud of gas or of fine solid particles which normally envelops a very hot star. A typical nebula embodies a star of type-O (temperature ~50 000 K) which is surrounded by a shell of gas; this has an emission spectrum containing the Balmer lines of H and prominent lines He I and He II. In Fig. 6.7, which is the spectrogram and microphotometer trace for the Orion Nebula (NGC 1976), the hydrogen and helium lines are quite apparent. Among

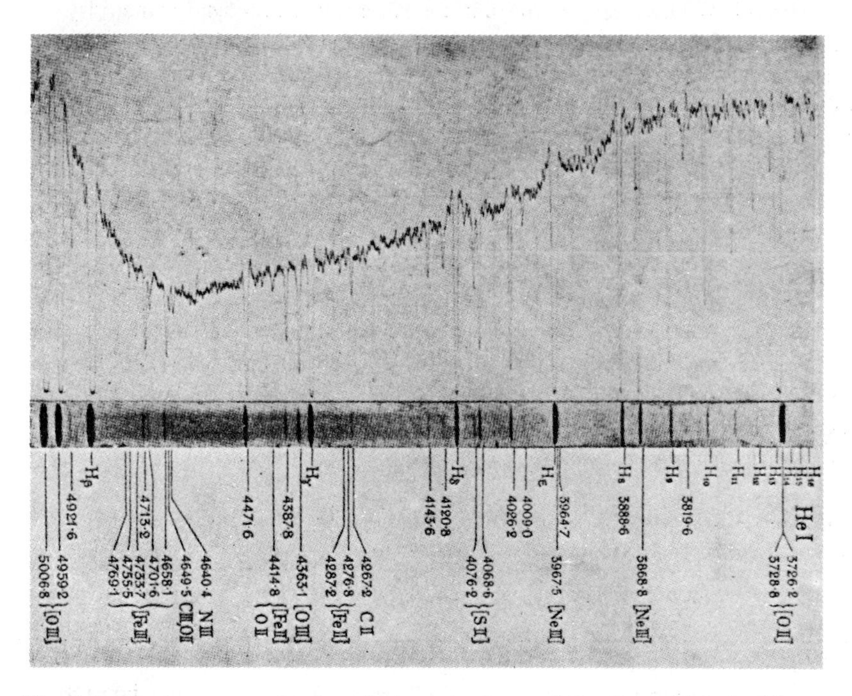

Fig. 6.7 Spectrogram and microphotometer trace of the Orion Nebula. The spectrogram was obtained using a spectrograph with a single prism employed in conjunction with a 120 cm telescope. (After Tcheng, Mao-Lin and Dufay [6.9] courtesy of Professor J. Dufay and Hutchinson and Co.).

weaker lines of the nebula are those attributed to C II, C III, and Ne III and sometimes also to O II and N III. In the near infrared region some hydrogen lines belonging to the Paschen series have been observed, and in one particular nebula (NGC 6572) a line from the neutral oxygen atom has been detected. Lines from O III and N III have also been observed in the spectra of nebulae and are often quite intense.

The unfamiliar conditions of extremely low density in nebulae (for example, the density of the Orion Nebula has been estimated at ~300 atoms/ml of that of air at normal temperature and pressure) give rise to spectra which are not normally encountered in the laboratory.

In addition to the already indicated lines, other lines are observed which were not readily identified initially. The wavelengths of these lines have been measured quite accurately and the values obtained did not correspond to any known lines. These unknown lines were at first attributed to an unknown element, nebulium. Since, however, no suitable gaps were available in the periodic table it appeared more probable that nebulium was a known element under unfamiliar conditions.

The problem was solved by Bowen, [6.14, 6.15] who showed that the frequencies were due to O, N, Ne, and S in various stages of ionization. Some of these lines have now been produced in the laboratory, although the majority have not yet been obtained, since their emission is forbidden by the selection rules for electric dipole radiation. They are consequently called forbidden lines, examples of which can be seen in some of the lines in the spectrogram in Fig. 6.7.

The selection rules for atoms for electric dipole radiations are;

(i) $\Delta J = 0, \pm 1, J = 0 \nrightarrow J = 0$.

(ii) $\Delta L = 0, \pm 1$ and $\Delta l = \pm 1$.

(iii) $\Delta S = 0$.

(iv) Even terms combine only with odd terms and vice versa.

However, for magnetic dipole and electric quadrupole radiation the selection rules are in some cases quite different, and although the normal conditions on earth are generally unsuitable for such transitions, in nebulae the reverse is true. Table 6.3 summarizes the selection rules for electric and magnetic dipole and for electric quadrupole radiation.

Table 6.3 Comparison of some of the selection rules for atoms for electric dipole, electric quadrupole, and magnetic dipole radiation

	Electric dipole radiation	*Electric quadrupole radiation*	*Magnetic dipole radiation*
ΔJ	0 ± 1	$0 \pm 1 \pm 2 (J' + J'' \geqslant 2)$	0 ± 1
Terms	$+ \longleftrightarrow -$	$+ \longleftrightarrow +$	$+ \longleftrightarrow +$
		$- \longleftrightarrow -$	$- \longleftrightarrow -$
ΔL	0 ± 1	$0 \pm 1 \pm 2$	0 ± 1
ΔS	0	0	0

The Einstein transition probability of spontaneous emission A_{nm} is given by (see Vol. 2, p. 333).

$$A_{nm} = \frac{64\pi^4 \tilde{\nu}^3_{nm}}{3h} |R^{nm}|^2 \tag{6.5}$$

A_{nm} gives the fraction per second of atoms in the initial state n undergoing transitions to the state m where n is the excited state. The mean life of state n is:

$$\mathcal{T} = 1/A_{nm} \tag{6.6}$$

For a permitted electric dipole transition $\mathcal{T}$ is approximately 10^{-8} s, that is the average lifetime of the state n is 10^{-8} s. If, however, no electric dipole transition is allowed from state n to state m, A_{nm} will be very small, and $\mathcal{T}$ will consequently be large. It has been calculated that for transitions which take place by virtue of the electric quadrupole $\mathcal{T}$ is of the order of 1 s, while for a magnetic dipole transition $\mathcal{T}$ is approximately 10^{-3} s. Under these conditions n is termed a *metastable state*.

Generally, if a transition from n to m cannot take place by the electric dipole mechanism, then the other possibilities are by electric quadrupole or magnetic dipole transitions. However, on earth the probability of most gaseous atoms radiating by such means is very unlikely, because even in the highest vacuum attainable the atoms would lose their excess energy by collision with other atoms, before they had the opportunity to emit. At extremely low densities prevailing in the gaseous nebulae atomic collisions are very infrequent, occurring for a given atom at an average interval of several hours or even days. Under these conditions emission may take place by magnetic dipole and electric quadrupole transitions giving rise to the so-called forbidden lines [6.16, 6.17].† If in any portion of the nebula the predominant ion of an element X is X^{n+}, then usually only forbidden transitions of X^{n+} appear. However, recombination of X^{n+} with electrons produces permitted transitions of $X^{(n-1)+}$. In certain cases, for example, H, He II, O III, and N III, permitted transitions do arise.

One particular well known forbidden line—the green auroral line at a wavelength of 5577.35 Å, which is emitted after dark in the high terrestrial atmosphere and which is of great intensity during polar aurorae—was reproduced in the laboratory by McLennan and Shrum in 1924. They diluted oxygen with an inert monatomic gas in the discharge tube and concluded from the spectrum that the green line was due to the forbidden transition $^1D_2-{}^1S_0$ of the neutral oxygen atom; the transition was considered to take place through the electric quadrupole moment.

Forbidden lines belonging to Fe II, Fe III, Fe VI, and Fe VII have also been detected. The solar corona has forbidden lines of Fe XI, Fe XIII, Fe XIV, Ca XII, Ca XV, Ni XIII, Ni XV, Ni XVI, and Mg X. Elements in gaseous nebulae detected by their permitted lines are relatively few in number and include H, He, C, N, O, and possibly Fe, Si, and Mg. The heavier atoms detected by their forbidden lines are, for example, F, Ne, S, Cl, K, Ca, Fe, and Ni.

† The concept of forbidden lines is based on the definition that lines which obey the selection rules for electric dipole radiation are the *permitted* ones whereas all the remaining lines are termed *forbidden*. This definition automatically classes any transition involving magnetic dipole or electric quadrupoles as forbidden transitions, even though they obey the selection rules for that type of interaction.

6.5 SPECTRA OF COMETS

A comet consists of three parts. (i) A nucleus of small dimensions generally of the order of a few kilometres or less. It is a collection of rocky and metallic particles coated with frozen ices of CH_4, CO_2, H_2O, and NH_3. The nucleus of the comet has been looked upon as a collection of meteorites. (ii) A coma or head surrounding the nucleus and extending to perhaps hundreds of thousands of kilometres. (iii) A tail directed away from the sun, the length of which may well be of the order of millions of kilometres.

The comet is in orbit with respect to the sun. However, its path is such that it spends most of its time at great distances from the sun. In recent years considerable popular interest occurs when its path in the orbit brings it sufficiently close to be seen from the Earth. When the comet's position in its orbit is such that it swings in towards the sun, its surface begins to sublime or evaporate, and, as a consequence, gas flows into space. Since the nucleus of the comet exerts only a weak gravitational effect, the outgoing molecules and atoms carry with them solid particles. Hence by these processes the gaseous and dusty clouds are created.

The brightness of the comet may be attributed to: (i) a process where the ultraviolet radiation from the sun dissociates molecules, and then the resulting species can absorb solar radiation and emit it; (ii) sunlight is scattered by dust in the comet.

The spectra observed from comets result from the gases in the head and tail. These gases are released from the solid particles under the influence of solar radiation. Cometary spectra consist mainly of a large number of strong emission bands. However, when a comet approaches the sun, the sodium D-lines are observed from the comet and may have a very high intensity when the comet–sun distance is small.

The band spectrum from the head and nucleus is very different from that occurring in the tail. The species observed from their band spectra in the heads of comets are C_2, CN, C_3, OH, NH, CH, CH^+, and NH_2, while in the tails CO^+, N_2^+, CH^+, and CO_2^+ have been identified. The identification of the species OH, NH, CN, C_2, and C_3 in comet 1940 I may be seen in Fig. 6.8.

The intensity distribution within these cometary bands differs from those found in the laboratory sources even at low temperatures. The irregular distribution is very evident in, for example, the CN band near 3880 Å and also less prominent for the CH, OH, and CH^+ bands. Instead of showing a smooth rotational intensity distribution, cometary spectra reveal complex band profiles with intensity minima in the branches. These minima appear even when the bands are incompletely resolved. The positions of the minima correspond perfectly to the wavelengths of strong absorption lines in the exciting sunlight, the absorption being due to the solar atmosphere. In addition, the radial velocity of the comet relative to the sun affects considerably the rotational profiles, since the positions of the minima are altered owing to the change in wavelength of the solar absorption lines in accordance with the Doppler principle.

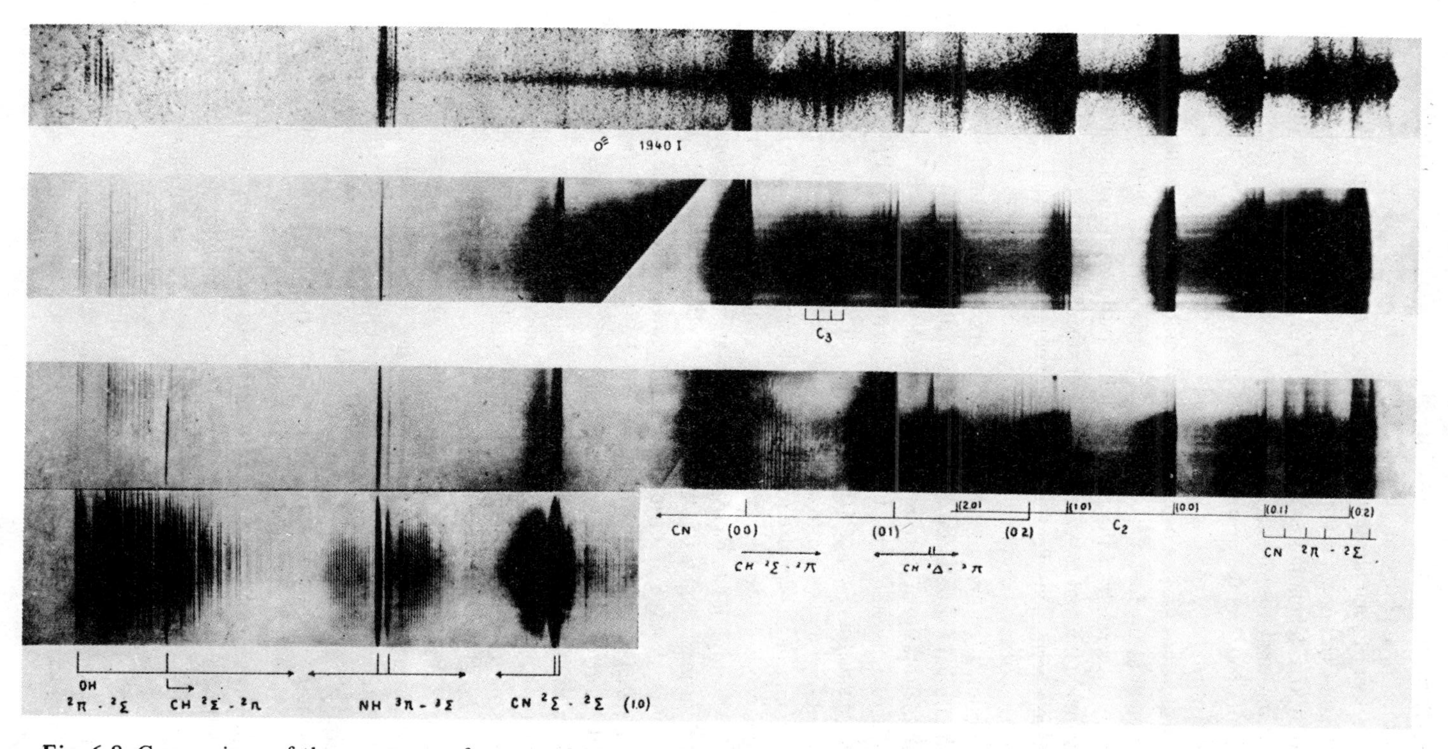

Fig. 6.8 Comparison of the spectrum of comet 1940 I with the spectra obtained from flames [6.19]. Spectrum of comet 1940 I obtained using an f/1 quartz spectrograph. (b) Spectrum of a rich acetylene flame with some addition of ammonia. (c) Spectrum of an acetylene flame with traces of ammonia. (d) Spectrum of an acetylene flame with ammonia. (After Swings and Haser [6.19], courtesy of P. Swings).

From a study of the rotational structure of a band (see p. 97), estimates of the effective rotational temperature may be made. However, allowance has to be made for the effect of solar absorption. When this is done it is found that CH, OH, NH, CN, and CH^+ have, on the whole, low rotational temperatures in the order of 200–400 K depending on the distance of the comet from the sun. A study of the rotational intensity distribution in a C_2 band, however, leads to much higher rotational temperatures of about 3000–4000 K. These differences in rotational temperatures may be appreciated from the following facts. (i) The rotational distributions are not brought about by collisions since collisions will be very infrequent at the low pressures prevailing. (ii) The distribution of rotational energy is a result of the continued absorption of sunlight, and whereas the low rotational temperature types may undergo transitions in the far infrared, since they have a permanent electric dipole moment, the C_2 molecule being homonuclear cannot dissipate its rotational energy in this manner. C_2 therefore has a high rotational temperature, while the CH, OH, NH, CN, and CH^+ species have a low rotational temperature. The comet–sun distance will have an important effect on the rotational distribution (see Fig. 6.9).

In Fig. 6.8 comparison is made between the spectrum of comet 1940 I and acetylene flame spectra. The spectrum of the comet was obtained by Swings [6.18] using an *f*/1 quartz spectrograph when the comet–sun distance was 0.87 A.U.† By comparing spectrogram (a) with those shown under (b), (c) and (d) the various bands due to OH, CH, NH, CN, C_2, and C_3 may be identified in the comet. In addition, it can be seen that the rotational and vibrational intensity distributions in the bands of OH, CH, NH, and CN are extremely different in the comet from those in the flames, while they are similar for the bands of C_2. The observed cometary lines within each band of OH, CH, NH, and CN correspond to electronic transitions from the lowest rotational levels. The details of the acetylene flames are as follows.

Spectrogram (b) results from a rich acetylene flame with some addition of ammonia. The main features are bands of OH, NH (Q branch), CN (violet system), C_3, C_2, and CH.

Spectrogram (c) is from an acetylene flame with traces of ammonia. The following bands are present in the inner cone of the flame: CH, NH, CN, CH, and C_2. In the outer cone of the flame are present bands due to OH, NH, CN, CH, and C_2.

Spectrogram (d) results from an acetylene flame with ammonia. The rotational structures of OH, NH, CH, and CN are clearly shown.

The abundance of molecules in comets can be estimated spectrophotometrically, although little work has been done along these lines. A very approximate estimate of C_2 molecules in the head of a comet gave 10^5 molecules per ml,

† A.U. is the astronomical unit, which is the Earth's average distance (92 960 000 miles) from the sun.

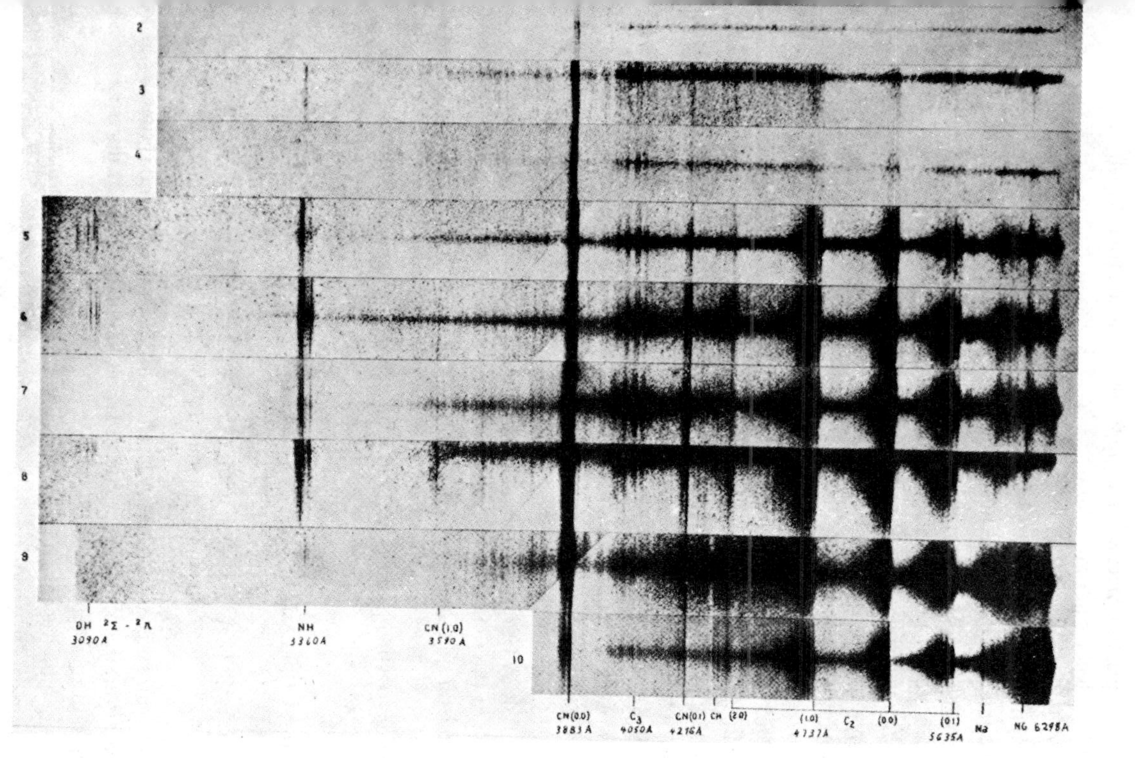

Fig. 6.9 Slit spectrograms of comet Cunningham (1940c-1941 I) at heliocentric distances of 2.24 to 0.48 A.U. The spectrograms were obtained as follows [6.19]: Spectrogram 1: McDonald Observatory nebular spectrograph, quartz prisms, f/1 camera. (b) Spectrograms 2 to 9: McDonald Observatory Cassegrain spectrograph, quartz prisms, f/1 camera, original dispersion 137 Å/mm at 3360 Å. (c) Spectrogram 10: same spectrograph as in (b) but with an f/2 camera, original dispersion 68 Å/mm at 3360 Å. (After Swings and Haser [6.19], courtesy of P. Swings).

while in the tail an estimate of CO^+ was found to be one molecule per ml. The effective vibrational and rotational temperature values may differ considerably and have little physical meaning. However, for systems at equilibrium these values should correspond. The lack of agreement serves to show that the prevailing densities are so low that thermal equilibrium in cometary systems cannot be established by collision processes. The upper limit for the total density is of the order of 10^6 molecules per ml.

The fundamental part of a comet is the collection of solid meteorites in the nucleus where almost the entire mass is concentrated, and from which the head and tail of the comet originate. The head and tail of a comet are formed by liberation of gas and dust from the nucleus, and since the nucleus has a small mass, that is it posseses little gravitational attraction, the head and tail need to be continuously replenished; otherwise they would disappear within a few days. As the comet approaches the sun the nucleus emits large quantities of gas and dust and eventually disintegrates. Sometimes the whole comet breaks up, and if fragments of it (meteorites) come under the gravitational attraction of the earth, then an opportunity of examining the fragments of former comets in the laboratory results. Molecules in a comet have a limited life in the radiation field of the sun, depending on the possibilities of photodissociation and photoionization and on the amount of solar radiation available in the spectral region where dissociation or ionization occurs.

The spectra of comet Cunningham (1940 c–1941 I) in the region 3000–6400 Å are given in Fig. 6.9. The spectrograms 1 to 10 inclusive were obtained at the following heliocentric distances: 2.24, 1.85, 1.47, 1.18, 1.03, 0.87, 0.75, 0.63, 0.50, and 0.48 A.U. respectively. The species detected are marked along the bottom edge of the figure. The resolution of the rotational structure of the OH and NH bands in spectrograms 5, 6, 7, and 8 enabled for the first time identification of these two radicals in comets. A study of comet 1948 I showed that the diameter to which radicals extend in the comet head varied with the heliocentric distance, e.g.:

$$r = 0.90\text{ A.U.}\begin{cases}\text{the diameter of CN is 62 000 km;}\\ \text{the diameter of } C_3 \text{ is 4140 km;}\end{cases}$$

$$r = 2.21\text{ A.U.}\begin{cases}\text{the diameter of CN is 166 000 km;}\\ \text{the diameter of } C_3 \text{ is 22 000 km.}\end{cases}$$

Over the comet–sun range 0.65–2.21 A.U. the profile of the CN band shows appreciable variation; the main systematic variation in profile of CN is due to its decrease in rotational temperature,

One interesting more recent study was of comet Bennet in 1970 by the Orbiting Astronomical Observatory when it was noted that :(a) a large cloud of hydrogen radiated at the wavelength of the Lyman α-line; (b) the far ultraviolet absorption of the hydroxyl radical is much stronger than had been suspected; (c) there appeared to be approximately equal numbers of hydrogen atoms and hydroxyl radicals, which was taken to indicate that ice was a major constituent of the comet's nucleus.

Not by any means are all the features of comets understood; even the formation of the ion tail is still open to speculation. One view is that, in addition to ionization brought about by solar radiation, the high-energy electrons in the solar wind also ionize the molecules in the coma. In addition the solar wind sets up a bow wave around the coma. Chaotic magnetic fields are created which act as a magnetic rake that selectively detaches ions from the coma. The un-ionized molecules and atoms are not so influenced and are left behind.

6.6 STELLAR SPECTRA

From a study of their spectra the surface temperature and chemical composition of the stars may be deduced. The majority of the stars show a continuous emission upon which are superimposed dark absorption lines. Some stars also show emission lines on this background.

In order to detect a particular absorption line two factors are important: (i) the amount of the species present; (ii) the f-value or oscillator strength for the involved transition. The value of this factor indicates the likelihood that the species (e.g. atom or ion) will actually absorb radiation of a particular frequency. The f-value is a measure of the Fourier coefficient of the particular harmonic and is employed to specify the strength of a particular line. Recently considerable attention has been paid to the measurement and computation of f-values especially with regard to its application in astrophysics. Two review articles cover the work on f-values [6.20, 6.21] up to 1963, and an excellent book by Kuhn [6.22] on atomic spectra should also be consulted.

The spectra of stars with very few exceptions can be grouped together into a limited number of spectral classes which form a sequence. The spectral sequence is:

```
                    R–N
                    /
      O–B–A–F–G–K–M
                    /
                   S
```

The sequence O to M is continuous while the R- and N-type stars, sometimes grouped together as C-type stars (i.e. carbon stars), follow on from group G, and S stars from group K. Initially it was considered that the division of stars into different spectral classes was to be related to their variation in chemical composition. However, it is now realized to be a division in terms of different surface temperatures.

The spectral sequence is also a colour sequence of the stars. Thus, stars of type O are bluish white in colour, M-type stars are red; those of intermediate types range from white through orange to red. The colour sequence is a temperature sequence. The intensity of molecular bands in the late sequence stars suggests that these stars are comparatively cool while the early type stars, whose spectra are due to neutral and ionized atoms, are much hotter. This also agrees with the colours of the stars. A summary of the spectral types, their colours, temperatures, and main spectral characteristics, is given in Table 6.4. It may be observed from Table 6.4 that the temperature is one of the main factors which governs the excitation and ionization in the atmospheres of the stars. It must be stressed that, although certain spectral features have been given to a particular class, they may well be met with in other classes; for example, the bands of TiO are dominant in class *M*, but they are also present in class *S* along with ZrO bands.

Each spectral class is sub-divided into groups called spectral types. These are identified by the numbers 0, 1, 2, 3, . . . placed after the capital letter of the class to which the star belongs. For example, *B*5 denotes a spectrum intermediate between *B* and *A* in appearance, while *F*8 indicates a spectrum resembling *G*0 closer than *F*0.

One particular striking feature in stellar spectra is that ionized helium lines may be taken to indicate that the temperature of the star is at least 30 000 K and is to be related to the high ionization potential of helium. In fact, only the very hottest stars, that is the *O*-type, contain large numbers of ionized helium atoms. The spectral characteristics of the *O*-type stars are lines of H, He I, He II, O III, N III, and C III. Thus the spectra of *O*-stars are relatively simple, and the reason for this is that all the other abundant elements which they contain, such as iron, are appreciably ionized owing to the very high temperature, and since their inner electrons are so tightly bound to the nucleus then very large amounts of energy are required to excite them. As a result the photon which is emitted when the excited electron falls back to the ground state corresponds to a very high frequency. Unfortunately, such frequencies cannot be detected on the earth's surface since they are absorbed by the atmosphere. In order to gain such information on hot stars telescopes are now being employed on satellites orbiting the earth above the earth's atmosphere.

In class *B*, lines of H and He I become stronger, while lines due to He II disappear. The hotter *B* stars show lines of O II, Si IV, and Si III while the colder *B* stars show Mg II, Si II, and C II. Near *A*2 the lines of H attain their maximum intensity. The lines due to metals gain prominence in types *A* to *G* while the H lines weaken. Molecular bands occur weakly in type *G* and increase in intensity together with the metallic lines from types *G* to *M*. Bands of TiO dominate *M*-type stars, bands of ZrO are shown by type *S*, while bands due to C_2 and CN are very prominent in the spectra of *R* and *N* types.

In Fig. 6.10 (b) and (c) the spectrum of an *A*2 star is compared with that of type *K*2. It will be observed that in the hotter star the Balmer hydrogen lines can be seen as can also the 3934 Å line of Ca II. In the spectrogram of the cooler star many lines characteristic of metals have made an appearance and bands can be observed. The lines on either side of the spectrograms (a), (b), and (c) are the iron arc reference lines.

Table 6.4 Colour, temperature, and the main spectral features of stars of the various types

Type	*Colour*	*Approximate temperature (K)*	*Example of such a star*	*Chief spectral features*	*Some other spectral features*
O	Bluish white	50 000	Alnitak	Absorption H; absorption He II; sometimes emission of He II at 4686 Å	Ionized C, N, O
B	Bluish white	20 000	Bellatrix	Absorption H; absorption He I	Ionized O, N, Si
A	White	10 000	Sirius	Hydrogen lines	Fe, Na, Ca II
F	Yellowish white	7000	Canopus	H and metallic lines	Lines of Ca II, Ca I; metallic lines
G	Yellow	5500	Sun, Capella	CH band at 4300 Å Metallic lines H lines still noticeable	CN, CH, C_2, Ca I, Ca II Fe I
K	Orange	4500	Arcturus	Metallic lines Ca I at 4226 Å	TiO, CN, CH, C_2, Ca II Fe I, VO, CO, Sr II
M	Red	3000	Betelgeuse	Bands of TiO Ca I at 4226 Å	MgH, SiH, AlH, ZrO, ScO, YO, CrO, AlO, BO, Ca II, Fe I, Ni I, Sr II, VO, CO, CN, H_2, H_2O
S	Very red	3000	—	Bands of ZrO	TiO, YO, LaO, CO, CN
R, N	Very red	3000	—	Bands of C_2, CN	NH, C_3, CH, CO, HCN, C_2H_2, $^{13}C^{16}O$

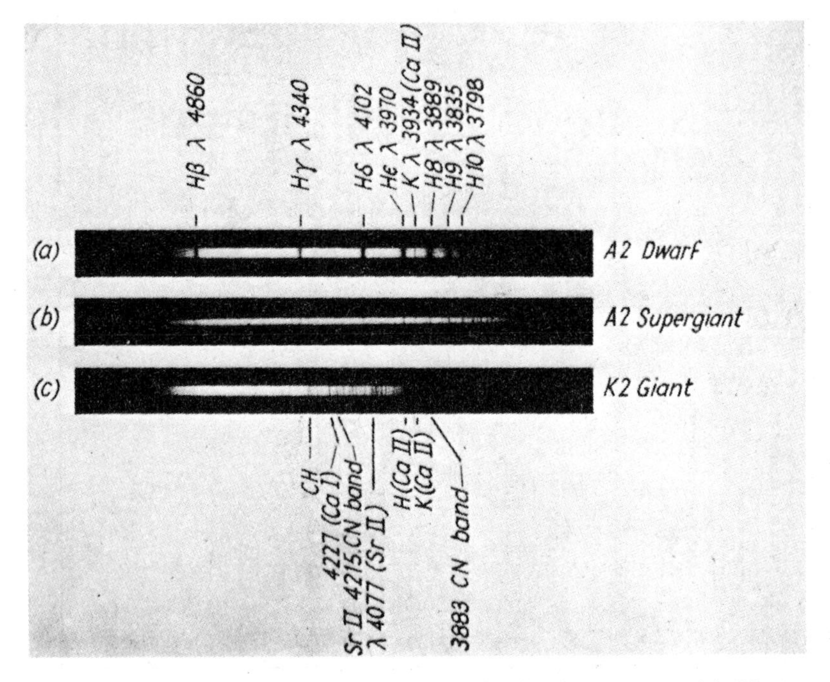

Fig. 6.10 Stellar spectrograms of (a) $A2$ dwarf, (b) $A2$ supergiant, (c) $K2$ giant. (*Courtesy of the Astronomer Royal.*)

The spectra of stars may be used to give an indication of stellar age, and one indicator is the strength of the lithium lines. Thus young stars such as the F and G stars in the Pleiades exhibit strong lithium lines whereas older stars such as the sun give only very weak, if any, lithium lines. The Ca II emission lines are also a fairly good indicator of age where the stronger the emission lines the younger are the stars.

If the spectral classes of the stars are plotted against their absolute magnitudes,† the points are found to lie in a definite pattern. The plot is called the Russell–Hertzsprung diagram and is illustrated in Fig. 6.11. The majority of the points fall along a curve running diagonally across the diagram; the hot blue stars lie at one corner and the cool red ones at the other. This group of stars is called the *principal series* or *dwarf sequence*, generally referred to as the main sequence, of which the sun is one of type $G2$. Probably more that 90 per cent of all the known stars lie in the principal series. It is possible for a star to be in the main sequence at one stage of its development and be a giant or white dwarf at

† The magnitude of a star expresses its brightness on a diminishing logarithmic scale. The absolute magnitude is the magnitude a star would have if it were placed at a distance of 10 parsecs, where 1 parsec is equal to 3.26 light years or 206 265 times the average distance of the sun from earth, i.e. 206 265 A.U.

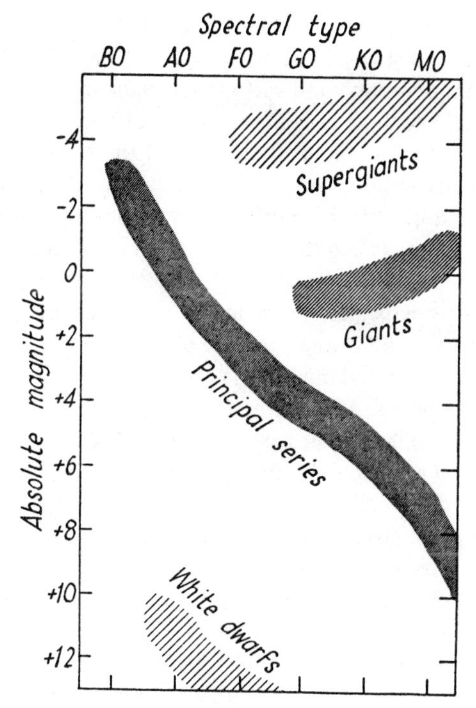

Fig. 6.11 The Russell–Hertzsprung diagram

another stage. Since the main sequence phase is the longest one in stellar evolution, the majority of stars lie in that sequence; this is the one in which nuclear reactions converting hydrogen into helium are supplying the energy radiated by the star.

Above the dwarf sequence lie two other groups of stars called the *giants* and *supergiants*, while below the main sequence is found the group referred to as *white dwarfs*. The masses of dwarf and giant stars differ little but, as with their luminosities, the mean densities of these types are vastly different. The mean density of giants is very small and in the highly rarefied atmospheres of such stars ionized atoms will tend to be more abundant than in the compressed atmospheres of dwarf stars which possess a greater density.

Thus, generally, an increased density will reduce the degree of ionization in stellar atmospheres resembling somewhat the effect of decreasing temperatures. The level of excitation of the atoms will, however, be unaffected. Consequently it might be expected that ionized lines would be stronger in the spectra of giants while neutral lines would gain more prominence in dwarf spectra.

However, since the spectral classification is based on the appearance of lines, giants tend to be slightly cooler than dwarfs of the same spectral class. While the increased density of a dwarf star is largely compensated by somewhat higher temperature, the compensation is imperfect for a few elements. Lines belonging to ionized atoms, for example, the 4215 Å and 4077 Å lines of Sr II and the 4233 Å line of Fe II, are strong in giant star spectra. The 4227 Å line of Ca I, on the other hand, is more intense in dwarf than in giant spectra. The CN bands at 3590, 3883, and 4216 Å are important for distinguishing between giant and dwarf stars of the types *G* and *K*. If the spectra of these stars are compared, then the CN bands in giants are more intense than in dwarfs of the same spectral class. The detailed reasons for these differences cannot be considered here.

A further difference between the spectra due to giant and dwarf stars is the width of the spectral lines. The lines from giant stars are generally sharp and narrow while those from dwarfs are broad and rather diffuse. In Fig. 6.10 (a) and (b) the spectra of an *A*2 dwarf and supergiant, respectively are compared. The broadening of the spectral lines in the case of the dwarf star is clearly seen. This broadening may be due to: (1) Greater rotation (see p. 205) in the hotter dwarf stars. (2) More collisional broadening in dwarfs due to the increased density. (3) Magnetic and electric fields may play a rôle in the line broadening phenomenon.

Such criteria as those given readily enable dwarf and giant stars of the same spectral type to be distinguished. The reason this is so important is that, once the spectral type of the star is known, and it is established whether it is a dwarf or a giant or super giant, then from the Russell–Hertzsprung diagram the absolute magnitude of the star may be obtained. By comparison of the absolute magnitude with the apparent magnitude (i.e. the apparent brightness based on a logarithmic scale) the distance of the star may be estimated. The accuracy of such a determination is about 25 per cent, the relative error being independent of distance, but difficulties sometimes arise owing to absorption by dark interstellar matter.

The great variety of spectra from stars does not necessarily imply any difference in chemical composition of the stellar atmospheres. The apparent absence of certain elements in a given type is probably due to the different conditions prevailing at that time. So far as is known the majority of stellar atmospheres– apart from those of types *R, N*, and *S*– are nearly identical in composition with only their temperatures and atmospheric pressures differing. The main constituent of stars is hydrogen. In the case of the sun 99.9 per cent by volume of the sun's atmosphere consists of H and He with hydrogen about five times more abundant than helium.

With these results, it is possible to establish the main processes of emission and absorption that give rise to the continuous background radiation on which these line spectra are superposed. At temperatures higher than about 4000 K, below which absorption by molecules begins to play an important part, the character of the continuous emission is fixed by the properties of hydrogen in the presence of free electrons. In the atmospheres of *O*- and early *B*-type stars, hydrogen is predominantly ionized and the continuous radiation (originally generated by thermonuclear processes in the deep interior of the star) is transferred through the atmosphere (that is the visible layers) mainly through Thomson scattering by free electrons. Shortward of the Lyman limit ($\lambda < 912$ Å), the residual neutral hydrogen in the ground state acts as an intense additional source of continuous absorption, so that one expects the intensity to be depressed below that of the corresponding black body in this extreme ultraviolet region, which has, however, not been observed so far except in the case of the sun (from rockets). For late *B* and *A* stars, free electrons become fewer (hydrogen only partially ionized), and the absorption is due to hydrogen in the second quantum level for the ultraviolet ($\lambda < 3646$ Å) and in the third quantum level for the visible ($\lambda < 8208$ Å). This effect is confirmed by the presence of a strong absorption edge (the Balmer discontinuity) at λ 3650 Å in the continuous spectra of *B*-, *A*-, and early *F*-type stars.

When the temperature is reduced further, for example, in solar-type stars, the Balmer discontinuity weakens, contrary to what would be expected from pure hydrogen absorption, and indeed the continuous spectrum then approximates remarkably closely to that of a black body (except in the ultraviolet, where intense line and band absorptions occur). This is due to the combination of neutral hydrogen atoms with free electrons supplied by the metals, which are mostly in the singly-ionized state, to form the H^- ion, as was first suggested by Wildt. The light from the sun is the recombination continuum of H^-, and the continuous absorption is due to the photoelectric dissociation of H^- into H + e, the cross-section of which varies only slowly over the near ultraviolet, visible, and near infrared regions. In giants, the electron pressure is lower than in dwarfs, owing to the lower surface gravity, and H^- is therefore, less abundant in comparison with H; the Balmer discontinuity is consequently stronger in giants than in dwarfs of the same spectral class. In most cases significant amounts of both line and continuous absorption occur in the same atmospheric layers, so that both must be taken into account in the interpretation of line intensities, which essentially depend on the *ratio* of one to the other.

The spectrum of the sun (our closest star of the *G*2 type) has received special attention since, on account of its brightness, good dispersion and resolution have been obtained. In addition to many atoms and ions (61 elements have been detected), the following molecular species have definitely been identified in the sun's atmosphere by means of their band spectra: CN, CH, C_2, NH, OH, SiH, MgH, TiO, CaH, ZrO, YO, AlO, and BH. Some evidence for MgO, ScO, MgF, and SrF also exists.

Stellar spectra are cut off at wavelengths less than 3000 Å by the ozone layer in the earth's atmosphere and by O_2 and N_2 at wavelengths less than about 2200 Å. However, with the availability of high-altitude rockets, which pass above the absorbing layers, it has been possible to photograph the extreme ultraviolet spectrum of the sun, and to examine this by recovery of the rocket. A series of such investigations has been carried out by Johnson et al. [6.23]. One of their spectrograms taken at an altitude of 115 km showed forty-five emission lines extending from 1892 to 977 Å which arise in the hot chromosphere and corona surrounding the visible disc of the sun. In Fig. 6.12 the microphotometer trace of this spectrum is given from 1850 to 950 Å. The main features of the solar spectrum in this region are an intense α-line of the hydrogen atom in the Lyman series and lines due to ionized C, Si, N, O, S, Fe, and Al. Except for the emission lines noted, the irregular appearance in the spectrogram from 1850 to 1550 Å is due to absorption by the sun's atmosphere.

The spectrograph employed to obtain the ultraviolet spectrum of the sun, from which the microphometer trace in Fig. 6.12 was obtained, is shown in Fig. 6.13, The spectrograph in a suitable housing was mounted on a rocket and

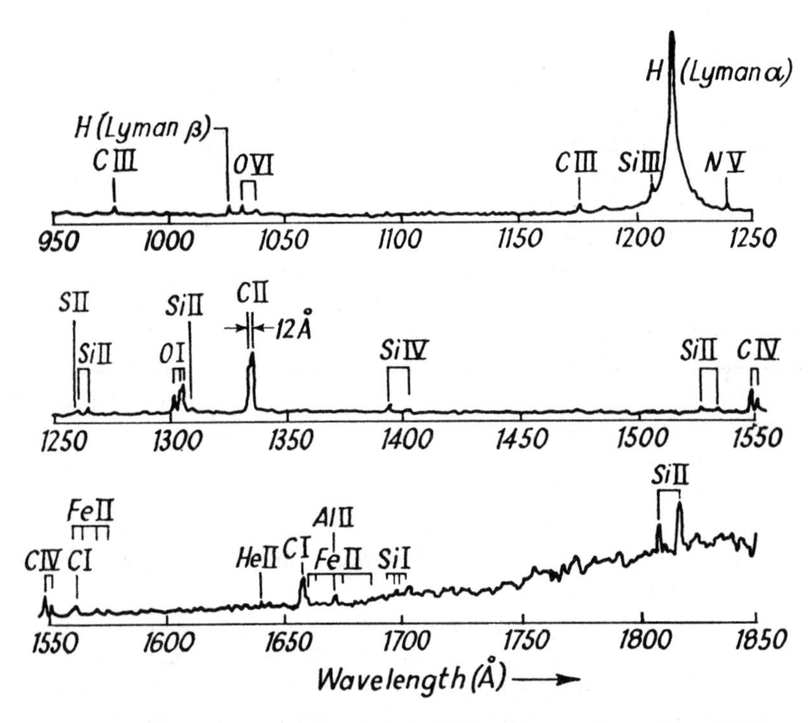

Fig. 6.12 Logarithmic densitometer trace of the solar spectrum from 950 to 1850 Å obtained at a height of 115 km above the earth's surface (after Johnson, *et al.* [6.23], courtesy of the Astrophysical Journal).

was designed to cover the region from 2500 to 500 Å. By means of a photoelectric sensing device the spectrograph housing was kept directed so that the sun's image fell on the slit. Sunlight collected by a spherical quartz collector mirror, M, was reflected on to the entrance slit, S, of the spectrograph. Light from this slit, located at a distance of 21.3 cm from the mirror, M, was incident on a concave grating, G. The grating, which, set at normal incidence, had a radius of curvature of 40 cm and 6000 lines/cm blazed for maximum first-order efficiency at about 1200 Å, focused the spectrum on to the photographic film, P. Since oxygen and water vapour absorb strongly in the vacuum ultraviolet region, the

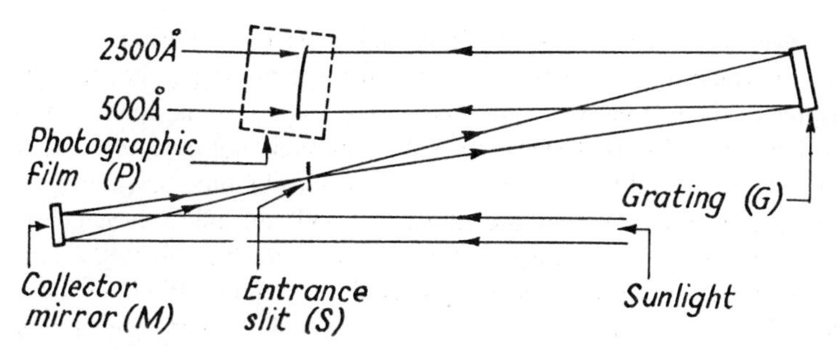

Fig. 6.13 Optical arrangement of the vacuum spectrograph employed to investigate the ultraviolet spectrum of the sun (after Johnson *et al.* [6.23], courtesy of the Astrophysical Journal).

spectrograph housing has to be either evacuated or, as was the case in this experiment, replaced by dry helium.

Spectra of the (sun's) corona and of the chromosphere were obtained in 1970 from a rocket flight and 25 new coronal lines were discovered. In addition, the Lyman alpha lines were detected and this established that un-ionized hydrogen atoms exist in the high temperature corona. By means of observations made on the ground during eclipses, from rockets, from satellites, and even from the moon, a detailed study is being made of the sun's outer atmosphere.

Of particular interest in recent work is the study of globular clusters. There are some 200 globular clusters in our galaxy and each consists of 100 000 to a million stars. It is thought that globular-cluster stars are about as old as the universe, and it is worth noting that the initial abundance of He in cluster stars is close to that formed in the 'big bang' model of the universe. Studies are made on cluster stars with a view to gaining evidence as to whether the universe has expanded from a tremendously hot primeval 'atom' that was formed in a unique event several billions of years ago.

Some of the features of cluster stars are that: (i) they are sufficiently cool so that helium atoms exist in their lowest state of excitation; (ii) they are strikingly deficient in heavy atoms and they contain only from one-tenth to a hundredth as many heavy atoms as stars of equal mass in the disc of the galaxy; (iii) the farther a cluster star is from the centre of the galaxy, the more deficient it appears to be in heavy elements.

Some recent works relating to stellar spectra worth considering are Chui, Warasila, and Remo [6.24] on stellar astronomy, Iben [6.25] on globular-cluster stars, Neugebauer and Becklin [6.26] on the brightest infrared sources, Pasachoff [6.27] on the solar corona, Rubin [6.28] on the dynamics of the Andromeda nebula, and Spinrad and Wing on [6.29] infrared spectra of stars.

6.7 SPECTRAL STUDIES OF THE INTERSTELLAR MEDIUM

The matter which is dispersed in space between the stars is known as interstellar matter. So far, in the interstellar medium in the gas between the stars in our galaxy, twenty-six different types of molecules have been detected. The number of such molecules is most dense where the dust is densest, and they have been discovered in the regions where the stars appear to be forming and in the outer atmosphere of cool stars. The spectral data from such molecules give an insight into the physical conditions in such regions. New techniques have been developed to gain information from the interstellar medium, and space vehicles have provided a means of overcoming the interference of absorption by the earth's atmosphere. One useful step which has been developed recently has been to employ a lightweight Michelson interferometer which can be flown successfully aboard a space vehicle with a view to studying emission spectra in the 10–1000 micron region. Such ventures have potential in studying the composition of interstellar medium, and studies should yield information on its heating and cooling mechanisms and its dynamic processes. This may lead to evidence on the formation of photostars, the nature of the galactic nuclei, and eventually to extragalactic objects such as quasars.

Many distant hot stars show in their spectra certain absorption lines with a shift differing from that shown by the remaining lines of the spectrum. This is illustrated clearly by certain spectroscopic double stars (see p. 206) which give Ca II lines with a constant velocity shift. In the spectrum of δ Orionis it was demonstrated that the H and K lines of Ca II gave a constant velocity shift, while the shift associated with the remaining lines went through a cycle of variation every few days owing to the orbital motion of the system. By the use of high dispersion spectrographs other stationary lines have been detected including those of Na I, K,I, Fe I, Ti II, and Ca I.

These stationary lines are really superimposed on the stellar spectrum and are due to the passage of light from the star through material dispersed in the space between the stars (interstellar material). Table 6.5 gives some of the identified interstellar absorption lines. Since all the atomic lines are resonance lines, the absorbing atomic species may be considered to be in the ground state

Although the atomic lines were readily identified, certain other lines caused more difficulty. However, Swings and Rosenfeld [6.30] and also McKellar [6.31] showed that the lines were due to interstellar molecules in their lowest rotational states. For example, the lines at 4300.3 Å, 3874.6 Å, and 4232.6 Å were shown, respectively, to belong to CH, CN, and CH^+. The observed absorption lines for these radicals in interstellar space are given in Table 6.5.

The lack of rotational energy of the interstellar molecules can be appreciated by a consideration of the conditions in interstellar space. Owing to the low density prevailing in these regions collisions are very infrequent, and excitation by radiation is either small or absent. Should any higher rotational energy levels become occupied, then depopulation can occur in the cases of CH, CN, and CH^+ by rotational energy emission.

Hydrogen is by far the most abundant element in the interstellar gas (as it is also in normal stars). However, its detection by optical methods is not possible except in emission from those clouds that are close enough to hot stars for the

Table 6.5 Interstellar absorption

Atom or ion	*Wavelengths* (Å)
Na I	3302.3, 3303, 5890, 5896
K I	7664.9, 7699
Ca I	4226.7
Ca II	3968.5, 3933.7
Ti II	3073, 3229.2, 3242, 3383.8
Fe I	3719.9, 3859.9

Radical	*Wavelengths* (Å)	*Remarks*
CH	4300.3	This is the $R(1)$ line belonging to the (0,0) band of the $^2\Delta$—$^2\Pi$ transition.
	3890.2 3886.4 3878.8	These lines belong to the (0,0) band in the $^2\Sigma$—$^2\Pi$ transition.
CN	3875.8 3874.6 3874.0	$P(1)$ $R(0)$ These lines belong to the (0,0) band $R(1)$ in the $^2\Sigma$—$^2\Sigma$ transition.
CH^+	4232.6 3957.7 3745.3	These lines belong to the $^1\Pi$—$^1\Sigma$ transition. The 4232.6 Å line is the $R(0)$ line of the (0,0) band, the 3957.7 Å is the $R(0)$ of the (1,0) band and the 3745.3 Å line is the $R(0)$ line of the (2,0) band.

hydrogen to be ionized. The ionization is caused by ultraviolet radiation emitted by the hot stars, and such regions where this process occurs are called H II regions. The hydrogen atom emission results from the proton capturing an electron and becoming an excited hydrogen atom. These atoms then lose their excess energy by emitting radiation some of which lies in the visible region. At a certain distance from a star, depending on its size, temperature, and on the density of the gas, radiation is no longer available to ionize the hydrogen, and from these regions (called H I regions) no optically detectable hydrogen atom emission can be observed.† H II regions are well defined volumes forming a sphere around a hot star and occupy only about one-tenth of the space occupied by the H I regions.

Neutral hydrogen in the H I regions can, however, be detected by radio astronomy from the 21.1 cm (1420 MHz) H I emission. This emission results from transitions between the $F = 1$ and $F = 0$ levels in the ground state of the atom ($n = 1$), where F is the total angular momentum quantum number of the whole

† It is from these H I regions that interstellar absorption originates.

atom. The two values of F arise since $S = \frac{1}{2} = J$ and $I = \frac{1}{2}$, where F takes the values:

$$J + I, J + I - 1, \ldots, |J - I|$$

The transition $F = 1$ to $F = 0$ corresponds to a reversal of the direction of the electron spin relative to that of the proton spin and takes place only very rarely, the mean life of an atom in the $F = 1$ state being approximately 11×10^6 years. Since, however, interstellar gas clouds are so extensive, detectable amounts of the 21.1 cm radiation are emitted.

From the Doppler velocity shift of the 21.1 cm line and on the assumption of galactic rotation† it has been found possible to determine the spatial distribution of the neutral hydrogen clouds and to confirm the spiral structure of the galaxy.

Optical studies of the light from distant stars are restricted owing to attenuation by interstellar dust. Radio waves, on the other hand, can penetrate the dust and wider regions become available for investigation. Radio astronomy has the additional advantage in extragalactic research of being less handicapped by red shifts brought about by the rapid recession of the universe. In fact, the use of the 21.1 cm line emitted from cool hydrogen clouds surrounding the intense extragalactic radio source in Cygnus showed that the recession velocity of these clouds deducted from the Doppler shift of this line was in agreement with the value of 16 700 km s^{-1} measured in the optical spectrum. This result confirms the view that the red shift observed in spectral lines of distant galaxies is a true velocity shift.

One big step forward in radio astronomy occurred in 1963 when the OH was detected by absorption at 18 cm. Most of the OH sources lie near the H II regions and young stars. The hydroxyl radical is observed in both absorption and emission. In fact, the OH gas is considered to be pumped into an excited condition and may be regarded as an interstellar hydroxyl maser. One of the features characteristic of OH emission is that the medium itself is very dense.

Another big step in interstellar space studies took place in 1967 when water was identified in microwave emission at 1.35 cm. Ammonia was also detected,

† The galaxy is composed of the sun, with its attendant planets, and of the stars nebulae, and gas clouds, and the stars. The boundaries of the galaxy are not defined, but it is estimated that the galaxy is about 30 000 parsecs across and about 3000 parsecs thick. The sun is near the edge of the disc at a distance of about 10 000 parsecs from the centre. An important property of the galaxy is that it is rotating – otherwise it could not exist – and the hypothesis of rotation is supported by extensive observations of stellar velocities in the line of sight. On its outer parts it rotates not as a solid body but in the same manner as the solar system, the outer planets rotating more slowly than the inner ones. The distribution of interstellar matter and hot, bright stars in the galaxy is now known to lie along the arms of a spiral near the central plane of the disc, the width of the arms being of the order of 500 parsecs.

and studies on these two types of molecule established that: (a) the clouds are quite dense compared with other known interstellar reactions; (b) the physical conditions of the molecules are quite different from those on earth.

Recently radio emission has been observed for the $J\ 1 \rightarrow 0$ transition of $H^{12}C^{14}N$ and $H^{13}C^{14}N$ at 88.6 and 86.4 GHz respectively. In addition formaldehyde has been detected in interstellar space by strong absorption at 6 cm and this molecule is considered to afford one of the most promising means for explaining the denser regions of the Milky Way.

The work on identification of interstellar molecules appears to be developing quite rapidly and a study of the species present gives direct information on the location and velocity of interstellar clouds and leads to information on the physical conditions and the dynamics within the clouds. Studies of spectral line width may be related to the temperature and turbulence or large scale motions within the cloud. Ammonia and carbon monoxide have proved most useful in temperature determinations.

Two important features which have emerged from the study of molecules in interstellar space are: (i) the formation of molecules from the number of atoms available in the interstellar clouds must be a most efficient process; (ii) there seems to be a tendency to favour the formation of organic molecules, and many simple organic molecules have been identified already, whereas a number of expected diatomic inorganic species have not been detected. Rank [6.32] has given a most useful account of interstellar molecules and the interstellar medium.

Table 6.6 lists the known interstellar molecules, where the molecules are listed in the sequence of their discovery. The wavelength column indicates the wavelength at which they were first discovered while an asterisk against the wavelength indicates that the species has since been detected at additional wavelengths.

About 90 percent of the interstellar medium consists of hydrogen atoms while atoms of helium make up the major constituent of the remaining 10 per cent. The proponents of the 'big bang' theory consider that nearly all the hydrogen and helium in the universe interstellar dates back to that event, which is considered to have taken place about 10 billion years ago.

Striking studies have been made of interstellar molecules containing different isotopes, and it has emerged that the interstellar isotope ratios appear to be the same as on the earth. This suggests that interstellar chemistry has changed very little since the earth was formed.

In recent years one of the most fascinating areas in interstellar studies has been the speculation as to how the molecules are formed. Various theories have been proposed and include: (i) formation of the molecules on the surface of dust particles; (ii) the molecules may be built up by atoms colliding and then sticking to one another in the gaseous phase; (iii) the molecules may come about as a result of evaporation or decomposition of dust particles when they encounter heating phenomena such as energetic particles or photons.

Thus, altogether the study of interstellar space by mainly molecular

Table 6.6 The known species in interstellar space, the year of their discovery, and the wavelength at which they were first identified. An asterisk indicates that the species has since been identified at additional wavelengths (after Turner [6.23], courtesy of Scientific American).

Year	*Molecule*	*Formula*	*Wavelength*
1937	Methylidyne (ionized)	CH^+	3,958 Å*
1937	Methylidyne	CH	4,300 Å*
1939	Cyanogen radical	CN	3,875 Å*
1963	Hydroxyl radical	OH	18.0 cm*
1968	Ammonia	NH_3	1.3 cm*
1968	Water	H_2O	1.3 cm
1969	Formaldehyde	H_2CO	6.2 cm*
1970	Carbon monoxide	CO	2.6 mm
1970	Hydrogen cyanide	HCN	3.4 mm
1970	X-ogen	?	3.4 mm
1970	Cyanoacetylene	HC_3N	3.3 cm*
1970	Hydrogen	H_2	1,060 Å
1970	Methyl alcohol	CH_3OH	35.9 cm*
1970	Formic acid	HCOOH	18.3 cm
1971	Carbon monosulphide	CS	2.0 mm*
1971	Formamide	NH_2CHO	6.5 cm*
1971	Carbonyl sulphide	OCS	2.5 mm*
1971	Silicon monoxide	SiO	2.3 mm*
1971	Methyl cyanide	CH_3CN	2.7 mm
1971	Isocyanic acid	HNCO	3.4 mm*
1971	Hydrogen isocyanide?	HNC	3.3 mm
1971	Methylacetylene	CH_3CCH	3.5 mm
1971	Acetaldehyde	CH_3CHO	28.1 cm
1972	Thioformaldehyde	H_2CS	9.5 cm
1972	Hydrogen sulphide	H_2S	1.8 mm
1972	Methanimine	CH_2NH	5.7 cm

spectroscopy provides a challenging area for both physicists and chemists which is rapidly developing.

6.8 PULSARS AND QUASARS

A pulsar is by definition an "object that emits radiation in rapid pulses", possessing a strong magnetic field. The radiation can be at different wavelengths, from radio waves to X-rays, and it is caused by "fast electrons spiralling outward along the lines of force in the nebula's magnetic field" [6.34] (the synchroton process).

It is believed that they were formed during the explosion of supernovae. One such pulsar was observed to be formed this way in 1054, the Crab nebula. Gravitational contraction can, if the object is rotating, cause a body to rotate much faster, and if the object posseses a magnetic field cause it to become much stronger. For example, if the gravitational collapse caused the sun to contract to the size of a neutron star, it would rotate at one thousand revolutions per second and gain a magnetic field of 10^{10} gauss. It is hypothesized that pulsars are neutron stars having the above properties. (Neutron stars are believed to be created from supernovae explosions). The pulsar in the centre of the Crab nebula spins at a rate of thirty revolutions per second.

It has been submitted that the synchrotron radiation is the result of the conversion of rotational energy since, if the rate at which a star is rotating and the rate at which it is slowing down are known, the rate at which energy is being given off can be calculated. The Crab nebula pulsar is slowing down at a rate of fifteen microseconds per year, thus losing energy 100 000 times faster than the sun. The mechanism that makes the conversion must be largely connected with the magnetic field. Now, consider a magnet spinning in a vacuum; it will emit electromagnetic waves with frequency equalling that of its frequency of rotation and with a speed depending on the strength of the field and its rate of rotation, provided that its axis of rotation does not lie on the axis of the field. If the Crab nebula had these properties its magnetic field at the surface would have to be 10^{12} gauss. This figure would be quite conceivable for a neutron star. Whether its environment is nearly a vacuum is another matter. However, even if electromagnetic waves could not survive, the cosmic equivalent of a unipolar inductor (a magnetized metal sphere attached at one of its poles of rotation to one terminal of an external circuit and at its equator to the other terminal, a stationary brush; the sphere is spun and the passage of the electrical current displayed by a galvanometer) would allow a torque equal to that created in a vacuum by the electromagnetic waves.

The fate of the Crab nebula must be one of obscurity amongst an abundance of cosmic rays in the galaxy, and as for its pulsar it will wane and eventually cease to radiate unless it gathers interstellar matter which heats it and thus allows it to emit X-rays [6.34].

Quasars are objects that look like stars but are such that they can radiate one hundred times as much energy as our galaxy.

There are three theories on the origin of quasars; the first is that quasars are in actual fact the centre of yet undeveloped galaxies, the second is essentially the same except that the galaxy will never develop, and the third is that they are explosions in the midst of already developed galaxies. The latter seems to have the most support since recently spectra of the photographs of quasars, surrounded by 'fuzzy nebulosities' [6.35], have shown the presence of stars. In particular this was done by Oke and Gunn who examined the spectra of B.L. Lacertae (owing to the discovery of radio waves emitting from it, B.L. Lacertae is now accepted as a quasar and not a variable star) using a telescope specially designed so that an obscuring disc mounted in the aperture concealed the quasar but not the 'fuzzy nebulosity'. B.L. Lacertae is about a billion light years away and consequently the nearest known quasar. It is less than one light year across in a galaxy stretching a distance of 100 000 light years.

If an explosion is the cause of the appearance of quasars then it is possible that quasars could appear often in a galactic lifetime and much more often in its childhood. However, we are ignorant of what would cause the explosions [6.35].

6.9 SPACE PROBES IN THE MID-SEVENTIES

The year 1974 brought a breakthrough in the understanding of our solar system, for it was in this year that we were presented with a volume of information on Jupiter, Venus, and Mercury by Pioneer 10 and Mariner 10 [6.36]. On December 4th, 1973, after 21 months and 2 days of anticipation by the N.A.S.A. ground crew, Pioneer 10 finally came within 286 Jupiter radii of Jupiter. Numerous experiments were put into operation, and the results were immediately flashed back to Earth. The radiation belts seemed to be 10 000 times stronger than the Van Allen belts of the Earth. There are three intensities: the first and the greatest penetrates to a distance of 15 Jupiter radii, the second to 35 Jupiter radii, and the third and the least to 100 Jupiter radii.

An ultraviolet photometer discovered helium to be present in the atmosphere of Jupiter probably in small quantities compared with the amount of hydrogen present. We now have obtained a list of constituents of the Jupiter atmosphere – hydrogen, helium, ammonia, methane, and deuterium, and acetylane and ethylene in lesser proportions.

An infrared radiometer was employed to measure the temperature of the side of Jupiter not visible from the Earth (Jupiter's phase angle is at most 12 degrees) and it was found to be similar to that of the sunlit side. The brightness temperature, which the infrared radiometer also measured, could only produce the conclusion that Jupiter must send forth twice the energy it admits from the sun. This could mean that, unlike the Earth, Venus, and Mars, which rely on the sun for their warmth, it has an internal heat source, possibly resulting from contraction in size due to gravity.

The ammonia clouds are a hindrance to observing the bright zones and dark belts of Jupiter, but Pioneer 10 [6.37] and subsequent missions now ensure a means of studying them and consequently a means by which we may understand the changing colours of the planet visible from the Earth. Jupiter's atmosphere is a mystery to all who behold it; perhaps the mission of Pioneer 11 will help to solve it. Failing Pioneer 11 there is always the N.A.S.A. planned voyage in 1977 to Jupiter, Saturn, and their satellites, Meanwhile Pioneer 10, after having given us close-up view of Jupiter and its satellites, speeds on and will continue to relay pictures of our solar system until the electrical power fails.

The cratered surface of Mercury resembles that of the moon, and until Mariner 10 arrived on the scene it was thought to possess even the same properties – no atmosphere and no magnetic field [6.38]. The estimation of Mercury's mass was now replaced by the more accurate figure of the ratio of the sun's mass to that of Mercury, 6 023 000 to 1. Surface temperature readings were recorded from 430°C above zero to 180°C below, that is a range of over 600°C which appears to be the widest range in the solar system.

Contrary to previous opinion Mercury was found to have an atmosphere one hundred billionth as dense as that of the Earth, composed mainly of helium at distances extending to 300 miles from the surface. The helium might have been present owing to the solar wind or to radioactive decay within the planet. Argon, probably again due to radioactive decay, and neon, due to the solar wind, were also present, but, except for a possible small amount near the surface, hydrogen was absent. A tail of helium stretching away from Mercury and the sun and a shock experienced by Mariner 10 on approach both indicated that Mercury had a weak magnetic field, perhaps only one thousandth that of the Earth. But where does this magnetic field originate, since Mercury rotates so slowly? It is a matter for speculation.

A shorter trip, but by no means a less important one, was that of Mariner 10 to Venus [6.39, 6.40]. Upon arrival Mariner 10 issued ultraviolet pictures of the so-called clouds of Venus, probably created by a photochemical process, which promise to yield information on the planet's wind velocities and direction. Further radio experiments using S-band signals at 2295 MHz were able to penetrate the cloud deck and should hint at its composition. The mass of the planet was pin-pointed much more accurately than the figures provided by Mariner 5. It also gave the magnetic moment to be one-third that given by Mariner 5. The magnetic field of the Earth was confirmed to be 2000 times stronger than that of Venus. Hydrogen is abundant in the Venusian atmosphere and could have been introduced during a cometry collision or have been gathered from the solar winds. The absence of both a magnetic field and deuterium (deuterium would have been present if there had been a cometry collision; Mariner 5 had recorded the presence of deuterium) made the latter theory more acceptable. Helium and atomic oxygen in the upper atmosphere were also detected to be in large proportions. It can be seen that the more precise and developed instruments of Mariner 10 over Mariner 5 enabled further and contradictory results. There is much to be learned from future missions.

Perhaps to the layman the most exciting question, and maybe the only one with which he is really concerned, is whether any life exists elsewhere in the universe. Certainly it may be argued that out of at least one hundred thousand million in our galaxy alone, why should we consider ourselves so privileged to be the only ones possessing life? The pictures relayed to the Earth from the Mariner 9 mission to Mars in the form of coded signals, which gave the brightness of each point in the picture, revealed a massive network of channels looking to all intents and purposes to have at one time contained flowing water. If this was so, since water is necessary for the formation of all life, the channels could have been the spawning grounds for the appearance of microorganisms which could have adapted themselves to survive the changing environment – from a dense atmosphere rich with water vapour to a thin, dry one. The early Mariners had thrown doubt on this theory, but the first pictures from Mariner 9 displayed evidence of internal activity, four volcanoes, one over 500 kilometres in diameter and between 15 and 30 kilometres high, canyons 80–120 kilometres wide and 5–6.5 kilometres deep, larger than any on the earth, and chaotic terrain. This may indicate that Mars once was active and simultaneously acquired an atmosphere that was capable of producing the channels.

Certainly Mars is no longer considered the dead planet, similar in appearance to the moon, as viewed by Mariner 4 in 1965. At this stage it is only possible to speculate: once an unmanned probe returns from Mars, in much the same manner as the U.S.S.R. probe from the moon, with specimens of martian soil we will have a much greater insight. This will probably not happen until 1980. For the time being we must content ourselves with the U.S. Viking mission in 1976 which will search for organic compounds and give the basic inorganic composition of surface minerals [6.13].

REFERENCES

6.1 Struve, O., and Elvey, C.T., *Astrophys. J.*, **88**, 364 (1938).

6.2 Fehrenbach, *J. Observateurs*, **41**, 41 (1958).

6.3 International Conference on Fourier Spectroscopy, March 16–20, Aspen, Colorado, 1970 Proceedings. Editors: Variasse, G.A., Stair, A.T., Jr. (U.S.A.F. Cambridge, Mass. Res. Laboratories), and Baker D.J., (Utah State University Electro-Dynamics Laboratories), *U.S.A.F. Systems Command – Special Reports,* No. **114**, (1970).

6.4 Thackeray, A.D., *Occ. Notes. R. Ast. Soc.*, **3**, 189 (1958).

6.5 Meinel, A.B., *Astrophys. J.*, **111**, 555 (1950).

6.6 Meinel, A.B., *Rept. Progr. Phys.*, **14**, 121 (1951).

6.7 Bobrovnikoff, N.T., *Rev. Mod. Phys.*, **16**, 271 (1944).

6.8 Kuiper, G.P., *Rept. Progr. Phys.*, **13**, 247 (1950).

6.9 Dufay, J., *Galactic Nebulae and Interstellar Matter* (translated by Pomerans, A.J.), Hutchinson, London (1957).

6.10 Barth, C.A., Hord, C.W., Pearce, J.B., Kelly, K.K., Anderson, G.P., and Stewart, A.I., *J. Geophys. Res.* (1971).

6.11 Herzberg, G., *Faraday Lecture,* 201 (1971).

6.12 Ingersoll, A., and Leovy, C., "The atmosphere of mars and venus", *Ann. Rev. Astron: Astrophys.*, **9**, 147 (1971).

6.13 Murray, B., "Mars from Mariner 9", *Scientific American,* Jan. (1973).

6.14 Bowen, I.S., *Astrophys. J.*, **67**, 1 (1928).

6.15 Bowen, I.S., *Rev. Mod. Phys.*, **8**, 55 (1936).

6.16 Mrozowski, S., *Rev. Mod. Phys.*, **16**, 153 (1944).

6.17 Herzberg, G., *Atomic Spectra and Atomic Structure*, Dover publications, New York (1944), p. 154.

6.18 Swings, P., *Vistas in Astronomy,* Vol. 2, Pergamon Press, London. (1956), p. 958.

6.19 Swings, P., and Haser, L., *Atlas of Representative Cometary Spectra,* University of Liège Astrophysical Institute.

6.20 Allen, C.W., *Astrophysical Quantities,* Athlone Press, London (1955).
6.21 Foster, E.W., *Rep. Progr. Phys.,* **27**, 469 (1964).
6.22 Kuhn, H.G., *Atomic Spectra,* Longmans, London (1969).
6.23 Johnson, F.S., Malison, H.H., Purcell, J.D., and Tousey, R., *Astrophys. J.,* **127**, 80 (1958).
6.24 Chui, H-Y., Warasila, R., and Remo, J.L., (Eds.) *Stellar Astronomy,* Vol. 1, Gordon and Breach, Science Publishers (1969).
6.25 Iben, I., "Globular-cluster stars", *Scientific American*, July (1970).
6.26 Neugebauer, G., and Becklin, E., "The brightest infrared sources" *Scientific American*, April (1973).
6.27 Pasachoff, J.M., "The solar corona", *Scientific American,* Oct. (1973).
6.28 Rubin, V., "The dynamics of the Andromeda nebula", *Scientific American,* June (1973).
6.29 Spinrad, H., and Wing, R., "Infrared spectra of stars", *Ann. Rev. Astron: Astrophys.,* **1**, 249 (1969).
6.30 Swings, P., and Rosenfeld, L., *Astrophys. J.,* **86**, 483 (1937).
6.31 McKellar, A., *Publ. Astron. Soc. Pac.,* **52**, 307 (1940).
6.32 *Molecular Spectroscopy: Modern Research* (editors: Rai, K.N., and Mathews, C.W.), Academic Press. (1972).
6.33 Turner, B., *Scientific American,* March (1973).
6.34 Pacini, F., and Rees, M.J., "Rotation in high energy astrophysics", *Scientific American,* **228**, 98 (Feb. 1973).
6.35 "Nearest quasar: an explosive birth", *Science News,* **105**, 222 (April 1974).
6.36 Frazier, K., "Our new view of the solar system", *Science News,* **105**, 235 (April, 1974).
6.37 Hunt, G., "Pioneer 10: the preliminary results", *New Scientist,* **125** (Jan. 1974).
6.38 "The strange and cratered world of mercury", *Science News,* **105**, 220 (April 1974).
6.39 Burgess, E., "Success at venus", *New Scientist,* 410 (Feb. 1974).
6.40 Burgess, E., "The first results", *New Scientist,* 540 (Feb. 1974).

7 Photoelectron spectroscopy

7.1 INTRODUCTION

7.1.1 Basic principles

In previous chapters it was seen that absorption and emission of energy in the range 1 nm to 1000 nm corresponded to changes in electronic energy involving transitions between atomic or molecular electronic energy levels. Such transitions involved an initial electronic energy level and a final electronic energy level, and the electromagnetic radiation absorbed or emitted in the process was exactly equal to the difference in energy between the two energy levels. Consider the absorption process, where the electron is excited by absorption of electromagnetic radiation from a lower to a higher energy level. If the energy of the electromagnetic radiation is greater than the difference between the initial electronic energy level and any higher energy level the electron will actually leave the atom or molecule and travel in free space with a velocity determined by the energy difference between the initial energy level and the energy of the electromagnetic radiation. The final state in this situation is not quantized and there are thus no selection rules that restrict the possible transitions, and, provided that the electromagnetic energy is large enough, all the electrons in an atom or molecule can be considered. The ejection of electrons from atoms or molecules by electromagnetic radiation in this way is known as the *photoelectric effect*, and the process is illustrated in Fig. 7.1.

Figure 7.1 shows the molecular orbital diagram for oxygen drawn to scale. It can be seen that the 1s oxygen atom orbitals are too low in energy for significant overlap to occur, so that in the O_2 molecule the two oxygen atom 1s orbitals remain essentially atomic in character and are known as *core* orbitals. The 2s

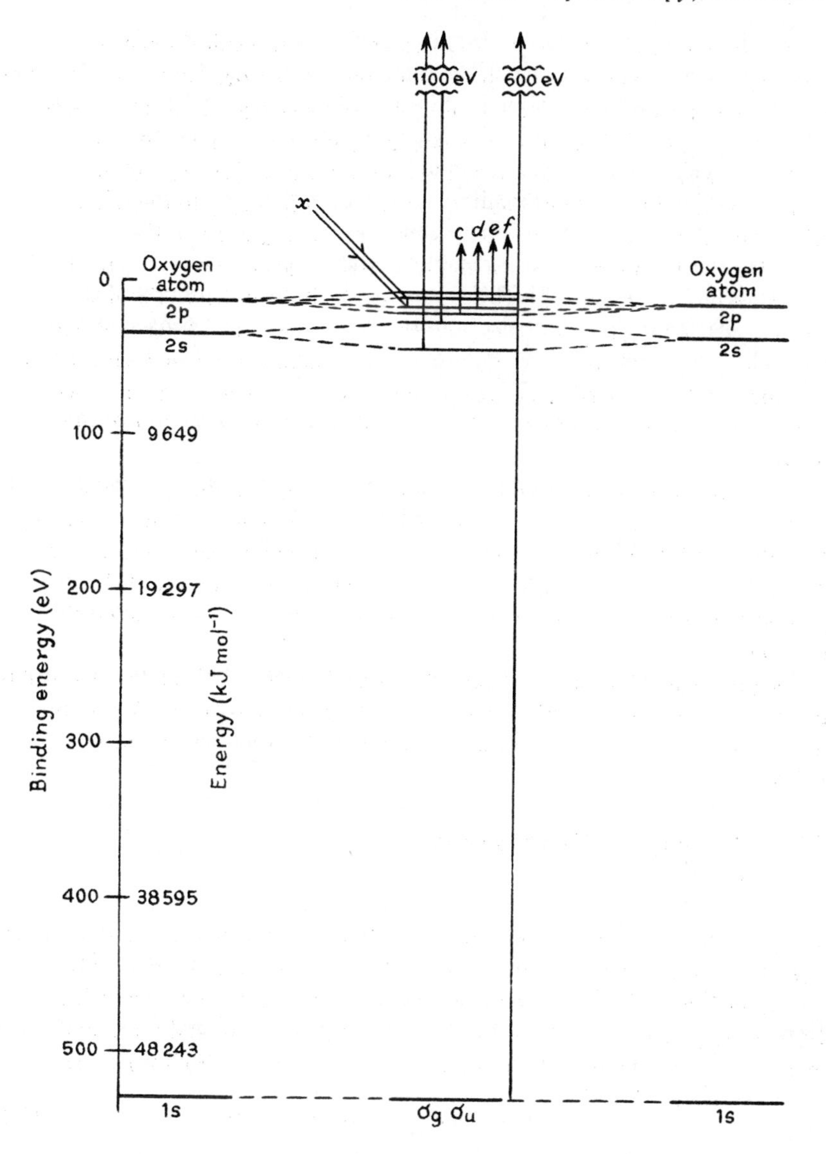

Fig. 7.1 Molecular orbital diagram for the oxygen molecule (O_2), showing a typical electronic transition (x), photoelectrons ejected from the valence orbitals using 21.21 eV radiation (c, d, e, f) and 1253.6 eV radiation [$r, s, t, u = (c - f) + 1232$ eV], and photoelectrons from the core orbital using 1253.6 eV radiation (v).

oxygen atom orbitals overlap to form bonding and antibonding molecular orbitals of σ character, and the 2p oxygen atom orbitals overlap to form bonding and antibonding molecular orbitals of both σ and π character. These orbitals that are essentially molecular in character are known as *valence* orbitals.

A typical electronic transition would be a transition of an electron in a π (bonding) orbital to a π (antibonding) orbital, corresponding to the $B^3\Sigma_u^- \leftarrow X^3\Sigma_g^-$ transition, and this transition is shown as the distance x in Fig. 7.1.

If electromagnetic radiation of energy in the ultraviolet range, say the He(I) resonance line at 58.4 nm (21.21 eV) were used, all the valence orbital electrons could be ejected to give photoelectrons of energy c, d, e, and f. As expected, these photoelectrons will have vibrational fine structure associated with the electronic states involved. This type of photoelectron spectroscopy is called *ultraviolet photoelectron spectroscopy*, or *molecular photoelectron spectroscopy*.

If electromagnetic radiation in the X-ray range, say the $MgK_{\alpha_1\alpha_2}$ radiation of energy 0.99 nm (1253.6 eV) were used all the core orbital electrons as well as the valence orbital electrons could be ejected to give photoelectrons of energy p, q, t, s, t, u, and y. This type of photoelectron spectroscopy is called *X-ray photoelectron spectroscopy* or *electron spectroscopy for chemical analysis (ESCA)*.

Electromagnetic radiation of variable energy has recently been made available by using synchrotron radiation. The use of a monochromator enables continuously variable electromagnetic radiation to be used from the u.v. through to the X-ray region.

7.1.2 The history of the photoelectric effect

The photoelectric effect was discovered by Hertz in 1887 when he observed that a spark gap illuminated with ultraviolet light could discharge more readily. Lenard in 1899 built an apparatus for measuring the velocity with which electrons were ejected from a metal surface by ultraviolet light, and Einstein in 1905 explained Lenard's observations using the quantum theory from which he derived the relationship:

$$\text{K.E. (ejected electrons)} = h\nu - B \tag{7.1}$$

where ν is the frequency of the ultraviolet light and B is the energy of the electronic energy level from which the electron was ejected measured as a binding energy (where the binding energy is zero when the electron has no interaction with the nucleus, as shown in Fig. 7.1. The Einstein relationship was verified experimentally by Millikan using ultraviolet light, and by Robinson and de Broglie using X-rays, during the first quarter of the twentieth century.

The early photoelectron spectroscopic experiments were concerned with fundamental information such as atomic binding energies. The information obtained from early experiments was limited by the comparatively wide lines in the photoelectron spectrum, which resulted from wide linewidths fromthe source of electromagnetic radiation used, and the relatively low resolving power of the electron energy analysers available.

Higher resolution provided additional information which was of considerable value to chemists, but since higher resolution was only achieved after the war, and its chemical applications only appreciated within the past decade, photoelectron spectroscopy is a relatively new chemical tool.

Higher resolution followed two different lines of development, one involving ultraviolet excitation and the other X-ray excitation. In the ultraviolet case rapid development became possible once a suitable monochromatic source of radiation could be found. The resolution problem was considerably reduced using ultraviolet radiation because the photoelectrons produced were of relatively low energy and thus gave reasonable linewidths even using electron energy analysers of relatively low resolving power. This is because resolving power is defined:

$$\text{Resolving power} = \frac{\text{Electron energy}}{\text{Width at half-height}} \qquad (7.2)$$

thus the width increases with increasing electron energy for a fixed resolving power. Turner [7.1] poineered the development of the monochromatic radiation provided by a helium discharge which provides a very intense line at 58.4 nm (21.21 eV), called He(I), with a linewidth of only 0.005 eV. Later, using lower pressures and by dissipating much more power in the helium discharge, nearly 100 percent of the He(II) line at 30.4 nm (40.8 eV) was obtained.

In the X-ray case rapid development followed the production of higher resolution spectrometers. Seigbahn [7.2, 7.3] in 1957 produced the first high-quality spectrum by using a high-resolution iron-free magnetic β-spectrometer in which an X-ray tube had been fitted [using $MoK_{\alpha_1\alpha_2}$ radiation of 0.071 nm (17 479 eV), and 0.713 nm (17 374 eV)]. This apparatus was used to look at the core level binding energies of copper, and in the following year a chemical shift was observed of 4.4 eV between the binding energies of the 1s and 2s core electrons in copper and cupric oxide [7.4]. After a thorough study of atomic binding energies of 232 levels in 76 elements Siegbahn and his coworkers [7.5] in 1964 published a paper entitled "Electron spectroscopy for chemical analysis" pointing out that chemical shifts were a general observation and could be related to the valence state of the atom whose core electron was ejected, and that the intensities of the peaks observed could be related to the number of atoms present, and the binding energies could be used to identify the type of atoms present. The term ESCA is now frequently used to describe X-ray photoelectron spectroscopy.

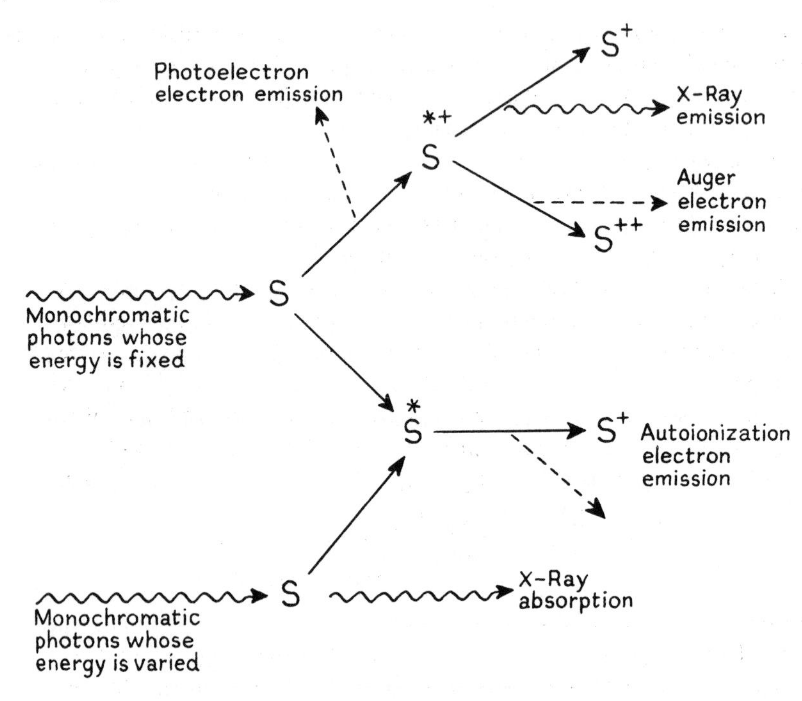

Fig. 7.2 Processes that can occur when a monochromatic beam of photons is directed at a sample. The sample atoms are represented by the symbol S, an excited sample atom by S^*, a singly ionized sample atom by S^+, and a doubly ionized sample atom by S^{++},

7.1.3 Other processes related to photoelectron spectroscopy

Before going on to discuss instrumentation and applications of photoelectron spectroscopy, it is important to point out that there are a number of processes closely related to photoelectron spectroscopy which often occur simultaneously with the photoelectron process.

Figure 7.2 shows the processes involved when a beam of monochromatic photons (ultraviolet radiation or X-rays) falls on a sample. The changes that occur to the atoms in the sample (one type of sample atom being represented by the symbol S) are also shown. Thus initially two things can happen; either the sample loses a photoelectron to produce an excited ion S^{*+}, or the sample undergoes an electronic transition to produce an excited state of the atom or molecule S^*. Although the former situation is the photoelectric process, the latter process cannot be dismissed, for in some cases the excited atom or molecule S^* can emit an electron by autoionization, and this electron will enter the electron energy analyser together with the photoelectrons. The ways in which the excited ion

S^{*+} relaxes is also important since it may emit an electron via the Auger process (to be described in Section 7.10.1), and these electrons will likewise enter the electron energy analyser together with the photoelectrons.

Fig. 7.2 also illustrates the process where the X-ray is absorbed and the electron excited from a core to a valence orbital. This process does not yield electrons and so causes no complications to the photoelectron spectra.

7.2 INSTRUMENTATION

7.2.1 The basic design of photoelectron spectrometers

The basic design of photoelectron spectrometers requires four components. Firstly a suitable source of electromagnetic radiation of high intensity is required. Secondly a sample chamber must be available designed to hold a sample in the required physical state while the sample is subjected to the electromagnetic radiation. The third and techiiically the most difficult, part consists of the electron energy analyser. There are a variety of possible designs for electron energy analysers, and both electrostatic and magnetic types may be used. In either case the analyser must be shielded from the earth's magnetic field, since this magnetic field will deflect the electron path in the analyser. The use of a shield made of paramagnetic material such as μ-metal cuts the force lines of the earth's field and is thus an effective shield for electrostatic instruments, but is unsuitable for magnetic instruments since it will also perturb the spectrometer magnetic field. Magnetic instruments may be shielded by rather elaborate Helmholtz coil arrangements around the spectrometer. All commercial instruments use electrostatic analysers because of their easier shielding and smaller size. Finally after energy analysis electrons whose energy has been measured must be detected by a suitable detection system. This is generally an electron multiplier which counts electrons with high efficiency down to very low energies (below 1 eV). Figure 7.3 gives a schematic diagram illustrating the basic

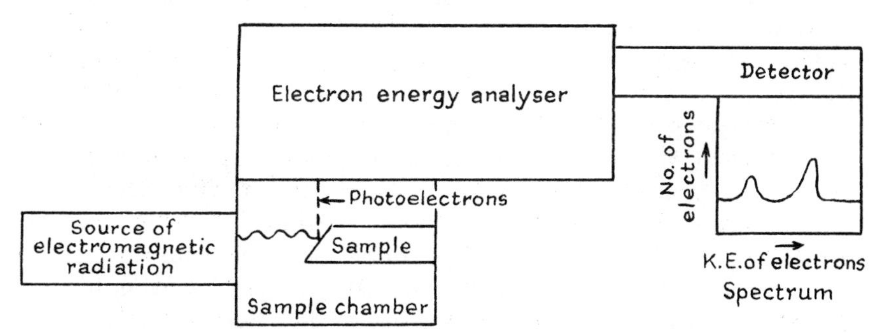

Fig. 7.3 Schematic diagram illustrating the basic design of a photoelectron spectrometer.

design of a photoelectron spectrometer. The whole system is under vacuum (generally less than 10^{-5} torr (1.33×10^{-3} N m^{-2}), so collisions between electrons are reduced to a minimum. Electrons that have lost energy by collision will have a whole range of energies and contribute to the background of the spectrum.

The spectrometer provides information concerning the number of electrons with a particular kinetic energy. The spectrum is obtained by adjusting the analyzer field so that electrons of a specific kinetic energy are brought to a focus at the exit slit. The information is generally presented as a spectrum obtained by plotting the count rate of electrons arriving at the detector against the photoelectronic kinetic energy usually plotted increasing from left to right, or after calibration, the binding energy plotted increasing from right to left.

7.2.2 X-ray photoelectron spectrometers

The main requirements for X-ray photoelectron spectrometers are high resolution and monochromatic X-radiation. The ways of achieving monochromatic X-radiation will be discussed more fully in Section 7.2.4. As explained in Section 7.1.2, the requirement for high resolution is more important the more energetic the electrons studied. Most commercial instruments use spherical (usually hemispherical) double focusing electrostatic analysers, in which there is a radial electrostatic field. For such systems the resolving power, as defined by Equation (7.2), is given by:

$$\text{Resolving power} = 2R/W \tag{7.3}$$

where R is the median radius, the average of the radius of the inner and outer spheres, and W is the width of the collector slit. Equation (7.3) applies to a point source and neglects aberrations. The aberrations may be reduced by fitting baffles in the analyser, and the resolution will then depend upon the analyser acceptance angle. Equation (7.3) is however a good approximation for a hemispherical analyser with equal source and collector slit width ($= W$). Equation (7.3) shows that the resolving power will be increased if the radius of the spheres is increased, and some instruments use large hemispheres of 36 cm radius. The problem with large hemispheres is that the larger the hemisphere the greater is the volume to be evacuated and the amount of metal to be machined. Many instruments get over this problem by using smaller hemispheres, and thus smaller resolving power, but slow down the electrons to be analysed, thus maintaining the same ΔE. Figure 7.4 illustrates such an instrument. In this instrument, manufactured by A.E.I., the photoelectrons are retarded by a retarding potential in the lens system which focuses the image of the source slit on the entrance slit of the hemispherical analyser (of 12.7 cm radius), the electrons entering the analyser with only 1/20th of their original energy. Using Equation (7.3) for $W = 0.635$ mm, the resolving power is 400, but since the electrons are slowed down by a factor of 1/20th, the effective resolving power in terms of the original energy of the electrons is 8000. Unfortunately this design produces a

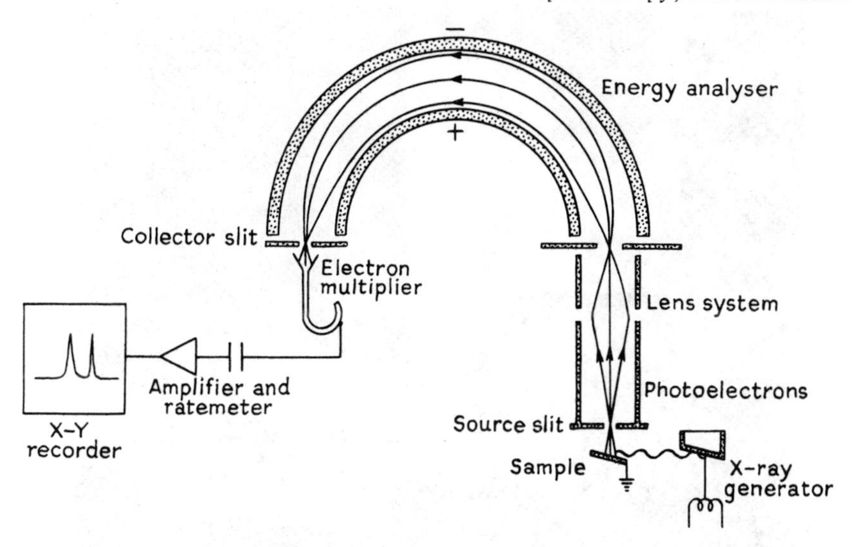

Fig. 7.4 Schematic diagram illustrating the basic design of an X-ray photoelectron spectrometer using a retarding lens system and a hemispherical electrostatic analyser.

considerable loss of sensitivity at low photoelectron energies. Figure 7.5 shows a typical complete X-ray photoelectron spectrometer of the type shown in plan in Fig. 7.4.

Most X-ray photoelectron spectroscopic studies are made on solid materials. The solid to be studied is usually attached to some type of plate. If the sample is in a solid strip it may be screwed on to the plate. If the sample is a powder (the most common situation) it may be held on to the plate by pressing on to a metal grid or double-sided sellotape attached to the plate. The plate is of small size (typically 0.5 × 1.5 cm) so that only small amounts of sample are required, especially since the sample thickness need be no more that 10 nm (this will be discussed in Section 7.4). The plate is placed close to the X-ray source, but is separated from the X-ray source by an X-ray transparent window, generally made made of aluminium, to prevent bombardment of the sample by electrons from the X-ray gun, and to allow the sample chamber and X-ray gun to be separately pumped. The X-ray beam and the electron analyser entry slit are generally at right angles as illustrated in Fig. 7.6.

Other sample states can be studied, but such studies of liquids and gases present considerable experimental problems. The easiest way to study gases is to freeze the sample plate and condense the gas as a solid on the sample plate. Likewise liquids may be studied by freezing the liquid on to the sample plate [7.6].

The study of samples in the gas phase presents problems because of the much lower concentration of molecules in the gas phase as compared with the compact solid phase. To obtain even a low count rate of photoelectrons (often less than

Fig. 7.5 A.E.I. ES 200 X-ray photoelectron spectrometer.

ten counts per second) the gas must have a pressure of a few tenths of a torr. A pressure as great as this in the whole system would be quite unacceptable, since all the photoelectrons would suffer many collisions before reaching the analyser, and it was pointed out above that the pressure should be less that 10^{-5} torr. To study gases therefore a differential pumping system is employed. The sample plate is generally replaced by a chamber into which the gas to be studied may be bled. The chamber has one opening into the sample chamber, a slit which faces the X-ray beam allowing X-rays to enter the gas sample and photoelectrons to leave. The slit causes different pressures to result in the chamber and in the sample chamber and instrument, the slit restricting the pumping out of the gas chamber allowing the gas pressure to be several tenths of a torr while that in the instrument is below 10^{-5} torr. Gas-phase spectra have the advantage of the absence of surface charging (Section 7.5) and generally give narrower spectral lines; they are however complicated by a dependence of the binding energy upon gas pressure. A full description of such spectra may be found in the book by Siegbahn and his coworkers [7.7].

The study of samples in the liquid phase is in its infancy, but liquid spectra have been obtained [7.8] by pumping a 'liquid beam' through the sample chamber using differential pumping.

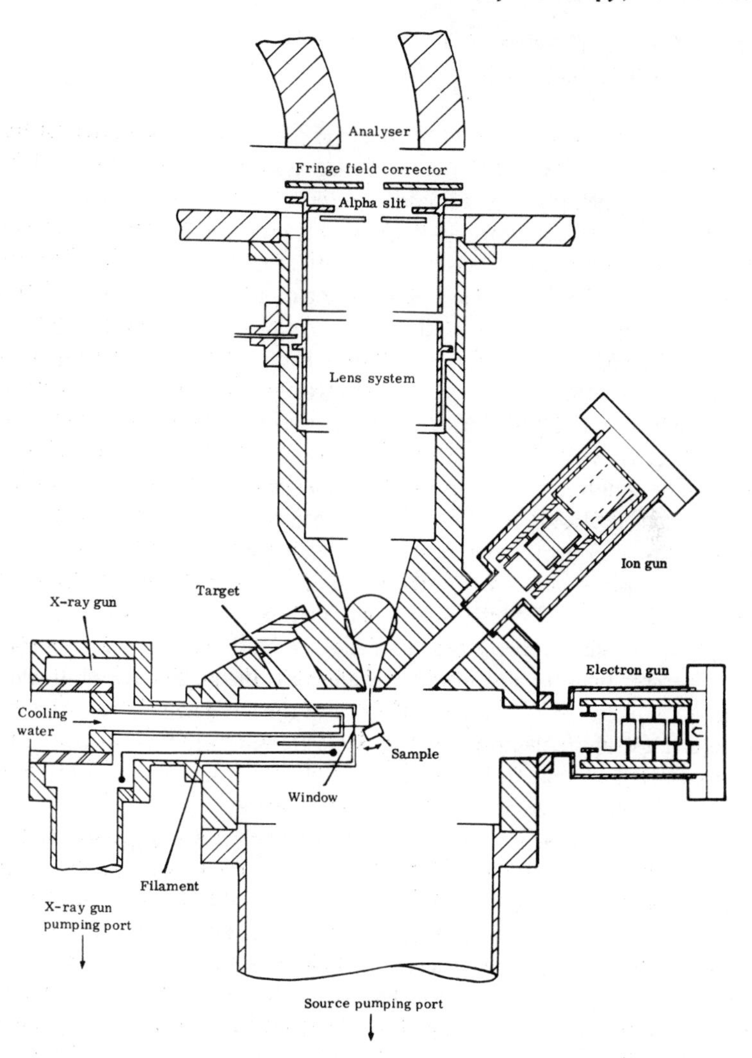

Fig. 7.6 Sample chamber design of an A.E.I. ES 200 X-ray photoelectron spectrometer. The plate on shich the sample is attached is a copper block. The copper block with the sample attached may be removed from the sample chamber through a vacuum lock for the fitting of a new sample.

The study of samples in molecular beams has also been carried out [7.9]. While the density of molecules is lower than in gas-phase studies, and thus the detection problems greater, the molecular beam study has the advantage that molecules can be studied under well defined physical and chemical conditions and can be directed so that no interference occurs with spectrometer components. This method may make the study of potentially unstable species possible.

7.2.3 Ultraviolet photoelectron spectrometers

The main requirements for u.v. photoelectron spectrometers are satisfactory resolution and monochromatic u.v. radiation. As explained in Section 7.1.2, the requirement for high resolution is less important because the photoelectrons produced are of relatively low energy. Thus a typical commercial instrument with a resolving power of 300 gives a linewidth at half-height of 0.017 eV for an electron energy of 5 eV. The ways of achieving monochromatic u.v. radiation will be discussed more fully in Section 7.2.4. Another advantage with u.v. excitation as compared with X-ray excitation is that the photoelectron intensity increases with the third power of the wavelength of the exciting electromagnetic radiation (Section 7.6), and so much greater photoelectron intensities may be expected with u.v. excitation. This means that the electron analyser design can be simplified. The hemispherical electrostatic analyser used in X-ray photoelectron spectrometers has a high electron transmission because there are many paths for the electron between the entrance and exit slits of the analyser. Ultraviolet photoelectron spectrometers generally use cylindrical electrostatic analysers with a $127° \, 17'$ sector (this angle is chosen because there is an electron trajectory refocusing property for this angle), and although the electron transmission is lower than for a hemispherical analyser, the greater u.v. photoelectron intensity compensates for this and the price of the analyser is lower, As a result, commercial u.v. photoelectron spectrometers are generally half the price of X-ray photoelectron spectrometers. Fig. 7.7 illustrates a typical 127° cylindrical electrostatic analyser u.v. photoelectron spectrometer.

Most u.v. photoelectron spectroscopic studies are made on gaseous materials. The low count rate experienced in X-ray photoelectron studies is compensated for by the increased photoelectron intensity in u.v. photoelectron spectrometers, and count rates of over one thousand counts per second are often observed. As in the X-ray photoelectron studies the gaseous samples are studied using a differential pumping system. The gas is bled into a chamber, generally a metal tube, with a slit to allow the photoelectrons to leave. The u.v. source unlike the X-ray source does not have to be separated from the sample by a window and can be directed down the metal tube in the opposite direction to the gas flow into the tube; sample vapour is prevented from entering the u.v. source by fitting a capillary tube between the source and metal tube. Figure 7.7 illustrates the source arrangement.

There have been a number of u.v. photoelectron spectroscopic studies of solids, but because of the small penetration of u.v. photons and the even smaller escape depth (only a few hundred picometres) of the photoelectrons, such studies are essentially surface studies. There have been a number of studies, mainly by physicists and metallurgists, of the photoelectron spectra of clean metal surfaces in order to obtain the valence band structure (usually called photoemission studies), and recently u.v. photoelectron spectroscopy has proved of considerable value for studying gases adsorbed on metal surfaces.

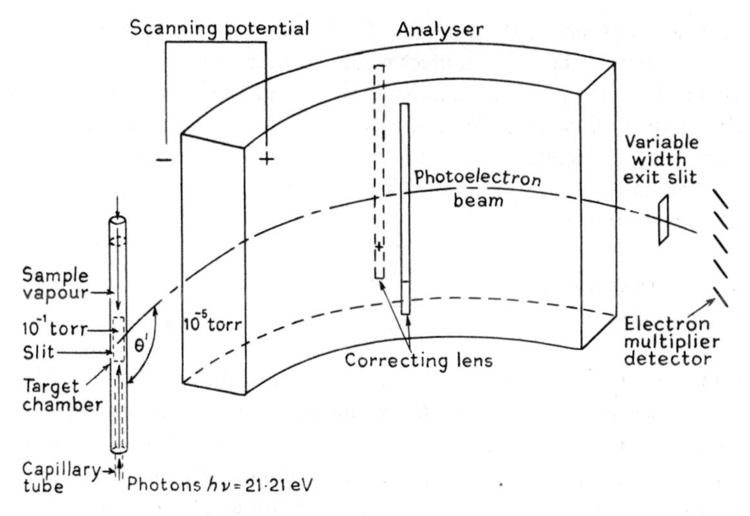

Fig. 7.7 Schematic diagram illustrating the basic design of a u.v. photoelectron spectrometer using a 127° cylindrical electrostatic analyser.

7.2.4 Linewidths of photoelectron sources sources

The first factor to consider when considering linewidths in photoelectron spectroscopy is the linewidth of the electromagnetic radiation excitation source. Ideally the excitation source should have a negligible linewidth, since the greater the linewidth of the excitation source the greater the minimum linewidth of the photoelectron peak. In general the greater the energy of the excitation source, the greater will be the source linewidth.

The linewidth of electromagnetic radiation depends upon a number of factors.

(i) *The natural width of the energy levels*

Emission of electromagnetic radiation occurs because an electron in a higher energy level falls to a lower energy level, the loss in electron energy appearing as emitted radiation. The linewidth of the emitted radiation will depend upon the width of the initial and final state electron energy levels. An electron in any energy level will have an uncertainty, ΔE, in its energy given by Heisenberg's uncertainty principle ($\Delta E \cdot \tau \approx h$), where τ is the time the electron spends in the energy level, i.e. the lifetime of the energy level. Taking ΔE as the peak width at half height, the broadening will be approximately $4 \times 10^{-15} \tau^{-1}$ eV or $2 \times 10^{-16} \tau^{-1}$ J. The width of energy levels increases appreciably with increasing energy because the lifetime is proportional to $(\text{energy})^{-3}$; thus valence orbitals (of 0–20 eV energy) have lifetimes of the order 10^{-8}–10^{-10} s giving widths of the order 5×10^{-5} to 5×10^{-7} eV, whereas core orbitals (of 100 000 to 40 eV energy) have lifetimes of the order 10^{-13}–10^{-18} s giving widths of the order 10^{-1} to 4×10^{3} eV. Thus u.v. sources would be expected to have linewidths of the order of less than 10^{-5} eV, and X-ray sources of an electron volt or more.

The lifetime of an energy level of a gas-phase molecule is affected by the pressure, since collisions can shorten the lifetime. This *pressure, resonance,* or *Lorentz* broadening is generally less than 10^{-6} eV and is thus insignificant compared with other effects.

(ii) *The width arising from motion of the emitter atoms or molecules*

The emitting atoms or molecules may move owing to thermal motion. This motion gives rise to *Doppler broadening* and the effect is proportional to the wavelength of the electromagnetic radiation and inversely proportional to the square root of the mass of the emitting atoms or molecules. For hydrogen u.v. sources some significant broadening arises from this effect, but for other sources of greater emitter mass and greater energy it is negligible.

(iii) *The monochromatic nature of the radiation used*

The He(I) source most frequently used for u.v. photoelectron studies (already discussed in Section 7.1.2) is very monochromatic, giving almost entirely the 58.4 nm (21.21 eV) line. Other sources such as the He(II) at 30.39 nm (40.8 eV), the Ne(I) at 73.58 nm (16.85 eV), the A(I) at 106.7 nm (11.62 eV), and the Lyman α line of atomic hydrogen at 121.55 nm (10.2 eV) may also be used, but they are not as monochromatic as the He(I) source and a number of weak lines having different energy from that of the main line are also excited.

X-rays are generated by directing a highly energetic beam of electrons at a metallic target to eject a core electron which is replaced by a transition within the core levels with the generation of an X-ray. This process gives the K_{α} line (corresponding to the transition of a 2p electron to fill a vacancy in the 1 s level) as the most intense line, but the spectrum is not monochromatic and a number of other lines are excited, together with a continuous background (called Bremsstrahlung radiation). A typical spectrum is shown in Fig. 7.8. The relative intensities of the lines vary considerably with the target atom, and additional lines arise from satellite peaks caused by the multiple ionization of the core shells in the X-ray process [7.10]. For the common target materials used in X-ray photoelectron studies, aluminium and magnesium, these other lines are of appreciably less intensity than the K_{α} line (Fig. 7.9). Unfortunately the K_{α} line is split into two lines by spin–orbit coupling in the 2p energy level giving $2p_{1/2}$ to 1s and $2p_{3/2}$ to 1s transitions of intensity ratio 1 : 2. The separation between the two lines falls with decreasing X-ray energy, and in the case of aluminium and magnesium the two lines overlap, increasing the width of the X-ray radiation (Fig. 7.10).

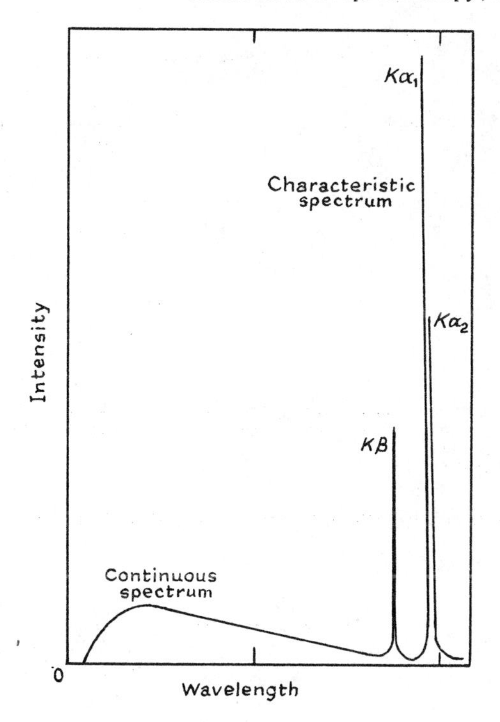

Fig. 7.8 Typical spectrum from an X-ray tube.

The overall effect of all the factors above is to give radiation of approximately 0.005 eV linewidth for the u.v. sources, and 0.85 eV for magnesium and 1.0 eV for aluminium X-ray sources. Heavier atom X-ray sources are little used, unless very deep lying core levels are studied, since as expected from the above the linewidth increases considerably with increasing atomic weight (since the orbitals involved are of higher energy). Thus chromium radiation is of about 18 eV width and molybdenum about 6.6 eV width.

The appreciable linewidth of X-ray sources can be reduced by monochromatization of the X-radiation, and some instruments have been fitted with a monochromator built to the design of Siegbahn and his coworkers [7.10]. The X-radiation is dispersed by Bragg reflection from a curved quartz crystal, which acts as a crystal diffraction 'grating'. The X-ray sonrce and sample and curved quartz crystal are situated on the Rowland circle, the circle which touches the crystal such that, if the source is located at any point on the circle, the diffracted radiation from the crystal comes to a focus on the same circle. Figure 7.11 shows the arrangement. For Al K_α radiation, reflection from the (0, 1, 0) crystal planes takes place at a Bragg angle θ of 78.5°. The resolving power of the crystal dispersion system is about 0.2 eV. Monochromatic radiation from the crystal dispersion system may be selected by means of a slit on the Rowland circle (before the radiation reaches the sample), at $\theta = 78.5°$ so that Al K_{α_1}

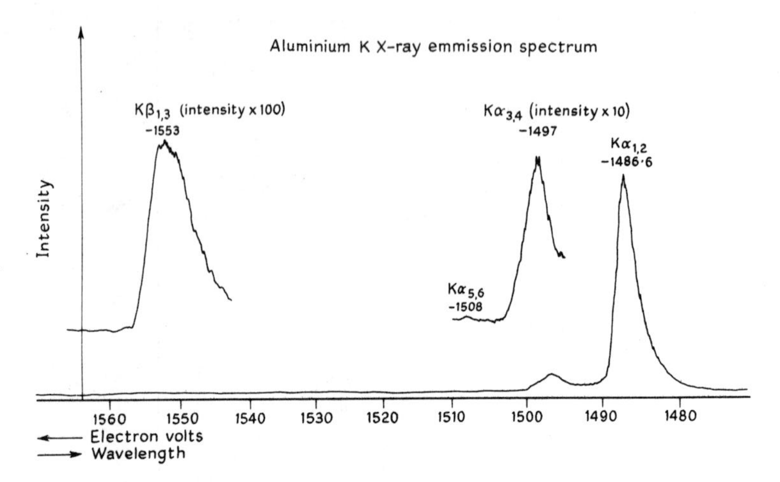

Fig. 7.9 X-ray emission spectrum of metallic aluminium [Urch, D.S., private communication].

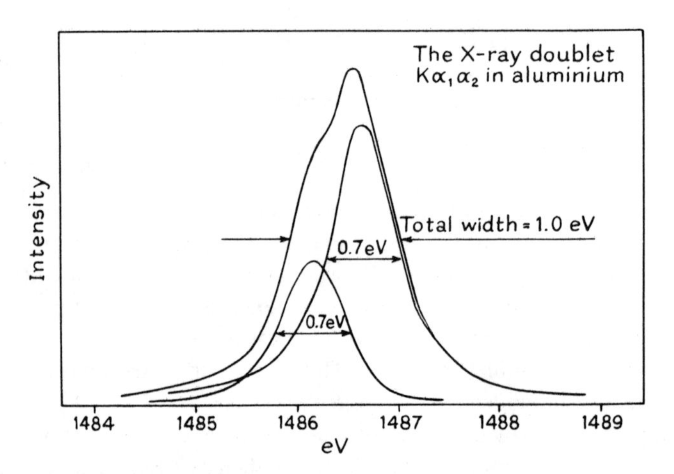

Fig. 7.10 X-ray emission spectrum of metallic aluminium, $K_{\alpha_\pi \alpha_2}$ doublet [7.10].

radiation is selected. This arrangement unfortunately results in appreciable loss in intensity, and this loss in in intensity can be reduced by placing the sample on the Rowland circle so that each point on the sample surface is exposed to monochromatic radiation, the wavelength of which varies across the surface, and the electron analyser is so arranged that it focuses the photoelectrons from the sample along a line the width of which depends upon the photoelectron energy and not linewidth. This method is known as dispersion compensation, and the whole system is illustrated in Fig. 7.11. The use of X-ray monochromatization

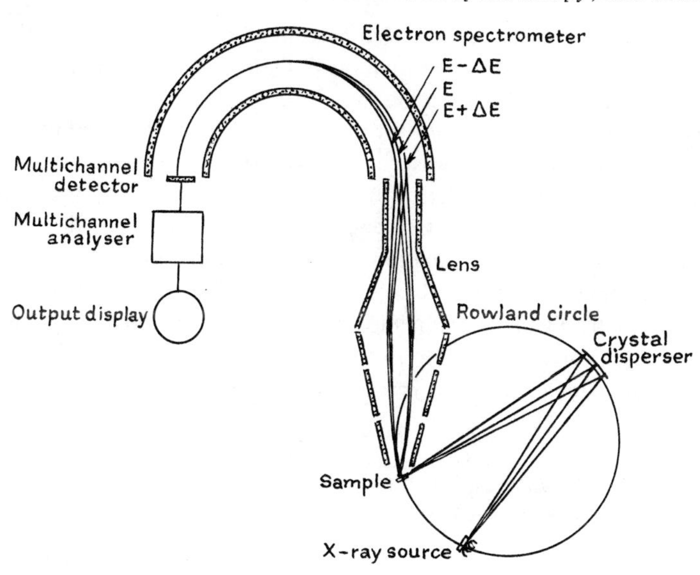

Fig. 7.11 Schematic diagram of X-ray photoelectron spectrometer fitted with monochromator with dispersion compensation [7.10].

produces not only narrow X-ray line widths but also removes completely the X-ray satellite lines and background Bremsstrahlung radiation. The loss of intensity may mean that special detection systems have to be used.

7.2.5 Linewidths in photoelectron spectroscopy

When the narrow linewidth radiation from u.v. sources, or from monochromatized X-ray sources, is used the photoelectron spectrum still exhibits significant linewidth due to the photoelectron process.

The linewidth of the photoelectron process (i.e. the range of photoelectron energies for monochromatic photon excitation) depends upon factors (i) and (ii) of Section 7.2.4, except that in the case of factor (i) only one energy level is involved in the photoelectron process. The natural width therefore depends upon upon the linewidth of the energy level from which the photoelectron is ejected (Section 7.10.2). As seen in Section 7.2.4 this is significant for core orbitals, but small for valence orbitals.

7.3 CHEMICAL INFORMATION FROM PHOTOELECTRON SPECTROSCOPY

It has been seen that photoelectron spectroscopy can yield values for the binding energies of all the electrons in an atom or molecule since, once the kinetic energy of the ejected electrons has been measured in the spectrometer, the binding energy can be obtained using Equation (7.1). In fact there are often more terms in Equation (7.1). Firstly there should be an additional term ($-E_r$) on the right-hand side of (7.1) to account for the recoil energy of the atom or molecule from which the photoelectron was ejected. However, the recoil energy is very small, approximately [mass(electron)/mass(molecule)] × K.E.(photoelectron), about 5×10^{-4} eV per photoelectron energy (in eV) for hydrogen, i.e. about 0.9 eV for Al $K_{\alpha_1 \alpha_2}$ radiation, and much less for heavier atoms (e.g. 0.1 eV for Al $K_{\alpha_1 \alpha_2}$ radiation on lithium). Recoil energy is thus ignored. Secondly, in the case of solid materials the attachment of the sample to the sample plate will give rise to a contact potential equal to the difference in work functions between the sample and the sample plate ($\phi_{cpd} - \phi_{spectrometer}$); thus (7.1) becomes:

$$\text{K.E.} = h\nu - B + \phi_{cpd} - \phi_{sp} \tag{7.4}$$

putting B on a scale with respect to the Fermi level, B', rather than with respect to the vacuum level, where $B = B' + \phi_{cpd}$:

$$\text{K.E.} = h\nu - B' - \phi_{sp} \tag{7.5}$$

In addition, for solid samples there is also a sample charging effect (to be discussed in Section 7.4) which introduces an additional correction term S:

$$\text{K.E.} = h\nu - B' - \phi_{sp} - S \tag{7.6}$$

Once the binding energy of a required orbital is obtained using the appropriate form of Equation (7.1) useful chemical information is obtained. If the orbital concerned is a valence orbital, important information about the molecule being studied is obtained since the binding energy reflects the molecular orbital energy. For any orbital x (valence or core orbital):

$$B_x = \begin{pmatrix}\text{Total energy of neutral} \\ \text{molecule}\end{pmatrix} - \begin{pmatrix}\text{Total energy of molecular} \\ \text{ion with vacancy in orbital x}\end{pmatrix} \tag{7.7}$$

B_x will equal the orbital energy if one assumes that the electrons outside the orbital from which the electron was ejected have the same energy in the molecular ion as in the neutral molecule. Such an assumption is known as *Koopman's theorem*, and it gives orbital energies that are consistently about 2–10 per cent too large because *reorganization* or *relaxation* energy has been ignored. Relaxation energy arises from the contraction of the electrons in the molecular ion leading to a gain in the total energy of the molecular ion and thus [Equation (7.7)] a lower value of B_x. Differences in the correlation energy and relativistic energy of the neutral molecule and molecular ion should also be included but these are generally very small. The Koopman's theorem approximation for valence orbitals will give the wrong prediction of the ordering of the molecular orbitals when there are large differences in relaxation energy between orbitals, and for core orbitals it will give chemical shifts that are difficult to compare when there are large differences in relaxation energies for different compounds.

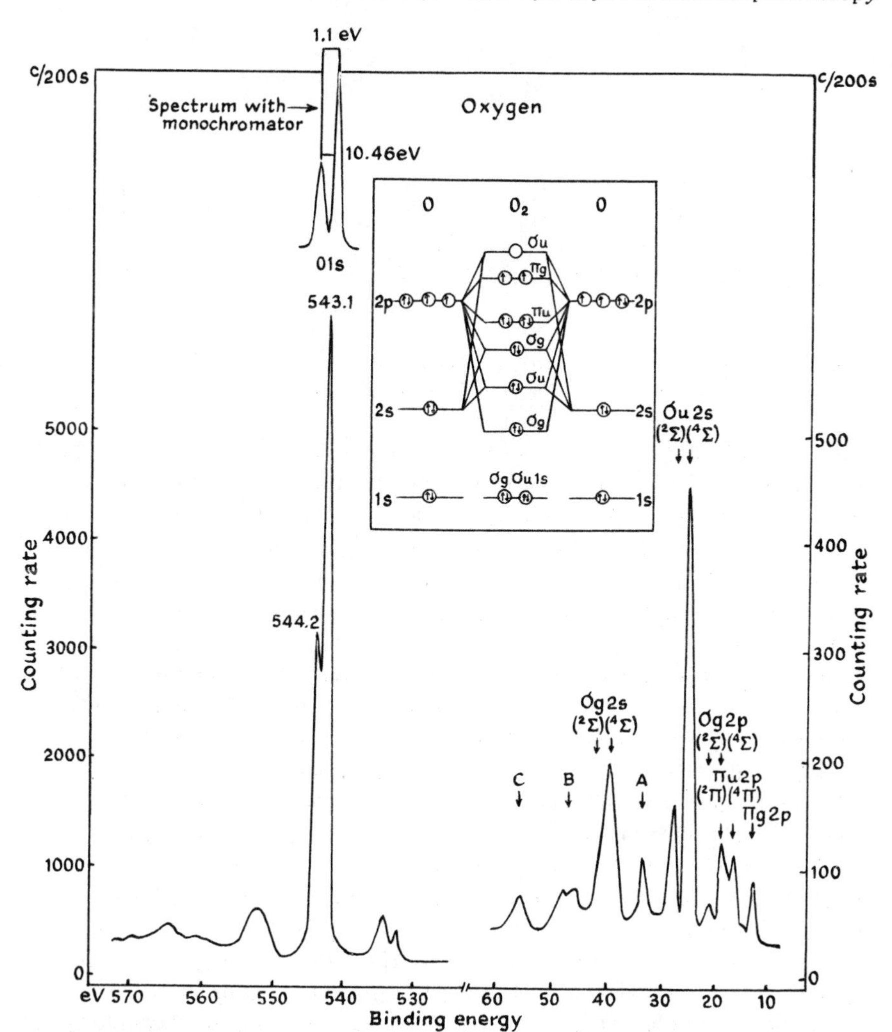

Fig. 7.12 Photoelectron spectrum of oxygen gas excited by Mg $K_{\alpha_1\alpha_2}$ radiation, and molecular orbital diagram for O_2. The insert shows the improved resolution of the O 1s core energy level peak achieved when the spectrum is excited by monochromatized X-ratiation [from [7.7], and Gelius, U., Basilier, E., Svensson, S., Bergmark, T. and Siegbahn, K., *J. Elect. Spec. Rel. Phen.*, **2**, 405 (1973)].

The approximation that the photoelectron spectrum from the valence band region will reflect the molecular orbital diagram provides considerable chemical information. Consider the case of oxygen. Figure 7.12 shows the photoelectron spectrum of the gas excited by Mg $K_{\alpha_1\alpha_2}$ radiation. It can be seen that the spectrum from the valence orbitals corresponds to the molecular orbital diagram

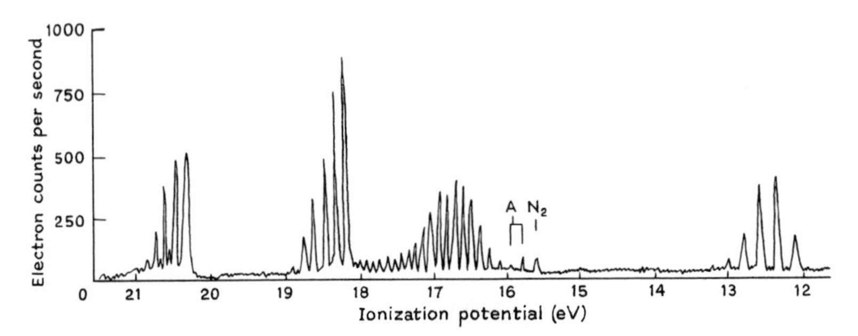

Fig. 7.13 Photoelectron spectrum of oxygen gas excited by He(I) radiation [Turner, D.W., *Chemistry in Britain*, 382 (1967)].

(drawn to scale in Fig. 7.1, and not to scale, but with electron occupancy, in the insert in Fig. 7.12). The peaks at A, B, and C are probably due to secondary processes to be discussed in Sections 7.10.3 and 7.10.4. When a source of lower energy but narrower linewidth is used [He(I) radiation] there is only sufficient energy to ionize the π_g, π_u, and $\sigma_g 2p$ electrons, but now each electronic energy level shows the vibrational structure clearly resolved (the spectrum is illustrated in Fig. 7.13). This vibrational structure gives information about the type of electronic energy level with which it is associated (see Section 7.7.1). Figures 7.12 and 7.13 show that photoelectron lines associated with one electronic energy level may be split into a doublet. This doublet arises through spin–orbit coupling and occurs whenever an electron is removed from a completely filled orbital for which the orbital quantum number $l > 0$ since the remaining electron has $s = \pm\frac{1}{2}$ and thus j $(j = l + s)$ for spin–orbit coupling has two values.

The chemical information immediately provided by the photoelectron spectrum of the valence orbitals seems immediately obvious, but what of the core orbitals. Thus Fig. 7.12 shows one peak [the doublet nature (two peaks separated by 1.1 eV) is explained in Section 7.10.5] for the photoelectrons ejected from the 1s core orbital. One might initially expect the binding energy of oxygen 1s electrons to be the same irrespective of the type of chemical compound considered, since this is a core orbital.

To examine the case of the core orbitals further, consider the molecule CO. There is a single peak for the photoelectrons ejected from the oxygen 1s core orbital, but the spectrum of CO gas, illustrated in Fig. 7.14(a), shows that the binding energy is somewhat less than in the case of O_2. If the CO gas is absorbed on to a clean polycrystalline tungsten surface at 300 K and the photoelectron spectrum [Fig. 7.14(b)] of the tungsten surface studied, *two* peaks separated by 3.4 eV are seen for photoelectrons ejected from the oxygen 1s core orbital. These two peaks may be explained as being due to two types of adsorbed CO, the strongly bound β-CO and the weakly bound α-CO. It will also be noticed that the binding energies of the CO on tungsten are lower than the gas-phase value. This is partly because the values given refer to B' and thus must be corrected for the work function (about (5.4 eV), but even so there is a significant shift.

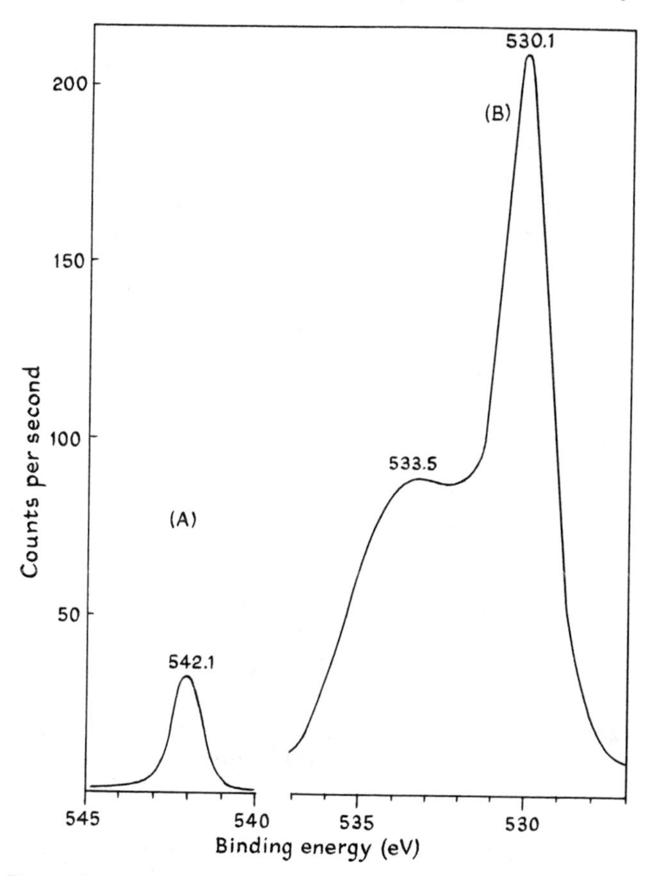

Fig. 7.14 Photoelectron spectrum of CO excited by Mg $K_{\alpha_1\alpha_2}$ radiation. (a) the spectrum of gas-phase CO [7.7]; (b) the spectrum of CO adsorbed on poylcrystalline tungsten at 300 K [Madey, T.E., Yates, J.T., Jr. and Erickson, N.E., *Chem. Phys. Lett.*, **19**, 487 (1973)] Note increase in intensity in adsorbed state.

Thus the binding energies of core electrons give rise to a *chemical shift*. Thus we see above that the binding energy of the oxygen 1s core electron varies according to the chemical environment of the oxygen atom, and when there are two types of oxygen atom present in the same system, two peaks are observed.

In fact, chemical shifts can be observed for the core electrons of all elements except hydrogen (whose single electron is involved in all componds involving hydrogen and is thus always a valence electron). This means that the photoelectron spectroscopy of core electrons (X-ray photoelectron spectroscopy, since X-ray excitation is required to excite core electrons) is a very powerful technique for investigating structure and bonding in chemistry. In addition, since the core electron binding energies are characteristic of a particular type of atom, the atomic contituents of a compound can be readily determined, and further, the

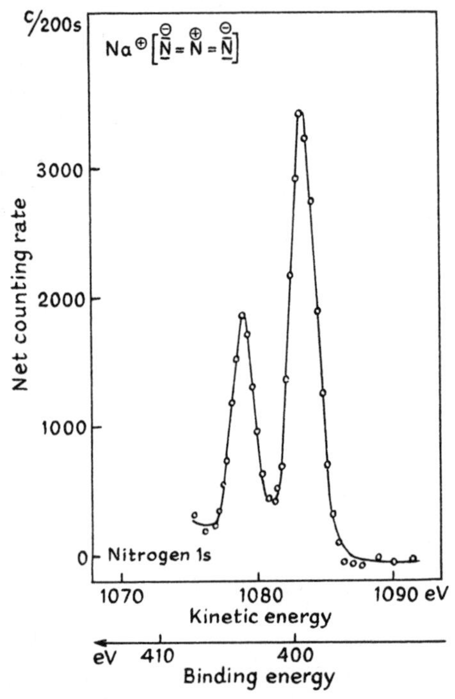

Fig. 7.15 Photoelectron spectrum of solid NaN_3 excited by Al $K_{\alpha_1\alpha_2}$ radiation, showing the region corresponding to photoelectrons ejected from nitrogen 1s core orbitals [7.10].

area under the peaks corresponds to the number of atoms present. It is these two features that lead to the name ESCA (Section 7.1.2). Thus in Fig. 7.15, which shows the spectrum of the N 1s core electrons in NaN_3, two types of nitrogen are seen in the ratio 2:1. The explanation of chemical shifts will be left to Section 7.9.2, but it may be noted here that writing the azide ion as $(\bar{N}=\overset{+}{N}=\bar{N})$ shows two types of nitrogen in the ratio 2:1, and that the nitrogen atom of highest binding energy corresponds to that of the greatest formal positive charge. Thus it appears that the chemical shift increases with increasing positive charge on the nucleus concerned, which seems reasonable since the greater the effective positive charge on the nucleus, caused by movements of valence electrons altering electron–nucleus interaction, the greater the attraction of the core electron for the nucleus and thus the higher the binding energy. The situation is not quite as simple as this but full discussion will be left to Section 7.9.2.

7.4 SOLID STATE SURFACE STUDIES

Photoelectron spectroscopic studies of solids are to a lesser or greater extent surface studies. This is because the penetration of electromagnetic radiation into a solid is limited by absorption of radiation in the solid, and the photoelectrons produced from atoms and molecules in the solid may suffer collisions with other solid atoms and molecules, thus losing energy. Electromagnetic radiation, especially X-radiation, penetrates deep into the solid, and the limiting factor is the escape depth of the photoelectrons. Figure 7.16 illustrates the situation. For sample atoms or molecules on the surface the photoelectrons produced suffer no collisions and enter the vacuum with energy e. Deeper below the surface, collisions with other molecules may occur, the electrons losing a variable amount of energy corresponding to the energy lost on collision, and entering the vacuum of the spectrometer with energy e' where $e' < e$. All electrons that suffer energy loss (e') will not give a single peak in the photoelectron spectrum and contribute to the spectrum background. The photoelectron peaks correspond to electrons ejected from the surface up to some maximum depth called the *escape depth* below which depth only e' electrons can be ejected. Clearly the greater the photoelectron energy the greater the chance that it may be ejected with energy e. Since the photoelectron energy depends upon the excitation energy [Equation (7.1)], the greater the excitation energy the greater the escape depth. In fact the escape depth depends approximately upon (photoelectron kinetic energy)$^{\frac{1}{2}}$. Typically the escape depth for He(I) excited photoelectrons is of the order of a few hundred picometres, and for Mg $K_{\alpha_1\alpha_2}$ X-rays of the order of two thousand to ten thousand picometres. The heavier the atoms that are present in the material the greater the chance of energy loss by collision and thus the smaller the escape depth.

The surface sensitive nature of photoelectron spectroscopy has both advantages and disadvantages. Many X-ray photoelectron spectroscopic studies are made in order to determine the nature of the bulk material, and when such studies are made care has to be taken to ensure that material at the surface is the same as the material in the bulk. Under the high vacuum conditions often used in photoelectron spectroscopy (10^{-6}–10^{-8} torr), oxygen, water, and hydrocarbons from the pump oil will condense on to the surface of the sample (of the order 1 monolayer/second at 10^{-6} torr) giving rise to additional peaks in the photoelectron spectrum. Such additional peaks may be used to advantage for calibration purposes (Section 7.5), but sample contamination prevents surface studies being made. At pressures of 10^{-10} torr or better, surface contamination becomes insignificant and photoelectron spectroscopy may be used for surface studies. The power of photoelectron spectroscopic studies of surfaces has already been illustrated in Fig. 7.14. Many studies are now being made of surfaces and of gases adsorbed on surfaces (fractions of a monolayer of adsorbed gases may sometimes be observed).

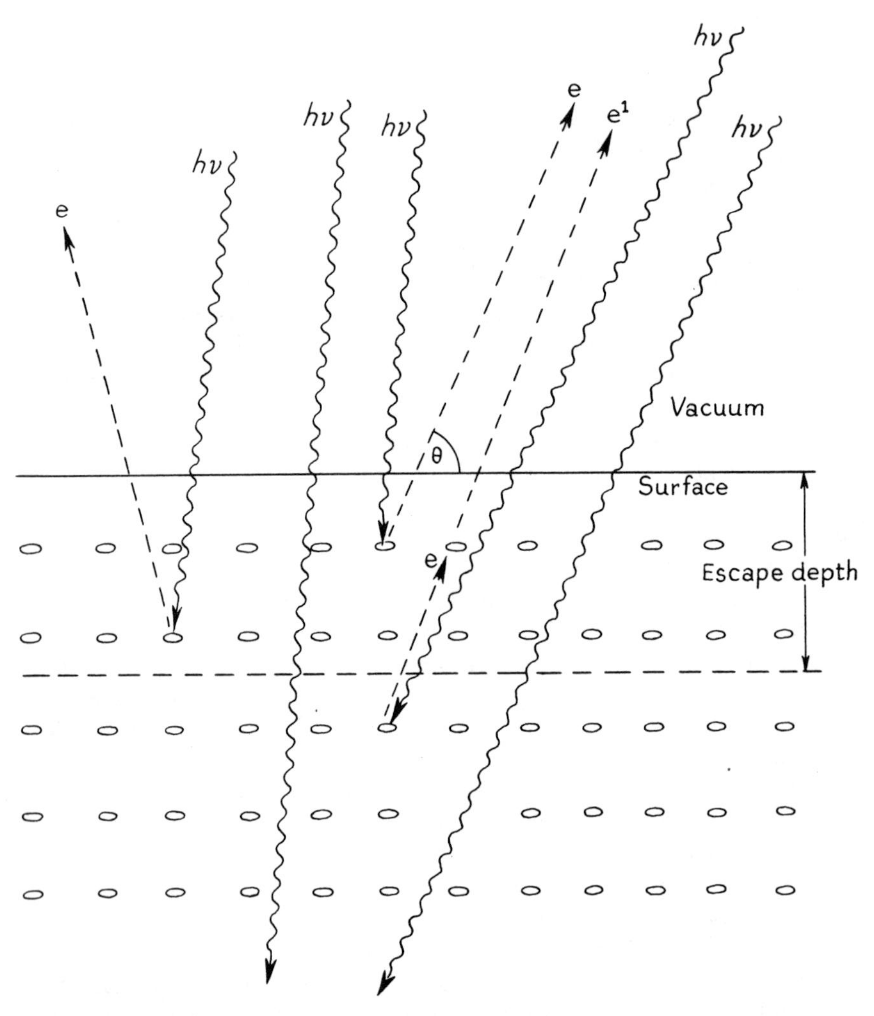

Fig. 7.16 Diagram illustrating the escape depth for photoelectrons. $h\nu$ represents electromagnetic radiation, and e represents a photoelectron which has suffered no energy loss from collision (elastic ejection), and e$'$ represents a photoelectron which has suffered energy loss from collision (inelastic ejection).

7.5 SURFACE CHARGING AND THE CALIBRATION PROBLEM

Having seen that photoelectron spectroscopy is a surface technique, one must consider what happens to the surface electrically when photoelectrons are emitted. Loss of electrons from the surface in the photoelectron process will give rise to positively charged ions on the surface which thus develops a positive charge which tends to attract escaping photoelectrons to the surface, thus reducing their kinetic energy. Whenever solid samples are studied, the kinetic energy of the electrons will therefore be reduced by an unknown surface charge. For electrically conducting solid samples this surface charge may be removed by earthing the sample, but for non-conducting samples surface charges will build up even though the sample is earthed. The surface charge in non-conductors will be reduced by bombardment of the surface by secondary electrons present in the sample chamber due to X-ray bombardment of the sample chamber walls, though in instruments fitted with monochromators the low X-ray intensity causes low secondary electron intensity and thus high surface charging. The surface charge in non-conductors may be reduced by bombarding the surface with an electron gun; however the exact charge on the surface can never be known accurately.

The surface charge thus introduces a correction term which was included in Equation (7.6). Equation (7.6) shows that for solid samples the terms ϕ_{sp} and S must be determined before the binding energy with respect to the Fermi level B' can be obtained from an experimental measurement of kinetic energy (K.E.) of the photoelectrons. While ϕ_{sp} is constant for a particular instrument, S varies from experiment to experiment. The sum of the terms ϕ_{sp} and S may be obtained for a particular experiment by calibration of the spectrum. Calibration is achieved by placing on the surface some material whose B' value is known accurately and which does not react with the sample being studied, and which is in electrical contact with the sample. For such a situation the Fermi levels of the calibrant material and the sample will equilibrate, and the calibrant material will take up the same surface charge, S, as the sample surface. The kinetic energy of photoelectrons from the energy level of the calibrant that is known, B'(calibrant), is measured, and hence $(\phi_{sp} + S)$ may be determined since B' and $h\nu$ [Equation (7.6)] are known. Fig. 7.17 illustrates the situation.

A number of materials have been used for sample calibration. These include the C1s peak from the hydrocarbon adsorbed on to the sample surface from the instrument pump oil, the Au $4f_{7/2}$ peak from metallic gold either deposited in small amounts on the surface or placed on the surface as gold powder, and various peaks from calibrants internally mixed with the sample. All methods of calibration seem to suffer from various drawbacks when applied to particular samples, and a full discussion of all the problems is beyond the scope of this chapter. The main point to note is that the assumption of equilibration of Fermi levels between sample and calibrant is at best approximate, and there can be no doubt that the best calibration is the use of an internal energy level that is chemically insensitive.

Calibration is a much simpler process for gaseous materials because the effect of space charges and contact potentials between the sample and spectrometer is much reduced. Equation (7.1) may be used with the addition of a correction term which may be evaluated by adding an inert gas for which a particular energy level is known accurately.

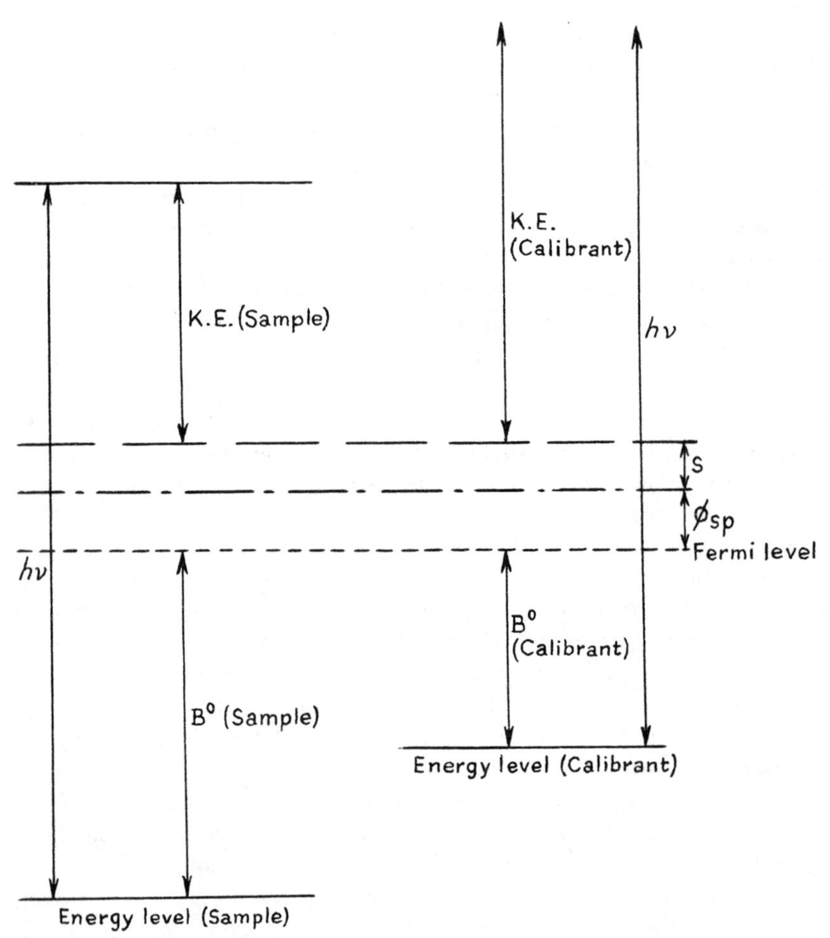

Fig. 7.17 Diagram illustrating the factors involved in the calibration of photoelectron spectroscopic studies of solids.

7.6 PHOTOELECTRON INTENSITIES

In Section 7.3 it was pointed out that the area under a photoelectron peak was proportional to the number of atoms present. This situation obviously applies only to core orbital photoelectrons when the photoelectron may be associated with a particular atom in the molecule. Nevertheless the intensities of photoelectron peaks from valence orbitals provide valuable information. Photoelectron intensities will depend upon a number of factors, which for the case of a homogeneous system of sample atoms or molecules of concentration C, where the photoelectrons after leaving the sample atom suffer no collisions before entering the spectrometer, are given by:

$$I = I_0 \alpha C k x \tag{7.8}$$

where I_0 is the intensity of the incident photon energy; α is the cross-section [units of $(\text{length})^2$] for photoionization of a given energy level of the atom or molecule for a particular incident photon energy, and is often known as the photoelectron cross section; k is an instrumental factor which has a particular value for the instrument used and the incident photon energy; C is the concentration of sample (units of number of atoms per volume); and x is the length of sample studied. These factors will now be discussed in more detail:

(i) *Instrumental factors* (k)

The design of the particular instrument being used will have a marked effect on the actual intensity observed. Thus instruments with high intensity sources and highly sensitive detection systems provide high intensity. Some instruments have different sensitivities to electrons of different kinetic energy [the retarding lens analyser is such a system (Section 7.2.2)] and all instruments will have a variable electron transmission depending upon the analyser design.

(ii) *Sample state* (C)

Equation (7.8) shows that the intensity increases with increasing sample concentration. Thus solid samples will give photoelectron peaks of higher intensity than gaseous samples. However, the ideal situation described by Equation (7.8) can never be realized in practice since, as soon as the sample has a sample concentration greater than one sample atom or molecule in the sample chamber, then there will be a finite chance of a collision between the photoelectron and another sample molecule. In Section 7.4 it was seen that such energy-losing collisions lose photoelectron peak intensity. In fact Equation (7.8) holds quite well for gaseous samples if the intensity data are recorded at several pressure values and an extrapolation is made to zero pressure. In the case of solid samples Equation (7.8) must be modified for the appreciable number of collisions that occur between photoelectrons and solid atoms. For a homogeneous solid of thickness x and mean escape depth d, Equation (7.8) becomes:

$$I = F\alpha Dkd(1 - e^{-x/d}) \tag{7.9}$$

where D is the density of the atom in the material under investigation. The thickness of the solid sample studied, x, depends upon the angle (θ in Fig. 7.16) at which the photoelectron leaves the surface, determined by the orientation of the sample with respect to the entrance slit of the spectrometer. It is clear from Fig. 7.16 that x represents a depth $x \sin \theta$ below the sample surface. By varying θ the actual depth below the sample surface studied may be varied. Thus when θ is very small only the first monolayer or so of the surface may be studied (e.g. if $d = 20$ Å and $\theta = 10°$, the effective depth studied is $d \sin 10° =$ approx 3.5 Å, about a monolayer). Angular variation studies are thus very useful for studying surfaces. In addition, if the surface studied is that of a single crystal, diffraction effects occur as θ is varied, a plot of intensity against θ giving rise to maxima corresponding to the different crystal planes. The angle between the photon beam and sample (θ') is fixed as θ is varied since α varies as θ' varies (see below).

(iii) *Photoelectron cross section* (α)

If a beam of monochromatic electromagnetic radiation of high energy (far u.v. or X-rays) is directed at a sample and the transmitted radiation measured by some detector, Beer's law is found to hold:

$$I' = I_0 \, e^{-\mu\rho x} \tag{7.10}$$

where I' is the intensity of the transmitted photon beam, μ is the mass absorption coefficient of the electromagnetic radiation, and ρ is the density of the sample material. The other terms in Equation (7.10) are defined above. The absorption of the electromagnetic radiation may be separated into two types: (a) photoelectric absorption, corresponding to the destruction of a photon and the excitation of a photoelectron; (b) apparent absorption due to scattering, the scattered photon being scattered out of the path of the detector. The mass absorption coefficient, μ, may thus be separated into two parts due to the two absorption processes described above:

$$\mu = \mathrm{a} + \mathrm{b} \tag{7.11}$$

where a is the mass photoelectric absorption coefficient, and b is the mass scattering absorption coefficient. Both a and b depend upon the atomic number of the sample atoms, Z, and the photon wavelength, λ, but they do so in different ways:

$$a = KZ^4\lambda^3 \tag{7.12}$$

$$b = K'Z^p\lambda^q$$

where K and K' are constants, p is a number between 1 and 2, and q is a number less than 1. In fact b may be ignored except for small values of Z and λ, and the mass absorption coefficient may be considered approximately equal to the mass photoelectron absorption coefficient (units of mass^{-1} length^{-2}), and thus proportional to the photoelectron cross-section (units of length^2):

$$\mu \approx a = \alpha(L/A) \tag{7.13}$$

where L is Avogadro's constant, and A is the atomic weight of the sample atoms. If Equation (7.13) is applied it can be seen that high photoelectric cross-sections may be expected for elements with high atomic number, and for electromagnetic radiation of long wavelength. Thus u.v. excitation will produce much more intense peaks than X-ray excitation, a valuable feature already discussed in Section 7.2.3 allowing simpler design for u.v. photoelectron spectrometers.

The photoelectron cross section, α, is approximately proportional to the square of the overlap between the orbital from which the photoelectron is ejected and the wavefunction for the outgoing free electron. For molecular orbitals the photoelectron cross-section may be expressed as the sum of the cross-sectional areas of the atomic orbitals which contribute to it. Several calculations of α have been made using this approach. It is found that α has an angular distribution variation; thus, for non-polarized incident electromagnetic radiation and a sample of randomly oriented molecules, the intensity is given by:

$$I \propto 1 + (\beta/2)(3/2 \sin^2\theta' - 1) \tag{7.14}$$

where θ' is the angle between the ejected photoelectron and the axis of the incident electromagnetic radiation (illustrated in Fig. 7.7 for a u.v. photoelectron spectrometer), and β is an asymetry parameter which may vary between -1 and $+2$ which depends upon the photoelectron energy and the nature of the orbital from which the photoelectron is ejected. Unfortunately, very little is known about the behaviour at the high kinetic energies in X-ray photoelectron studies, though a number of studies have been made in u.v. photoelectron studies. For u.v. photoelectron studies of molecular orbitals the values of β can help to assign the spectrum. Thus non-bonding orbitals give the highest β values, and bonding orbitals the lowest β values. Overlapping bands may sometimes be resolved by making the sometimes erroneous assumption that β is invariant over the vibrational structure.

(iv) *Satellites*

The presence of satellites (Sections 7.8.4, and 7.10.3, 7.10.4, and 7.10.5) complicates the situation by taking intensity away from the parent peak. Thus the intensity described above applies to the sum of the parent peak and satellite peaks.

7.7 VALENCE ENERGY LEVEL STUDIES

7.7.1 Vibrational structure in gases

In Section 7.3 it was seen how the u.v photoelectron spectrum of valence orbitals showed vibrational fine structure. The origin of this vibrational fine structure will now be discussed in more detail. Figure 7.18 shows the potential energy curve for some diatomic molecular ion with a vacancy in one of its orbitals together with the potential energy curve for the corresponding diatomic neutral molecule. Equation (7.7) shows that the lines drawn connecting some vibrational state in the ion with the ground vibrational state of the neutral molecule (assuming that the ground vibrational state is the most populated state in the neutral molecule) correspond to the binding energies and thus the peaks in the photoelectron spectrum. The Franck–Condon principle states that the internuclear distance should not alter during an electronic transition, thus the vertical line in Fig. 7.18 will correspond to the transition of greatest probability and

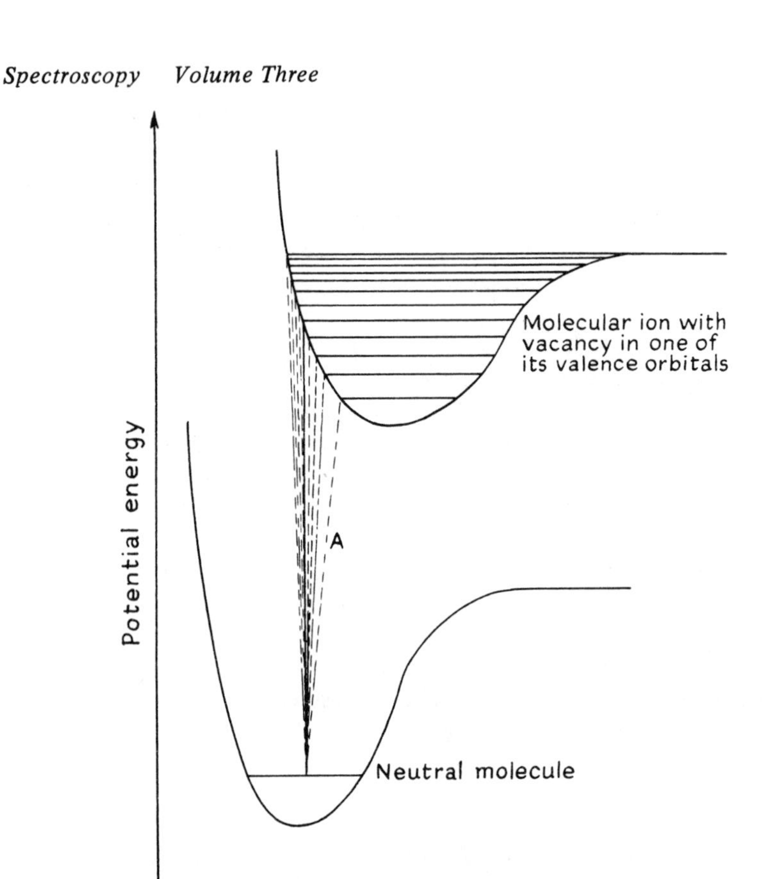

Fig. 7.18 Potential energy diagram showing the potential energy curves for the neutral molecule and the molecular ion formed in the photoelectron process. The lines drawn correspond to the peaks seen in the photoelectron spectrum. The vertical transition is shown by a solid line, and other transitions by broken lines.

thus greatest intensity, such a transition being known as a *vertical transition*, the binding energy in the photoelectron spectrum being called the *vertical ionization potential*. The intensity of the peaks corresponding to the other lines in Fig. 7.18 diminishes as the lines depart from the vertical. The transition from the ground vibrational state in the neutral molecule to the ground vibrational state of the molecular ion (shown as A on Fig. 7.18) is called the *adiabatic transition*, the binding energy in the photoelectron spectrum being called the *adiabatic ionization potential*.

The vibrational fine structure observed in a u.v. photoelectron spectrum can be valuable in giving information about the type of electronic energy level with

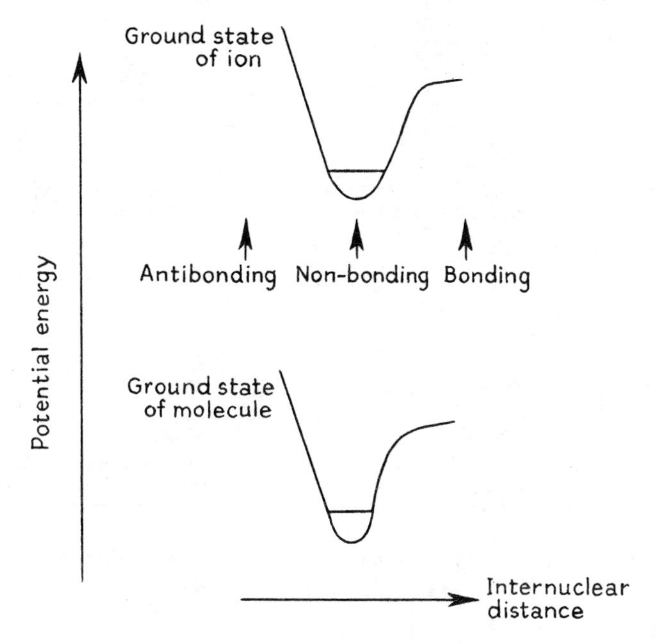

Fig. 7.19 Diagram showing how the potential energy curve for the molecular ion moves according to whether the photoelectron comes from an antibonding, non-bonding, or bonding orbital in the neutral molecule.

which it is associated. This point is illustrated by Fig. 7.19 which shows how the potential energy curve for the molecular ion moves according to whether the photoelectron comes from a bonding, non-bonding, or antibonding ortibal in the neutral molecule. The movement occurs because the removal of a bonding electron will weaken the bonding in the ion, the reverse being true of an anti-bonding electron. The removal of a non-bonding electron has no effect. The vibrational fine structure will reflect the different positions of the potential energy curves for the ion and the neutral molecule as shown in Fig. 7.20. Thus removal of a non-bonding electron will cause the vertical and adiabatic ioniz-ation potentials to become identical, giving a high intensity peak with little fine structure, since there is a much greater difference in the slope of lines of the type drawn in Fig. 7.18 in this case than in other cases because of the great change in slope of the potential energy curve near the bottom of the curve. Removal of a moderately strongly bonding or antibonding electron results in a vertical transition which intersects the potential energy curve for the ion at a point of only slightly changing slope, and appreciable vibrational fine structure results. As the electron removed becomes more strongly bonding or antibonding the potential energy curve for the ion moves more with respect to the neutral molecule. The vertical transition then intersects the potential energy curve for the ion at points corresponding to increasingly large vibrational excitation in the

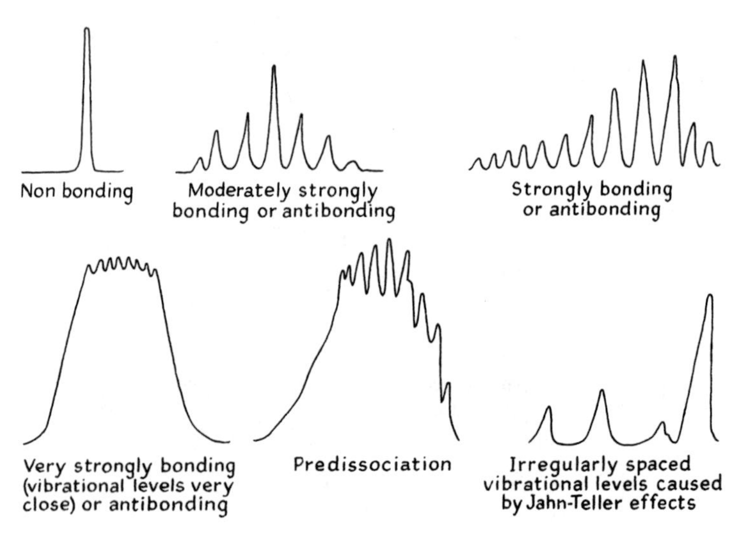

Fig. 7.20 Interpretation of vibrational fine structure in u.v. photoelectron spectroscopy in terms of the nature of the electronic energy level from which the photoelectron was ejected.

ion. Since the larger the vibrational excitation the closer packed are the vibrational energy levels and the smaller the change in slope of the lines drawn to connect these vibrational levels with the ground state vibrational energy level of the neutral molecule, then the spectrum changes as shown in Fig. 7.20.

The vibrational fine structure may display a distorted shape due to secondary features to be discussed in Section 7.8.

7.7.2 Valence bands in solids

The use of u.v. photoelectron spectroscopy for the study of valence bands in solids was mentioned in Section 7.2.3. In that section and in Section 7.4 it was made clear that such studies will be surface studies and thus it is important to ensure that the surface state reflects the bulk situation. The valence bands of metals and semiconductors have been studied by u.v. photoelectron spectroscopy for some years by physicists and metallurgists using u.h.v. condititions with the radiation entering the vacuum system through a LiF window, thus limiting photon energies to less than 11.6 eV (the cut-off point of the LiF window). Such studies have been called u.v. photoemission studies. Recently there have been a number of studies of the valence bands in solids using the full u.v. range with He(I) and sometimes He(II) excitation, and some of these studies have been carried out under u.h.v. conditions.

In solid materials the discrete molecular orbitals of the gaseous state may interact as a result of intermolecular interaction. As the intermolecular interaction increases, the discrete molecular orbitals become split into a number of closely spaced energy levels called *bands*. Those bands that are occupied by electrons are called the valence bands and those that are unoccupied are called the conduction bands. The Fermi level separates the valence and conduction band. The photoelectron spectrum of a solid material will therefore not show a single peak with associated fine structure due to vibrational excitation from excitation from a single molecular orbital, but will show a continuous peak due to the excitation of electrons from a continuous valence band the width of which is equal to the width of the valence band. The intensity of the photoelectron peak would be expected to reflect the density of states (the number of energy levels between E and $E + \mathrm{d}E$) of the valence band together with the photoelectron cross section (α) of the valence band. Equation (7.12) shows that the value of α in the u.v. photoelectron studies would be expected to be much larger than for X-ray photoelectron studies, but this apparent advantage from using u.v. excitation is offset by appreciable theoretical problems. The problem with u.v. studies is that one might expect a dependence of the spectrum upon the reciprocal lattice vector which would mean that the spectrum would not fully reflect the density of states, and although it is possible to consider a mechanism which gets over the dependence upon the reciprocal lattice vector, the situation raises doubts about correlating the u.v. photoelectron spectrum with the density of states. X-ray photoelectron studies use much higher photon energies giving photoelectrons of appreciable greater energy than valence bands, and thus little dependence upon reciprocal lattice vector, leading to a density of states that is constant. This gives X-ray photoelectron spectroscopy a considerable advantage over u.v. photoelectron spectroscopy for the study of valence bands. Further, since there is little fine structure to be resolved in such studies, the narrower photon linewidth of u.v. photoelectron spectroscopy offers no advantage. The greater surface sensitivity of u.v. photoelectron spectroscopy (Section 7.4) may make it more useful in surface studies.

There are now in increasing number of X-ray photoelectron spectroscopic studies of valence bands and most of these have given good agreement with calculations of densities of states. Figure 7.21 shows the X-ray photoelectron spectrum of the valence band of aluminium metal compared with the calculated density of states curve [7.11].

7.7.3 Ultraviolet photoelectron spectra and their interpretation

In Section 7.3 it was explained how the u.v. photoelectron spectrum of the valence orbitals could be approximated to the molecular orbital diagram, though the limitations of such an approximation were also pointed out. Very many u.v. photoelectron spectra have been interpreted by comparing the photoelectron spectrum with calculations of the molecular orbital energies using this approximation, and more than half the compounds studied have been organic molecules of varying complexity. It is beyond the scope of this chapter to discuss the

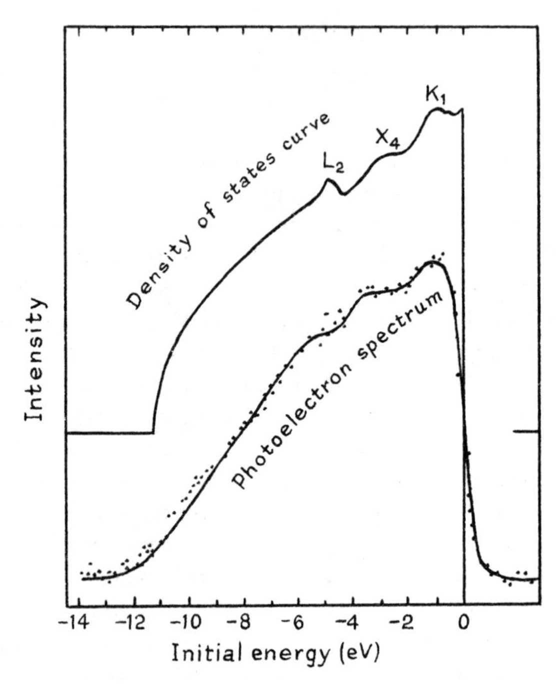

Fig. 7.21 X-ray photoelectron spectrum of the valence band of aluminium compared with the calculated density of states curve [7.11].

various molecular orbital calculations that have been made, but it should be pointed out that there are an increasing number of cases where calculations based on Koopman's theorem do not fit the observed spectrum.

Bearing in mind the care that must be taken in the relation of the photoelectron spectrum to the molecular orbital diagram, there are often features of the spectrum that give immediate information. Thus the vibrational fine structure gives information about the type of orbital from which the photoelectron was ejected (Section 7.7.1). The examination of the photoelectron spectrum of a series of related compounds may allow an empirical correlation to be made between some factor characterizing the chemical change and the change in binding energy of some band in the photoelectron spectrum. Thus the dihedral angle between the planes of the two rings in biphenyls may be correlated with the change in binding energy between two sets of photoelectron peaks assigned to different π-electron molecular orbitals (Fig. 7.22). There are many other examples [7.12, 7.13].

The ability of u.v. photoelectron spectroscopy to give high intensity photoelectron spectra of gaseous species has enabled transient species to be detected. The free radicals may be prepared in a flow system where some suitable gas is photolysed or pyrolysed just before entry into the spectrometer to yield a steady-state concentration of free radical at the point of photoionization.

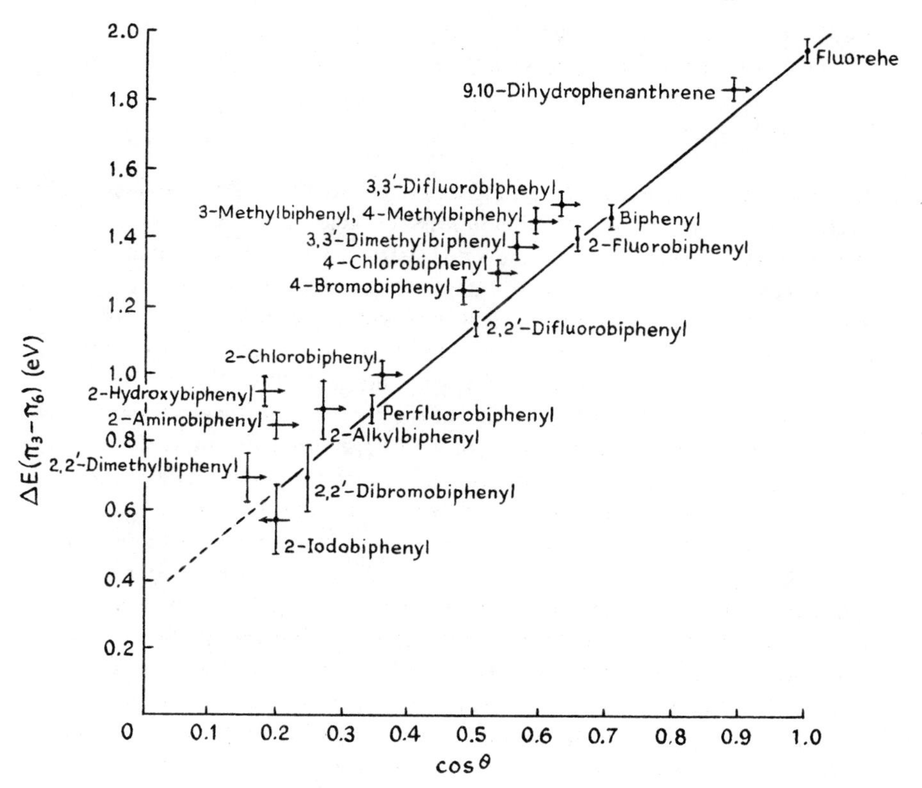

Fig. 7.22 Correlation diagram between the angle between the planes of the two rings in biphenyls (θ) and the difference in binding energy of two peaks assigned to different π-electron molecular orbitals [Maier, J.P., and Turner, D.W., *Faraday Discussion Chem. Soc.*, 54 (1973)].

Radicals such as CS, NF_2, SO_3F, N_2H_2, and CH_3 have been studied in this way [7.14, 7.15].

The decomposition of a gas sample while the photoelectron spectrum is being studied, or complications from the presence of impurities, may be monitored by *in situ* mass analysis of the ions formed in the photoelectron process. Such a mass analysis is achieved by carrying out an ion time-of-flight measurement by reversal of the electrostatic field with the provision of a pulsed ion accelerating field within the ionization region [7.16].

7.8 ADDITIONAL STRUCTURE IN ULTRAVIOLET PHOTOELECTRON SPECTRA

7.8.1 Autoionization

The process of autoionization was described in Section 7.1.3. The mechanism of electron ejection in the autoinization process is similar to that of the Auger process (Section 7.10.1) except that the initial state before electron ejection is an excited neutral atom or molecule and the final state a singly charged ion. Autoionization can occur only if the incident electromagnetic radiation is of just the correct energy to promote an electron from an orbital of higher energy to one of lower energy thus giving an excited atom or molecule. When autoionization occurs difficulties arise in interpretation of the photoelectron spectrum due to the detection of autoionization electrons by the spectrometer and, in the case of u.v. photoelectron studies, from interaction between the ground state of the molecule and the autoionization state. When the probability of ionization by the autoionization process becomes more likely than direct ionization, and the excited state formed in autoionization (S^* in Fig. 7.2) has a sufficiently long lifetime, appreciable interaction may occur between the ground-state molecule and the excited autoionization state. Since the excited autoionization state will have dimensions different from those of the ground state, the potential elergy diagram for the inital state in Fig. 7.18 will have to be modified, and thus the vibrational structure in the photoelectron spectrum will be different. The interaction also leads to an enhancement of the photoelectron cross-section (α) leading to an appreciable enhancement of intensity. The interaction is very sensitive to the photoelectron velocity, and it is only large when the velocity is small and there is appreciable time for the interaction to occur. Autoionization may therefore be reduced by increasing the photon energy, thus increasing the photoelectron velocity and thus reducing the interaction.

Figure 7.23 shows the u.v. photoelectron spectrum of oxygen gas using Ne(I) excitation (16.85 eV). Comparison with Fig. 7.13 shows how the vibrational fine structure in the Ne(I) spectrum is affected by interaction between the ground state and the excited autoionization state (about 17 eV above the ground state).

7.8.2 Predissociation

The phenomenon of predissociation was mentioned in Section 7.7.1. In u.v. photoelectron spectroscopy it is possible that the molecular ion produced in the photoelectron process predissociates. It is clear from Fig. 7.18 and Fig. 7.20 that, if predissociation of the molecular ion occurs, part of the vibrational fine structure of the photoelectron spectrum will be lost. Figure 7.20 illustrates the typical appearance of a photoelectron spectrum where predissociation has occurred.

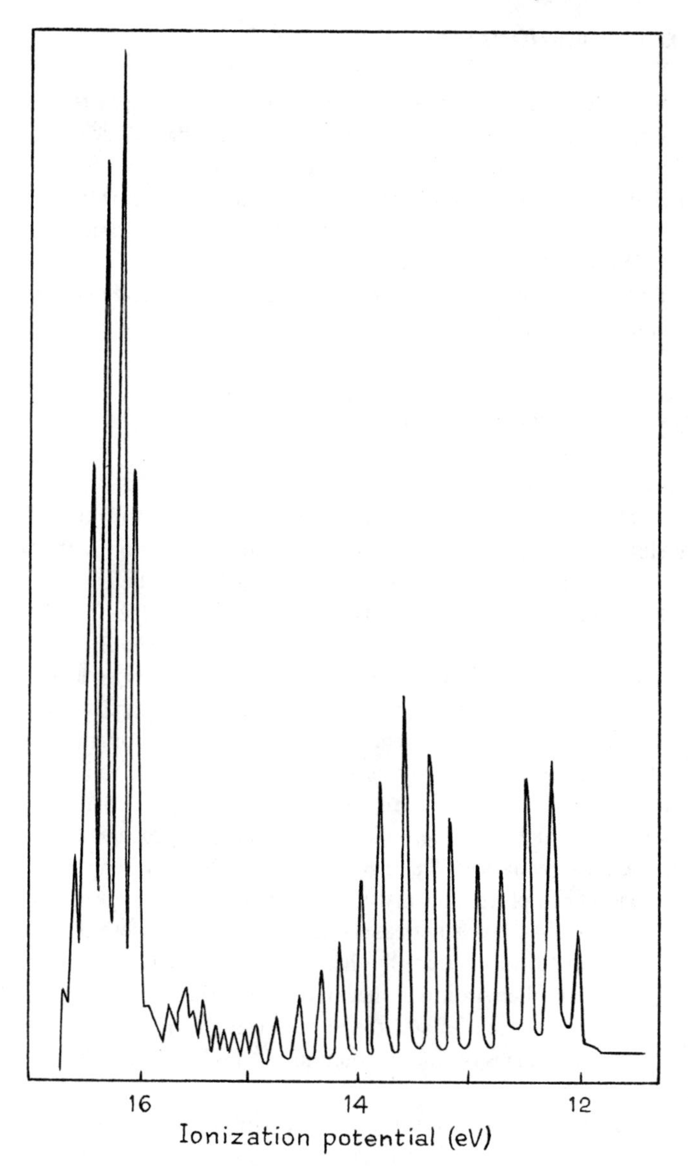

Fig. 7.23 Photoelectron spectrum of oxygen gas excited by Ne(I) radiation [Price, W.C., *Molecular Spectroscopy*, Institute of Petroleum (1968)].

7.8.3 Jahn–Teller effects

The u.v. photoelectron spectrum of non-linear molecules shows irregular vibrational structure on bands from molecular orbitals that are degenerate. This is the Jahn–Teller effect [7.17] which causes the molecular ion to have vibrational levels that are irregularly spaced. The vibrational envelope in the photoelectron spectrum will therefore be irregular, and in extreme circumstances may exhibit double or even triple maxima.

It should be noted that molecular orbitals that are degenerate will have an orbital quantum number $l > 0$, and will thus be liable to spin–orbit interaction (Section 7.3).

7.9 CORE ENERGY LEVEL STUDIES

7.9.1 Chemical shifts

In Section 7.3 it was seen that the binding energies of core electrons showed a chemical shift, and that this chemical shift showed a correlation with the charge on the nucleus of the atom concerned. It is found that the chemical shift observed has the following general features. (i) The chemical shift is the same irrespective of the particular core energy level studied. (ii) The size of the chemical shift depends upon the charge on the atom concerned. For an atom *in the same chemical environment* the shift increases with increasing positive charge on the atom ($+q$). This sensitivity of chemical shift to q varies roughly as the inverse of the valence-shell radius, thus increasing towards the top and right-hand side of the periodic table. (iii) The size of the chemical shift depends upon the chemical environment *for an atom of constant charge* q. Features (ii) and (iii) above need some explanation. There are a number of cases when the charge on the atom (q) has been plotted against the binding energy giving very good correlation (e.g. Fig. 7.24), but other cases (e.g. Fig. 7.25(a)) where the correlation is poor. The reason for this is that the chemical environment must be considered. Of course the chemical environment has partially been considered in the sense that the charge on the atom reflects the nature of the surrounding atoms (as in Fig. 7.15), but the changes on the surrounding atoms themselves have not been considered.

Certainly the best way to interpret chemical shifts is to use Equation (7.7) to calculate the binding energies. Many calculations have been carried out, some using the full Equation (7.7) and others just calculating the core orbital energy in the neutral molecule using Koopman's theorem. The reader is directed to a review by Shirley [7.19] for a critical discussion of the various calculations that can be made. The valuable feature about the interpretation of core energy level photoelectron spectra is that much valuable chemical information may be obtained without detailed calculations. For the purposes of this chapter it is useful to discuss a simple and widely applicable model that explains how the chemical shift depends upon the charge on the atom concerned and upon the

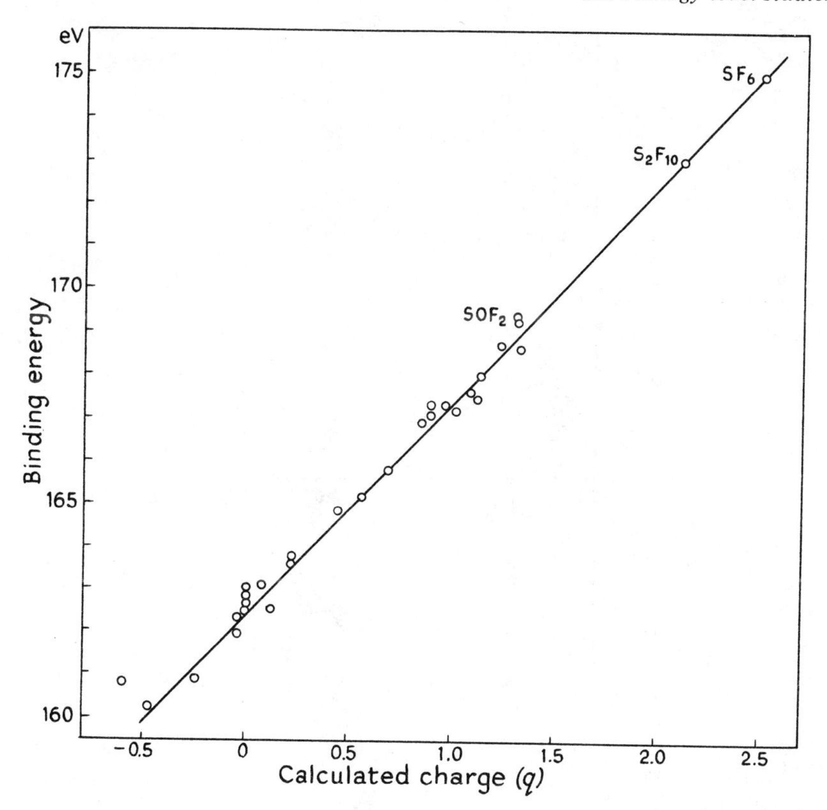

Fig. 7.24 Correlation between experimental S 2p chemical shifts for sulphur compounds and atomic charge [7.18].

charges on the surrounding atoms. This model is the potential model developed by Siegbahn and his coworkers [7.18]. In this model the atom from which the core electron is ejected may be considered as a charged conducting hollow sphere. The sphere has a charge q_i on it equal to the charge on the nucleus (the number of protons) less the charge from the electrons. When the atom is on its own q is zero, but when the atom is in a chemical bond the valence electrons may move to or from the atom, and thus in or out of the sphere giving rise to a charge $\mp q_i$ on the sphere. Using the classical analogy of a charged conducting hollow sphere of radius r_i, then the potential inside the sphere has a constant value:

$$\text{Potential energy} = q_i/r_i \tag{7.15}$$

This explains feature (i) above, since the change in potential energy that occurs in a chemical environment (thus giving rise to the charge q_i on the atom) is constant within the sphere and so identical for every core orbital. One may be tempted to consider the difference in binding energy between the atom of

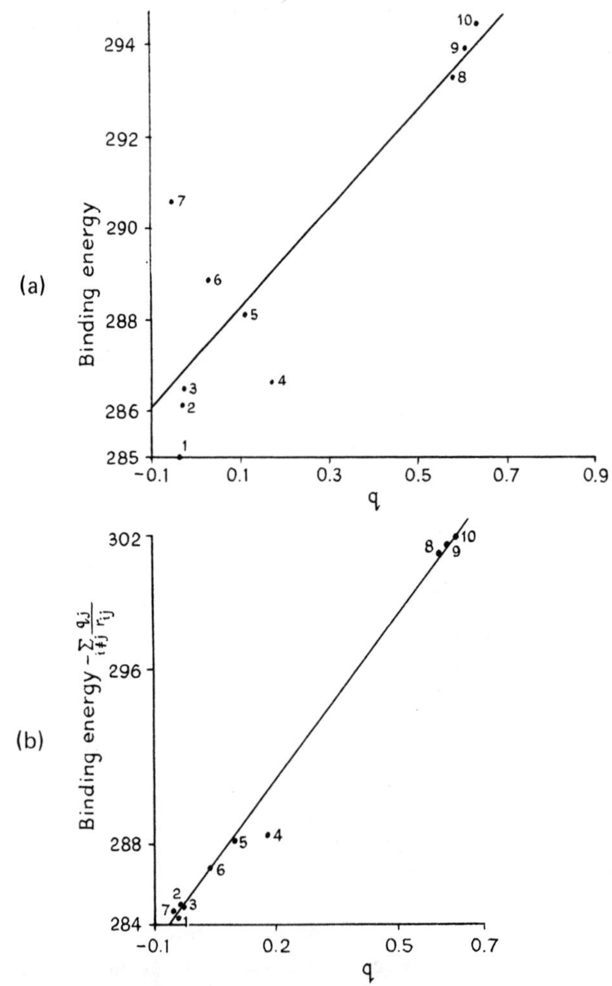

Fig. 7.25 Correlation between experimental C 1s chemical shifts for *tert*-butyl alcohol and three of its fluoro derivatives (particular carbon atoms numbered 1 to 10) and atomic charge (from CNDO M.O. calculations). In (a) binding energy is plotted against charge. In (b) binding energy-molecular potential is plotted against charge. The data is taken from Carver, J.C., Gray, R.C. and Hercules, D.M., *J. Amer. Chem. Soc.*, **96**, 6851, 1974 and the diagrams reproduced by permission of the authors.

charge q_i in a chemical compound and the neutral atom (i.e. the chemical shift) as being given by Equation (7.15). It is however necessary to consider what has happened to the rest of the molecule. Clearly the situation of a single sodium ion in a vacuum will be vary different from a sodium ion in a sodium chloride lattice. In the former case the electron removed from the sodium atom to give the ion is at infinity, but in the latter case the electron is on the chlorine atom (giving a chloride ion) on the adjacent chloride ion in the crystal lattice. If the surrounding atoms are reduced to point charges of charge q_i then the potential energy they will cause at the original atom will be:

$$\text{Potential energy} = q_j/r_{ij} \tag{7.16}$$

The overall effect will be the sum of (7.15) and (7.16), where (7.16) is summed for all the other atoms in the sample, thus:

$$\text{Potential energy} = \frac{q_i}{r_i} + \sum_{i \neq j} \frac{q_j}{r_{ij}} \tag{7.17}$$

Equation (7.17) will represent the difference in binding energy between an atom of charge 0 and an atom of charge q_i in some chemical environment containing other atoms of charge q_j. Equation (7.17) may be modified to compare the difference in binding energy measured experimentally and the binding energy of some reference compound (e.g. the calibrant; Section 7.5), ΔE, for a series of compounds containing atoms of charge q_i:

$$\Delta E_{\mathrm{A}} = k_{\mathrm{A}} q_i + \sum_{i \neq j} \frac{q_j}{r_{ij}} + K \tag{7.18}$$

where K is a constant determined by the choice of reference level, and k_{A} is a constant for the particular core level chosen. Using this model good agreement with experimental shifts and calculated charges may be obtained. The charges may be calculated by simple methods based on electronegativities or more complicated methods based on charges obtained from molecular calculations. The relationship between Equations (7.18) and (7.7) may be found explained elsewhere [7.20].

It is clear from Equation (7.18) that correlations between the binding energy and q_i will be successful only if the second term (sometimes called the potential term V) is proportional to the value of q_i. This was true in Fig. 7.24, but not Fig. 7.25(a). Inclusion of the second term in Fig. 7.25(a) (the circles) gives a good correlation (Fig. 7.25(b)).

The potential term V will include the charges on every other atom in the sample studied, but it is clear that as r_{ij} increases the effect falls off. In gases V will include all other atoms in the gas molecule. In covalent solids a summation over the molecule is generally sufficient because intermolecular distances are much greater than intramolecular distances (V is then known as a *molecular potential*). In ionic solids the summation has to be made appreciably far from the atom concerned, and V becomes the *Madelung potential*, or *self potential* (the Madelung potential at the particular site at which the atom concerned is situated). In order to preserve electroneutrality the signs of q_i and Σq_j are different, and thus the first two terms of Equation (7.18) tend to cancel one another. This effect is especially marked for the large V term found in ionic compounds. Table 7.1 illustrates this effect for lithium compounds. Here

Table 7.1 Chemical shifts in lithium salts

Compound	$(B_c^{qc} - B_c^0)$ eV (*measured*)	(ϕ/R) eV	$(\lambda_c + \phi/R)$ eV	q
Lithium acetate	5.9	−3.90	16.6	0.36
Lithium bromide	7.4	−9.162	11.44	0.65
Lithium carbonate	5.6	−6.58	13.92	0.40
Lithium chloride	6.7	−9.79	10.51	0.64
Lithium fluoride	7.6	−12.53	7.77	0.98
Lithium hydroxide	6.2	−10.68	9.82	0.63
Lithium hydride	5.6	−11.34	9.16	0.61
Lithium iodide	6.9	−8.32	12.68	0.54
Lithium nitrate	6.8	−9.15	11.35	0.59
Lithium oxalate	5.8	−13.7	6.8	0.85
Lithium oxide	1.7	−18.14	2.36	0.73
Lithium perchlorate	7.5	−2.07	18.43	0.47
Lithium peroxide	6.0	−13.1†	7.4	0.82
		−13.2†	7.3	
		−11.0†	9.5	0.63
		−11.0†	9.5	
Lithium sulphate	6.9	−9.53†	10.97	0.63
		−9.53†	10.97	
		−10.6†	9.90	0.70
		−10.6†	9.90	

† Refers to one lithium ion in the unit cell
[Reproduced by permission from Povey, A.F. and Sherwood, P.M.A., *J. Chem. Soc. Faraday Trans. II*, 70,1240 (1974)].

$(B_c^{qc} - B_c^0)$ represents the experimentally measured difference in binding energy between the metal (which one may assume has $q_i = 0$) and the lithium salt, ϕ/R is the term V, and λ_c is the difference in binding energy between the metal and a Li^+ ion at infinity (modified by correction terms). The term q_i corrects for covalent bonding giving Li^{q_i+} ions. It can be seen that for a completely covalent situation the series iodide, bromide, chloride, fluoride would have increasing binding energy [i.e. $(B_c^{qc} - B_c^0)$ would increase], whereas the reverse is true of a completely ionic situation where Table 7.1 shows that the V term predominates (column 3). The actual situation mixes ionic and covalent bonding and a mixed series results.

7.9.2 Correlations between chemical shifts of core energy levels and chemical shifts in n.m.r., n.q.r., and Mössbauer

It has been seen above that the chemical shift of core energy levels is directly related to the molecular charge distribution, and thus the chemical shift will be related to practically every parameter of chemical interest. It may thus seem tempting to correlate the chemical shift of core energy levels with the chemical shifts observed in n.m.r., n.q.r., and Mössbauer studies. Such correlations may be very successful as illustrated by Fig. 7.26, or very unsuccessful as illustrated by Fig. 7.27. In fact there is no general relation between the four different spectroscopic methods, though in a particular case a correlation may exist.

In n.m.r. spectroscopy (see Vol. 1, Chapter 2) the diamagnetic screening constant shows a chemical shift that is related to the chemical shift of core energy levels [7.21]. A correlation may be expected when the average screening constant is predominantly made up of this term. However, the average screening constant is given by the sum of the diamagnetic screening constant and the paramagnetic screening constant, and unfortunately the paramagnetic screening constant is the dominating term in most cases. However, the chemical shift of the core energy levels together with appropriate structural data for a molecule may be used to calculate the diamagnetic screening constant, and thus allow the proportions of diamagnetic and paramagnetic character of the average screening constant to be determined. The paramagnetic screening constant shows no relation to the chemical shift of the core energy level because the former depends solely on the electron structure of the excited states of the molecule, while the latter may be expressed (via Koopman's theorem) in terms of the charge distribution of the neutral molecule.

In n.q.r. spectroscopy transitions are observed between nuclear energy levels that are split by the interaction of their quadrupole moment with the electric field gradiant at the nucleus caused by the distribution of electrons about the atom concerned. The more asymmetrical the molecular charge distribution the greater will be the splitting between the levels. The asymmetry of the molecular charge distribution caused by changes in the valence shell electrons will thus give rise to a chemical shift. The p orbitals in the valence shell provide by far the largest electric field gradient at the nucleus of the atom concerned. Thus changes in the valence shell p orbital electron density give rise to appreciable chemical shifts in n.q.r. Such changes also give rise to changes in the overall molecular charge distribution and thus appreciable chemical shifts in the core energy levels. For such a situation good correlation between n.q.r. and energy level shifts results, an example of this case being shown in Fig. 7.26. Changes in the s, d or f valence shell orbital electron density on the other hand give rise to changes in the overall molecular charge distribution and thus appreciable chemical shifts in the core energy levels, but little change in the electric field gradient at the nucleus and thus small n.q.r. chemical shifts. For this situation poor correlation between n.q.r. and core energy level shifts would be expected.

In Mössbauer spectroscopy (see Vol. 1, Chapter 5) the chemical shifts (or isomer shifts) depend upon the s electron density at the nucleus. The chemical shifts of the core energy levels depend upon the *total* molecular charge distribution which is not always the same as the s electron density at the nucleus. In fact an increase in the s electron density at the nucleus may correspond to either

(a)

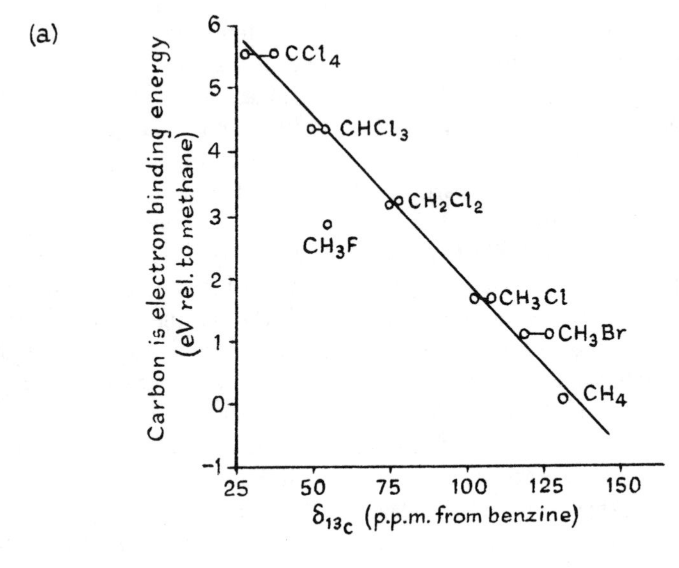

(b)

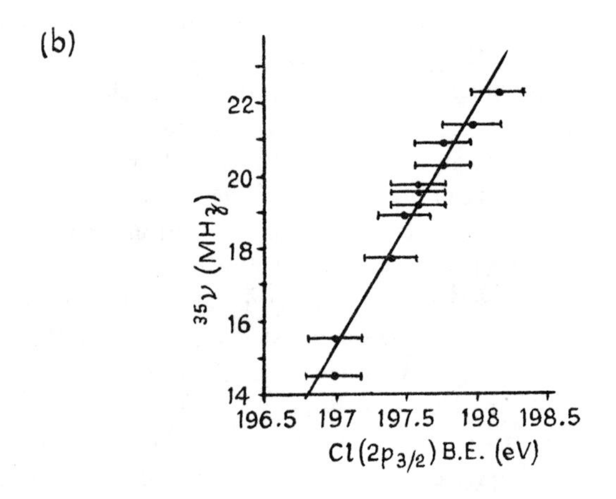

(c)

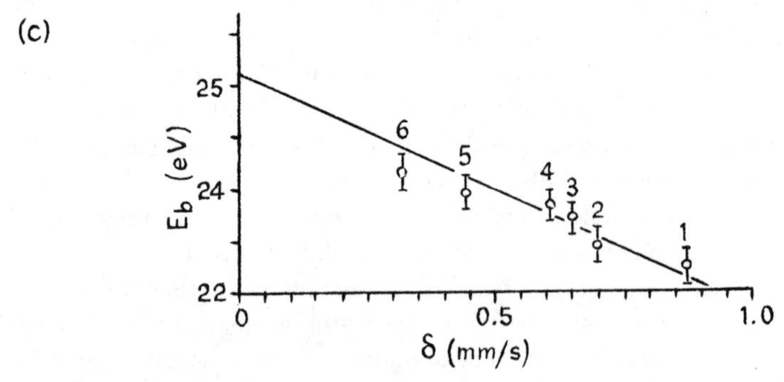

Fig. 7.26 Correlations between chemical shifts of core energy levels measured by X-ray photoelectron spectroscopy with chemical shifts in n.m.r. (A), n.q.r. (B), and Mössbauer (C). (A) shows a plot of C 1s binding energies against ^{13}C

an increase or a decrease in the total electron density. Thus an increase of the s electron density of the valence electrons leads to an increase in the s electron density at the nucleus and of the total electron density; thus Mössbauer and core electron chemical shifts move in the same direction. For such a situation good correlation between Mössbauer and core energy level shifts results, an example of such a case being shown in Fig. 7.26 [the sign of the Mössbauer chemical shift is negative in this case. On the other hand an increase of the p, d, or f electron density of the valence electrons leads to a *decrease* in the s electron density at the nucleus because the p, d, or f electrons shield the nucleus from s electrons, but the total electron density will *increase*. For such a case the Mössbauer and core chemical shifts move in different directions. For such a situation poor correlation would be expected between Mössbauer and core energy level shifts, an example being shown in Fig. 7.27. A comparison of the direction of Mössbauer and core energy level shifts may help in understanding the types of bonding present. Thus, for the case of a metal atom attached to a ligand, σ donation from ligand to metal would increase the metal s electron density of the valence electrons and give a good correlation between the two chemical shifts, while π acceptance by the ligand would decrease the d or p metal electron density and the two chemical shifts would go in opposite directions.

7.9.3 Applications of X-ray photoelectron spectroscopy

The main application of photoelectron spectroscopy has been for the study of structure and bonding by the examination of chemical shifts. There have been many studies of structure and bonding in many very different compounds, and for a full discussion the reader is directed to the series of review articles by Orchard et al. [7.13].

A few examples illustrate the immediate structural information that may be obtained from X-ray photoelectron spectroscopic studies of core energy levels. Thus the four different carbon atoms in ethyl trifluoroacetate are readily revealed in the gas-phase spectrum (Fig. 7.28). The spectrum is readily interpreted in terms of the positive charges on the atoms since these would be expected to fall off from the CF_3 end to the C–H end as the electronegativity of the substituents falls.

Valuable structural and bonding information may be obtained in the study of

n.m.r. chemical shifts [Block, R.E. *J. Mag. Res.*, **5**, 155 (1971)]. (B) shows a plot of ^{35}Cl n.q.r. frequencies against Cl $2p_{3/2}$ binding energies for a series of square-planar platinum chlorides [Clark, D.T., Briggs, D. and Adams, D.B. *J. Chem. Soc. Dalton,* 169 (1973)]. (c) shows a plot of Sn 4d binding energies against Mössbauer isomer shifts (δ) for a series of tin compounds [Barber, M., Swift, P., Cunningham, D. and Frazer, M.J. *Chem. Comm.*, 1338 (1970)].

(a)

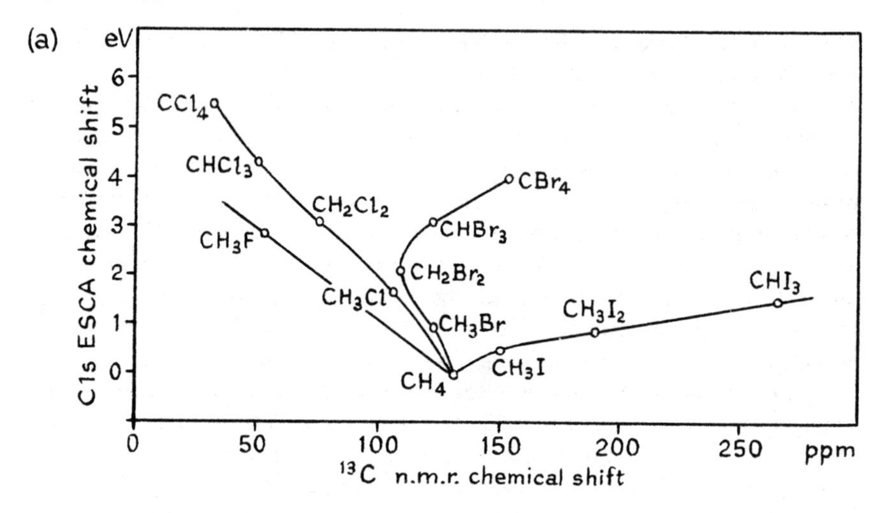

(b)

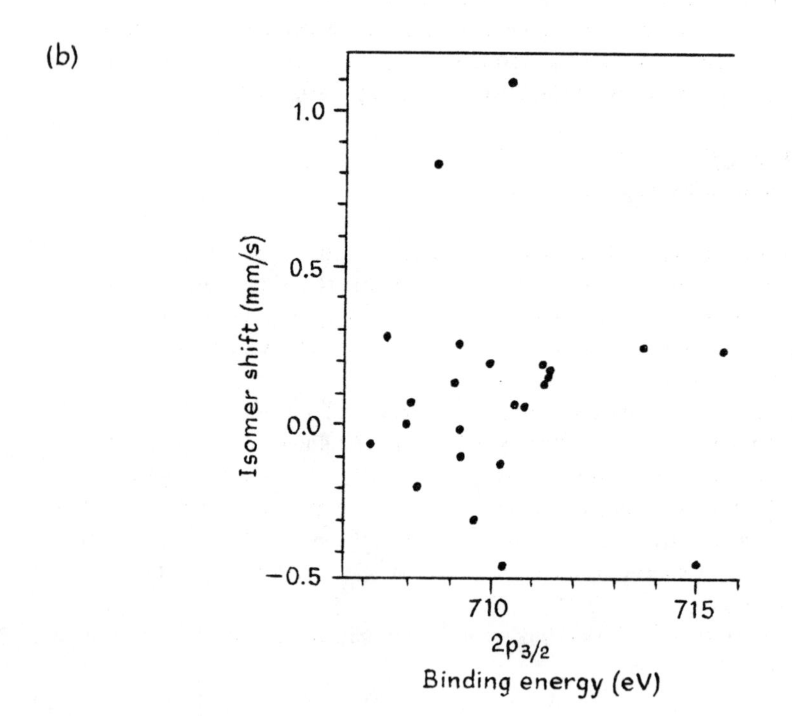

Fig. 7.27 Correlations between chemical shifts of core energy levels measured by x-ray photoelectron spectroscopy with chemical shifts in n.m.r. (A) and Mössbauer (B). (A) shows a plot of C 1s binding energies against ^{13}Cl n.m.r. chemical shifts for a number of halomethanes [Gelius, U., Johansson, G., Siegbahn, H., Allan, C.J., Allison, D.A., Allison, J. and Siegbahn, K. *J. Elect. Spec. Rel. Phen.*, **1**, 285 (1972/73)]. (B) shows a plot of Mössbauer isomer

transition-metal complexes. For example, the bonding in carbonyl compounds has been known to occur by a synergic process in which π-bonding occurs causing electron density to move from metal to carbon monoxide, and this movement is opposed by σ-bonding causing electron density to move from carbon monoxide to metal. X-ray photoelectron spectroscopy [7.22] shows that the net effect is to cause an increase of electron density on the carbon monoxide; the electron density on the carbon atom increases since the C 1s binding energy falls with respect to free carbon monoxide. In another example the question of the nature of the interaction of the ligand N_2 can be understood. Thus in complexes of the type $MX(N_2)L_2)$, where M is a metal (e.g. Re, Ir, Fe), X = Cl or H, and L is a ligand such as $Ph_2PCH_2CH_2PPh_2$, the N 1s core spectrum [7.23] shows two peaks of equal intensity suggesting that the nitrogen atoms in N_2 are inequivalent in line with a complex of the type $N \equiv N{-}M$. Calculations show both nitrogen atoms to be negatively charged, one more so than the other, consistent with a π and σ bonding situation as in the carbonyl case with the overall electron density moving in the same way.

The differentiation on non-equivalent atoms of the same element is clear in the cases above where widely separated photoelectron peaks are observed. In many cases however the separation is very small and the photoelectron peaks overlap. The ease with which overlapping peaks can be distinguished obviously depends upon the width of the excitation and the natural linewidth of the energy energy level from which the photoelectron is ejected (Sections 7.2.4 and 7.2.5). Figure 7.12 shows how the former can be improved by using monochromatized X-ray radiation. Overlapping peaks may be distinguished by a curve resolution technique whereby the spectrum may be split up into separate peaks by assuming that the photoelectron peak has a Gaussian–Lorentzian shape [7.7] and a peak width at half height obtained by comparison with photoelectron peaks for the same core level where only one type of atom is known to be present. Figure 7.29 illustrates a typical case. Spectrum (A) shows the N 1s region for a compound for which there is only one type of nitrogen atom. On the basis of the peak width at half height in this spectrum, spectra (B) and (C) may be resolved into two peaks consistent with the two types of nitrogen atom present in these two compounds. In spectrum (B) the best fit corresponds to two peaks of intensity ratio 1:1 corresponding to the three ring N atoms and the three NMe_2 nitrogen atoms. In spectrum (C) the best fit corresponds to two peaks of intensity ratio 1:2 corresponding to the three ring N atoms and the six NMe_2 nitrogen atoms.

Curve resolution procedures should be treated with some care for there is seldom a unique resolution possible since a number of combinations of numbers

shifts (δ) for a series of iron comppunds against Fe $2p^3_{3/2}$ binding energy [Johansson, L.Y., Larsson, R., Blomquist, J., Cederström, C., Grapengiesser, S., Helgeson, U., Moberg, L.C. and Sundbom, M., *Chem. Phys. Lett.*, **24**, 508 (1974)].

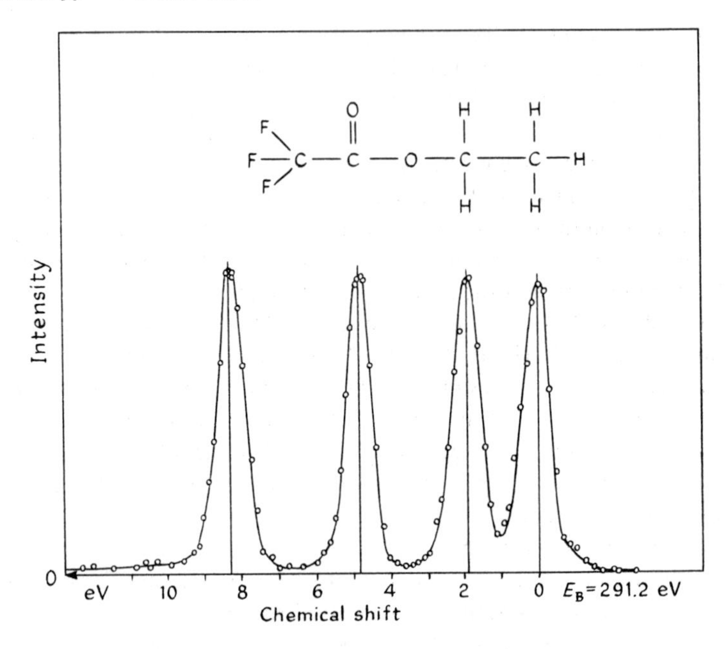

Fig. 7.28 Photoelectron spectrum of ethyl trifluoroacetate gas excited by monochromatized X-radiation showing the region corresponding to photoelectrons ejected from carbon 1s core orbitals [Gelius, U., Basilier, E., Svensson, S., Bergmark, T. and Siegbahn, K., *J. Elect. Spec. Rel. Phen.*, 2, 405 (1973)].

of peaks and intensity can often be chosen. The chosen resolution is usually shown to be consistent with a reasonable chemical structure. The assumption that the peak width at half height is constant for a particular core energy level is only approximately valid, and in Section 7.10.2 it will be explained how the width depends upon the chemical environment.

The relation of the area under the photoelectron peaks to the number of atoms present, together with the characteristic binding energy for a particular atom, allows a chemical analysis of the sample to be made. Quantitative studies may be made but, since the actual intensity depends upon a number of factors (Section 7.6), the analysis of the percentage composition of a particular compound in an unknown mixture by measuring the intensity of some core electron peak requires comparison of the intensity of the unknown sample with a calibration curve of percentage composition against intensity for known mixtures on the particular spectrometer used. Since escape depths differ for different compounds containing the same atom, calibration curves should be used for each type of compound studied [7.24]. Metal ions may be detected with high sensitivity by scavenging the ions from solution, by electrochemical deposition of the ions on to an electrode [7.25], or be chelating fibreglass surfaces [7.24].

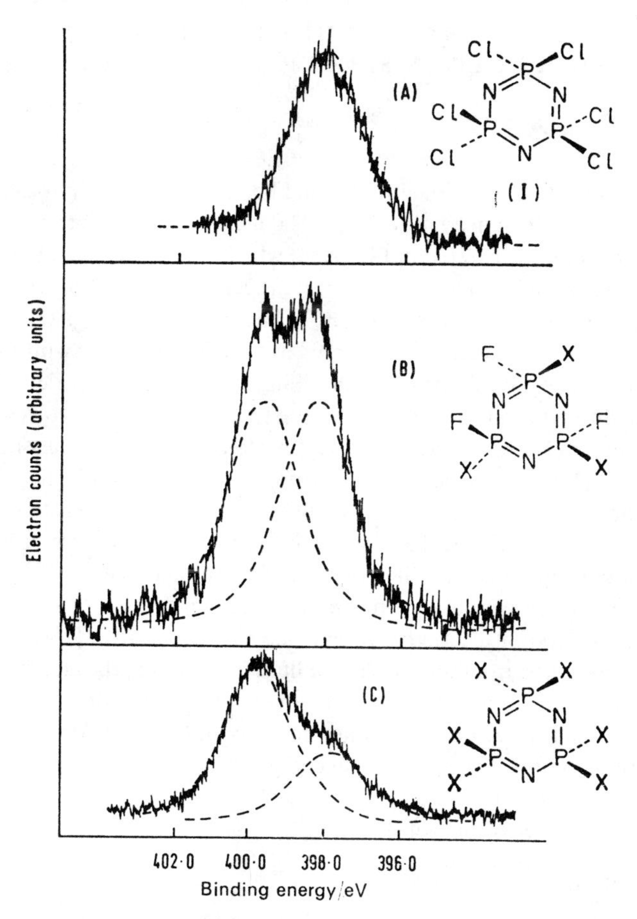

Fig. 7.29 Photoelectron spectrum of some solid dimethylamino-substituted cyclotriphosphazenes [X = $N(CH_3)_2$] excited by Al $K_{\alpha_1\alpha_2}$ showing the region corresponding to photoelectrons ejected from nitrogen 1s core orbitals. The spectra (B) and (C) are deconvoluted to show photoelectron peaks due to non-equivalent atoms [Green, B., Ridley, D.C. and Sherwood, P.M.A., *J. Chem. Soc. Dalton*, 1042 (1973)].

Analytical applications have included the analysis of smog particles [7.26], lunar soil [7.27], and the revelation that 'poly-water' is really a mixture of inorganic ions and organic material [7.28].

7.10 ADDITIONAL STRUCTURE IN X-RAY PHOTOELECTRON SPECTRA

7.10.1 Auger lines

The process by which Auger electrons could be produced in a photoelectron spectrum was described in Section 7.1.3. The mechanism of electron ejection in the Auger process is illustrated in Fig. 7.30, which shows the possible ways in which the ion formed in the photoelectron process (S^{*+}) may fill the core level vacancy to give S^+ (X-ray emission) or S^{++} (Auger emission) (Fig. 7.2). In Auger electron emission two electrons in energy levels higher than that with the core level vacancy are involved, one electron filling the vacancy and the energy gained (the difference in energy between the two energy levels involved) being given to another electron which is thereby ejected. In the X-ray emission process the core level vacancy is filled by an electron in an energy level higher than that with the core level vacancy, the energy gained being emitted as electromagnetic radiation. Figure 7.31 shows that the relative probabilities of these processes vary with atomic number, and the presence of Auger lines will clearly only be important for elements of low atomic number. A useful study of Auger lines in X-ray photoelectron spectra has been made by Wagner [7.29]. Auger lines may be distinguished from photoelectron lines by changing the energy of the electromagnetic radiation which will shift the photoelectron lines according to Equation (7.1) but leave the Auger lines unchanged, since the Auger lines depend only upon the difference in the energy of two energy levels.

7.10.2 Dependance of linewidth upon chemical environment

In Section 7.2.4 it was seen that the width depended upon the lifetime of the core energy level. This value depends upon how long the vacancy caused in the core level by the photoelectron process remains unfilled, i.e. the lifetime of the molecular ion S^{*+} (Fig. 7.2), which depends upon the filling of the core level vacancy to give S^+ or S^{++} via X-ray or Auger emission. The electrons that may be involved in filling the vacancy in the core level will be the intra-atomic electrons and the valence electrons from nearest neighbour ligands [7.30], and this suggests that the lifetime will show a dependence upon chemical environment. If the main process involves only the intra-atomic electrons, then one would expect the lifetime to decrease (and thus the width to increase) as the binding energy increases, since usually the greater the binding energy the larger the number of intra-atomic electrons and so the larger will be the number of electrons available to fill the vacancy in the core energy level. If the main process involves an appreciable number of valence electrons from nearest neighbour ligands, then one would expect the internuclear distances in the ion to be different from that of the neutral molecule, and vibrational fine structure would be

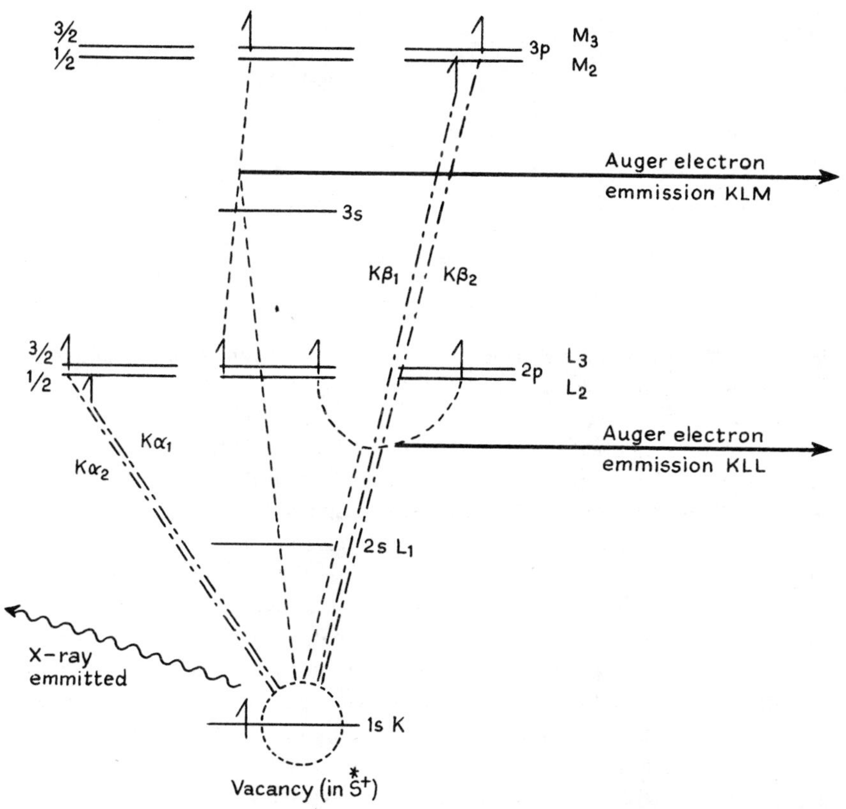

Fig. 7.30 Energy level diagram for the ion formed in the photoelectron process with a vacancy in its core 1s level. Some possible Auger and X-ray emission processes are illustrated which allow the ion to relax by filling the 1s vacancy.

expected as illustrated in Fig. 7.18. Vibrational structure will not normally be seen for core levels since the internuclear distance of the ion and neutral molecule are generally the same, the identical situation to the removal of a non-bonding valence electron explained in Section 7.7.1. As explained in Section 7.7.1, the greater the difference in internuclear distance in the ion and neutral molecule the greater will be the width of the vibrational envelope and thus the greater the width of the core energy level. Figure 7.32 shows a dramatic example in the gas-phase spectrum of trithiapentalene which has two different types of sulphur atom (S_1 and $S_{2,3}$) with different linewidths for each type of atom. The spectrum shows the S 2p region, with two lines due to the S $2p_{1/2}$ and S $2p_{3/2}$

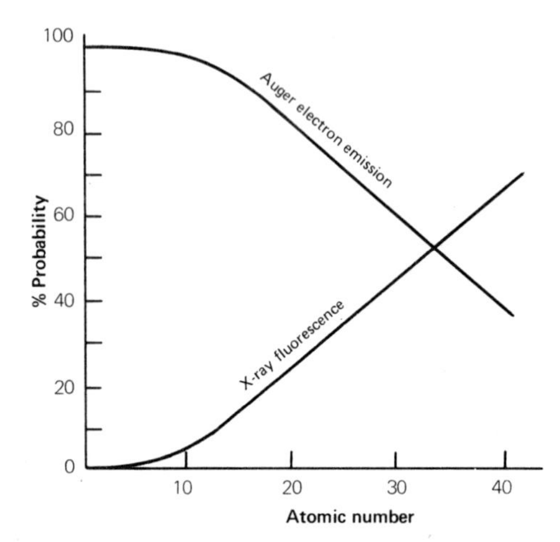

Fig. 7.31 Graph showing the probability of Auger electron emission and X-ray fluorescence as a function of atomic number.

levels since the S 2p level is split by spin–orbit coupling (Section 7.3), with intensity proportional to $(2j + 1)$, i.e. 2 and 4 respectively. The sulphur atoms $S_{2,3}$ give ions in the photoelectron process with appreciably different internuclear distance from that of the neutral molecule, and thus the S 2p peaks are of appreciable linewidth. The S_1 sulphur atom, on the other hand, as a result of its symmetrical position, gives an ion of similar internuclear distance to that of the neutral molecule and thus a narrow line. The difference in width between the two types of sulphur atom raises a further problem since binding energy calculations (Section 7.9.1) calculate the binding energy for the adiabatic process (Section 7.7.1), and the peak maximum for the $S_{2,3}$ peak corresponds to the vertical process (for the S_1 peak the vertical and adiabatic processes coincide) and so binding energy calculations should be compared with the difference for the S_1 and $S_{2,3}$ peaks for the adiabatic process (Fig. 7.32).

It is a general observation that the width of photoelectron peaks in solid materials is larger than that of gaseous samples. This solid-state effect is caused partially by differences in the surface charge S over the surface in the case of insulators (Section 7.5), and partly by the greater number of vibrational modes (phonons) possible in the solid state (the maximum number of vibrations in a solid sample is $3n$ where n is the number of atoms in the solid). This latter effect is important since, if there is any difference in the internuclear distances in the ion and neutral molecule, caused by relaxation involving valence electrons from nearest neighbour ligands, then vibrational structure will be seen that gives a photoelectron linewidth that increases as the energy range of the phonons increases [7.30]. Since the energy range of the phonons increases in going from the metal, where only acoustic phonons are possible, to the metal compounds, where both acoustic and optical phonons of rather different energy are possible, the photoelectron linewidth increases.

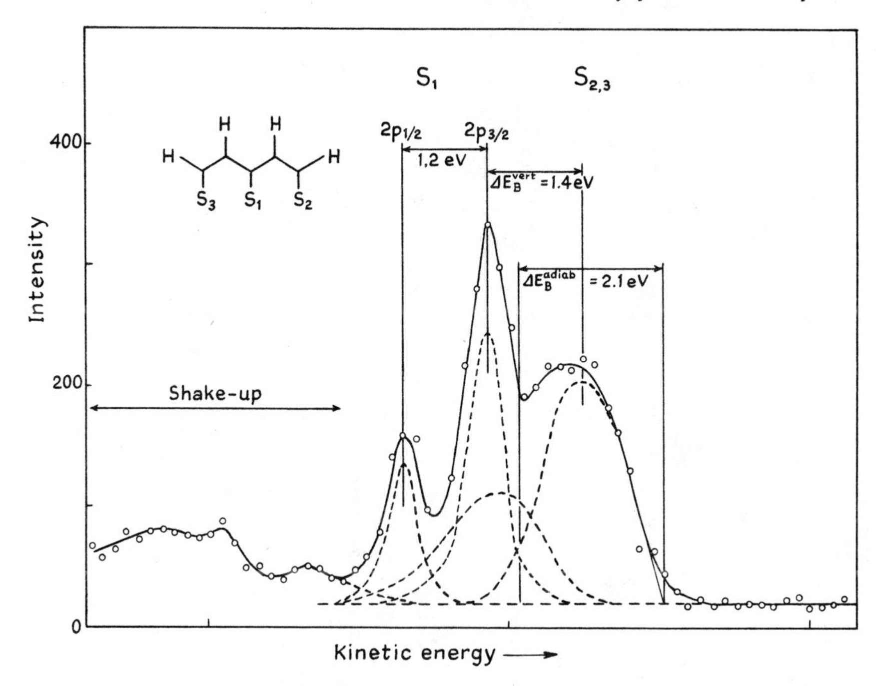

Fig. 7.32 Photoelectron spectrum of trithiapentalene gas excited by monochromatized X-radiation showing the region corresponding to photoelectrons ejected from sulphur 2p core orbitals [Gelius, U., Basilier, E., Svensson, S., Bergmark, T. and Siegbahn, K. *J. Elect. Spec. Rel. Phen.*, 2, 405 (1973)].

7.10.3 Shake-up and shake-off processes

The main peak in the photoelectron spectrum due to core energy level excitation corresponds to a process whereby the ion formed is in its ground state (S^{*+} in Fig. 7.2). Other states of the ion are possible where there is valence electron excitation (a *shake-up process*) or ionization (a *shake-off process*). In most cases only one valence electron is excited or ionized, though two valence electron excitations and ionizations have been observed. When shake-up and shake-off processes occur, the photoelectron loses energy to excite or ionize the valence electron and Equation (7.1) must be modified:

$$\begin{matrix}\text{K.E. (ejected electrons affected} \\ \text{by shake-up or shake-off)}\end{matrix} = h\nu - B - \Delta E \qquad (7.19)$$

where ΔE is the energy used in exciting or ionizing the valence electron. Electron shake-up and shake-off are very common processes and the photoelectron

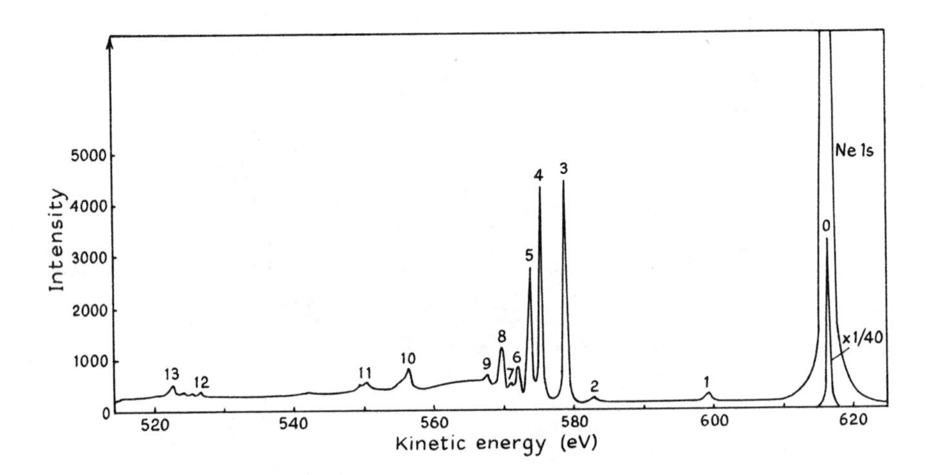

Fig. 7.33 Photoelectron spectrum of neon gas excited by monochromatized X-radiation showing the region corresponding to photoelectrons ejected from neon 1s core orbitals and the associated shake-up and shake-off structures [Gelius, U., Basilier, E., Svensson, S., Bergmark, T.,and Siegbahn, K. *J. Elect. Spec. Rel. Phen.*, **2**, 405 (1973)].

spectrum of every core energy level should have some shake-up and shake-off structure to the low kinetic energy side of the main peak. Figure 7.33 illustrates the photoelectron spectrum of the 1s core level in neon gas with associated shake-up and shake-off peaks. The two most intense peaks are for example due to shake-up processes where a 2p electron has been 'shaken-up' to a 3p orbital. Many other examples can be found discussed in the review by Orchard et al. [7.13]. Apart from being interesting in their own right, the presence of shake-up and shake-off features may provide valuable chemical information. Thus there is a definite correlation between the presence of shake-up satellites in transition-metal complexes and paramegnetism [7.31], and for example this may be applied to distinguish isomeric materials with different magnetic behaviour [7.32 [7.32].

The intensity of shake-up and shake-off processes is generally much less than that of the main peak but in some cases may have an intensity of $\frac{1}{4}$ to $\frac{1}{2}$ that of the main peak, in which case care must be taken not to confuse the peak with a main peak.

7.10.4 Discrete energy loss processes

Photoelectrons ejected from core energy levels may interact with other atoms and molecules (rather than with the valence electrons of their own atoms as seen above). Such inelastic processes lead to energy-loss electrons (e') which have been seen usually not to give a single peak in the photoelectron spectrum but contribute to the spectrum background (Section 7.4). However, if the interaction causes transitions of specific energy in the other atoms and/or molecules, then the photoelectrons will lose a discrete quantity of energy, and in this case they will give a single peak in the photoelectron spectrum.

The transitions of specific energy will obviously depend upon the possible transitions of the system which will depend upon the physical state of the sample. In a gas, for example, valence electron excitation in other atoms may occur. Figure 7.33 shows such a case (peak labelled 1) and as expected the intensity of the peak is pressure dependent and becomes more intense than the shake-up peaks as the pressure is raised. In a solid plasmon excitation of the conduction electrons of conductors and semiconductors may occur. The displacement of the electron gas representing the conduction electrons from the positive core gives rise to a restoring force and the resulting oscillation is quantized and known as a plasmon. Plasmon excitation thus gives rise to discrete energy loss peaks corresponding to both bulk and surface excitation. Figure 7.34 illustrates such a spectrum from aluminium.

7.10.5 Multiplet lines

Most studies of core energy level photoelectrons are made for closed shell system, i.e. systems which have all their occupied orbitals completely filled. When open shell systems are studied, i.e. systems with one or more unpaired electrons in the valence shell, ionization of a core electron will result in more than one final state. The simplest case is the case of an s core orbital (for which $l = 0$); the electron remaining in the s orbital when the photoelectron has been ejected has spin, $s = \mp \frac{1}{2}$ and can couple with the spin, s_V, of the unpaired electrons in the valence orbitals to give two final states corresponding to $(s_V \pm \frac{1}{2})$, which are separated by an energy ΔE given by:

$$\Delta E = (2s_V + 1)K \tag{7.20}$$

where K is the average exchange integral between the core s orbital and the valence orbitals with unpaired electrons. The intensity of the two photoelectron lines that will be seen corresponding to the two states separated by ΔE will be proportional to the multiplicity of the states $[2(s_V \pm \frac{1}{2}) + 1]$. Figure 7.12 shows such a splitting for the O 1s core orbital in gaseous oxygen for which $s_V = 1$,

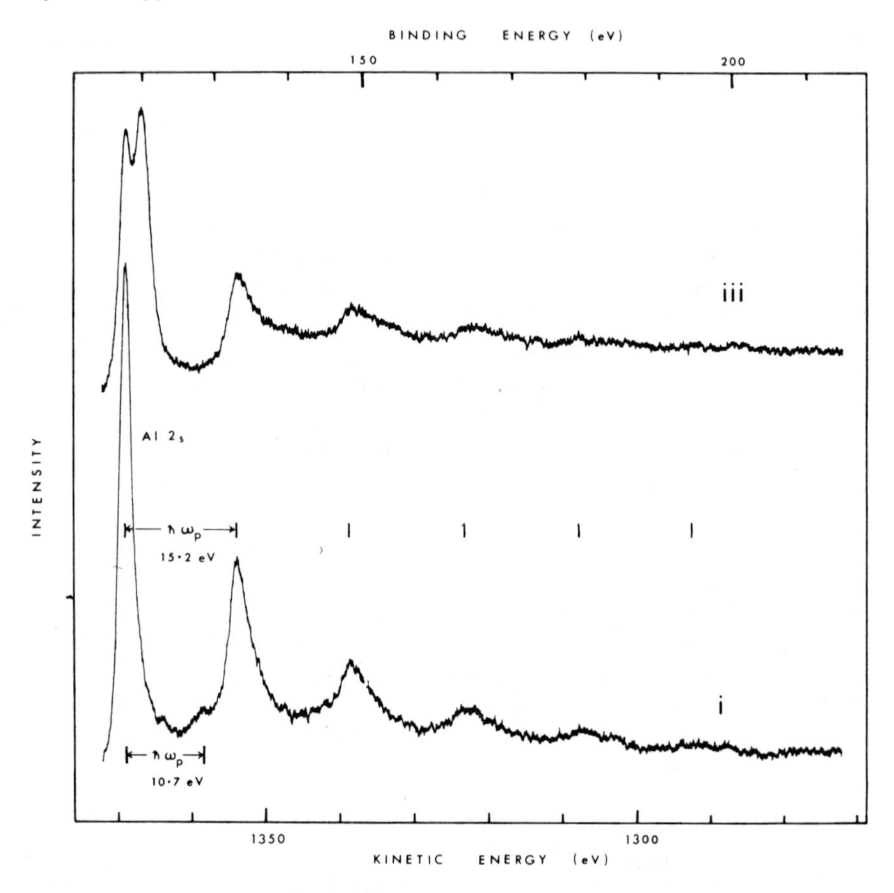

Fig. 7.34 Photoelectron spectrum of aluminium metal excited by Al $K_{\alpha_1\alpha_2}$ radiation, showing the region corresponding to photoelectrons ejected from aluminium 2s core orbitals with corresponding plasmon loss peaks (15.2 eV = bulk plasmon, 10.7 eV = surface plasmon) [Barrie, A., *Chem. Phys. Lett.*, **19**, 109 (1973)].

giving two lines separated by 1.1 eV with the expected intensity ratio 2 : 1. The multiplet splitting ΔE increases as K and s_v increase, and splittings as large as 6 eV may be seen for the core s orbitals of transition-metal compounds (where s_v may be very large). The amount of unpaired spin density on atoms in free radicals can be related to the multiplet splitting for the atom concerned [7.19, 7.33], giving results that are less sensitive but less ambiguous that e.s.r.

When core orbitals other than s orbitals are studied, $l > 0$, and the situation becomes more complex because orbit–orbit, spin–orbit (Section 7.3), and electrostatic (due to the angular dependence of Coulombic interactions between electrons bound in different orbitals) splitting give rise to a number of states which may interact with one another via configuration interaction [7.19].

Unresolved exchange interaction may appreciably broaden the photoelectron spectrum from core orbitals of open-shell transition metal compounds, making the differentiation of non-equivalent atoms of the same element impossible, and making curve resolution procedures unreliable.

7.10.6 Additional lines in the exciting radiation

When the radiation used is not obtained from an instrument fitted with a monochromator, the various X-ray lines generated by typical sources (Section 7.2.4) may give rise to spurious lines due to photoelectrons excited by secondary X-ray lines. Apart from the various X-ray lines emitted from a pure target (Figs. 7.8 and 7.9) the target material may contain impurities which also give rise to their own characteristic X-ray lines.

7.10.7 Sample decomposition

Care must always be taken to ensure that the sample studied has not decomposed. Decomposition may occur owing to the high vacuum conditions in the spectrometer, thus hydrates may lose water, halides may lose halogen, oxides may lose oxygen, and transition metal complexes may lose ligands. Reaction may occur with residual gases in the spectrometer. Further decomposition may occur in the X-ray beam (especially serious from the high intensity beams of non-monochromatized spectrometers). All types of decomposition may be reduced by cooling the sample, and many samples have to be studied at low temperature to prevent decomposition. For solid samples decomposition occurs on the sample surface, and because of the surface sensitivity of the technique decomposition of solids may especially serious.

REFERENCES

7.1 Turner, D.W. and Al-Joboury, M.I., *J. Chem. Phys.*, **37**, 3007 (1962).
7.2 Nordling, G., Sokolowski, E.,and Siegbahn, K., *Phys. Rev.* **105**, 1676 (1957).
7.3 Sokolowski, E., Nordling, C. and Siegbahn, K., *Arkiv Physik,* **12**, 301 (1957).
7.4 Nordling, C., Sokolowski, E. and Siegbahn, K., *Arkiv Physik,* **13**, 483 (1958).
7.5 Hagström, S., Nordling, C. and Siegbahn, K., *Phys. Lett.,* **9**, 235 (1964).
7.6 Kramer, L.N. and Klein, M.P., *J. Chem. Phys.,* **51**, 3620 (1969).
7.7 Siegbahn, K., Nordling, C., Hohansson, G., Hedman, J., Heden, P.F., Hamrin, K., Gelius, U., Bergmark, T., Werme, L.O., Manne, R. and Baer, Y., *ESCA Applied to Free Molecules,* North-Holland, Amsterdam (1969).
7.8 Siegbahn, H. and Siegbahn, K., *J. Electron Spec. Related Phenomena,* **2**, 319 (1973).

7.9 Khodeyev, Y.S., Siegbahn, H., Hamrin, K. and Siegbahn, K., *Chem. Phys. Lett.,* **19**, 16 (1973).
7.10 Siegbahn, K., Hammond, D., Fellner-Feldegg, H. and Barnett, E.F., *Science,* **176**, 245 (1972).
7.11 Baer, Y. and Busch, G., *Phys. Rev. Lett.,* **30**, 280 (1973).
7.12 Baker, A.D. and Betteridge, D., *Photoelectron Spectroscopy, Chemical and Analytical Aspects,* Pergamon Press, Oxford (1972).
7.13 Orchard, A.F., et al., *Chemical Society Specialist Periodical Reports, Electronic Structure and Magnetism of Inorganic Compounds,* Vols. 1, 2, and 3 (1972, 1973, 1974).
7.14 Jonathan, N., Morris, A., Okuda, M., Ross, K.J. and Smith, D.J., *Faraday Discussions Chem. Soc.,* **54**, 48 (1973).
7.15 Cornford, A.B., Frost, D.C., Herrong, F.G. and McDowell, C.A., *Faraday Discussions Chem. Soc.,* **54**, 56 (1973).
7.16 Ames, D.L., Maher, J.P., Watt, F. and Turner, D.W., *Faraday Discussions Chem. Soc.,* **54**, 277 (1973).
7.17 Murrell, J.N., Kettle, S.F.A. and Tedder, J.M., *Valence Theory,* Wiley, London (1972).
7.18 Siegbahn, K., Nordling, C., Fahlman, A., Nordberg, R., Hamrin, K., Hedman, J., Johansson, G., Bergmark, T., Karlsson, S-E., Lindgren, I. and Lindberg, B., *Nova Acta Regiae Societatis Scientiarum Upsaliensis,* Ser IV, **20** (1967).
7.19 Shirley, D.A., *Advances Chem. Phys.,* **23**, 85 (1973).
7.20 Gelius, U., *Physica Scripta,* **9**, 133 (1974).
7.21 Basch, H., *Chem. Phys. Lett.,* **5**, 337 (1970).
7.22 Clark, D.T. and Adams, D.B., *Chem. Comm.,* 740 (1971).
7.23 Leigh, G.J., Murrell, J.N., Bremser, W. and Proctor, W.G., *Chem. Comm.,* 1661 (1970); Folkesson, B., *Acta Chem. Scand.,* **27**, 287,1441 (1973).
7.24 Hercules, D.M., *J. Electron Spec. Rel. Phen.,* **4**, 219 (1974).
7.25 Brinen, J.S., *J. Electron Spec. Rel. Phen.,* **4**, 377 (1974).
7.26 Araktingi, Y.E., Bhacca, N.S., Proctor, W.G. and Robinson, J.W., *Spec. Lett.,* **4**, 365 (1971); Novakov, T., Mueller, P.K., Alcocer, A.E. and Oteras, J.W., *J. Colloid Interface Sci.,* **39**, 225 (1972).
7.27 Bremser, W., *Chem.-Ztg.,* **95**, 819 (1971).
7.28 Davis, R.E., Rousseau, D.L. and Board, R.D., *Science,* **171**, 167 (1971).
7.29 Wagner, C.D., *Anal. Chem.,* **44**, 967 (1972).
7.30 Citrin, P.H., *J. Electron Spec. Rel. Phen.,* **5**, 273 (1974).
7.31 Yin, L., Adler, I., Tsang, T., Matienzo, L.J. and Grim, S.O., *Chem. Phys. Lett.,* **24**, 81 (1974).
7.32 Matienzo, L.J., Yin, L.I., Grim, S.O. and Swartz, W.E., Jr., *Inorg. Chem.,* **12**, 2762 (1973).
7.33 Davis, D.W., Martin, R.L., Banna, M.S. and Shirley, D.A., *J. Chem. Phys.,* **59**, 4235 (1973).

A Appendix

The object of this chapter is to indicate how vibrational and rotational wavefunctions and energy levels and the various spectroscopic selection rules may be derived in terms more mathematical than in the main body of the text. However, this chapter still only indicates the route along which this information emerges; for a fuller treatment the reader is referred to references [1–6] at the end of the chapter.

A.1 ABSORPTION AND EMISSION OF RADIATION

By using time-dependent perturbation theory, the rate of absorption of radiation via an electric dipole process from a lower state l to an upper state u can be readily calculated. For a molecular system bathed in radiation of energy density ρ of frequency ν_{lu} corresponding to the transition $l \rightarrow u$, this rate is given by $n_l B_{ul} \rho$. B is called *Einstein's coefficient of induced absorption* and is equal to:

$$B_{ul} = \frac{8\pi^2}{3h^2} \left| \int \psi_l^* \mu \psi_u \, \mathrm{d}\tau \right|^2 = \frac{8\pi^2}{3h^2} R_{ul}^2 \tag{A.1}$$

where the integral R_{ul} is called the *transition moment*. n_l is the fraction of molecules in the lower state l and conventionally the first subscript of B_{ul} corresponds to the final state and the second to the initial state. μ is the dipole moment. If n is the number of molecules per litre of sample irradiated, then the change in intensity of the light beam after it has passed through a length δx is given by:

$$-\delta I = \frac{8\pi^2}{3h^2} R_{ul}^2 \rho h \nu_{lu} n \delta x \tag{A.2}$$

The energy density is simply related to the intensity of the radiation by the velocity of light, $I = c\rho$, and thus:

$$-\delta I = \frac{8\pi^2}{3h^2} R_{ul}^2 \frac{I}{c} h\nu_{lu} n \delta x \qquad \text{(A.3)}$$

If C is the concentration in moles dm^{-3}, then:

$$n = CN \qquad \text{(A.4)}$$

where N is Avogadro's number, and:

$$\delta I = -\frac{8\pi^3}{3h^2} R_{ul}^2 \frac{I}{c} h\nu_{lu} NC\delta x \qquad \text{(A.5)}$$

This represents the amount of light absorption over the entire absorption band due to the transition $l \rightarrow u$. Beer's law $I(x)/I(0) = \exp[-\epsilon(\nu)\cdot C\cdot x]$ may be written in differential form as:

$$\delta I = -\epsilon(\nu) IC\delta x \qquad \text{(A.6)}$$

and the terms compared between Equations (8.5) and (8.6). The integrated absorption intensity $A = \int \epsilon(\nu)\,d\nu$ is simply obtained as:

$$A = \frac{8\pi^3 N}{3hc} R_{ul}^2 \nu_{lu} \qquad \text{(A.7)}$$

For electronic transitions, R_{ul} represents the transition dipole moment between the two electronic states; for rotational and rotational-vibrational transitions the dipole moment needs to be expanded; for the simple case of a diatomic system as:

$$\mu = \mu_0 + \left(\frac{\partial \mu}{\partial x}\right)_{x=0} x + \frac{1}{2}\left(\frac{\partial^2 \mu}{\partial x^2}\right)_{x=0} x^2 + \ldots \qquad \text{(A.8)}$$

where x is the displacement of the system from equilibrium $(r - r_e)$. (At the equilibrium position $\mu = \mu_0$, the permanent dipole moment.) The application of these equations to electronic and vibration transitions will be considered below.

Decay from an excited state to a state of lower energy may occur by two radiative processes, *spontaneous* and *stimulated* emission. Stimulated emission is entirely analogous to the induced emission process introduced above, in that the rate of emission of quanta is proportional to the radiation density of the transition frequency ν_{lu} and to the fraction of molecules in the excited state. The rate of emission by this route is thus equal to $n_u B_{lu} \rho(\nu_{lm})$ where B_{lu} is *Einstein's coefficient of stimulated emission.* Spontaneous emission occurs via a pathway which is independent of the power density of the radiation and occurs at a rate $n_u A_{lu}$ where A_{lu} is *Einstein's coefficient of spontaneous emission*. This is equal to:

$$A_{lu} = \frac{64\pi^4 \nu_{ul}^3}{3h} |\int \psi_u{}^* \mu \psi_l \, d\tau|^2 \quad (A.9)$$

At any time for a system in equilibrium the number of molecules in the states l and u must be constant, i.e. the rate at which molecules enter the state u must equal the number leaving it. Thus:

$$n_u A_{lu} + n_u B_{lu} \rho(\nu_{lu}) = n_l B_{ul} \rho(\nu_{lu}) \quad (A.10)$$

For non-degenerate states l and u the Boltzmann distribution gives a relationship between n_l and n_u:

$$n_u = n_l \exp\left(-\frac{h\nu_{lu}}{kT}\right)$$

and for a system in thermal equilibrium with the radiation, it can be shown that for non-degenerate levels:

$$B_{lu} = B_{ul} \quad (A.11)$$

and

$$\frac{A_{lu}}{B_{lu}} = \frac{8\pi h}{c^3} \nu_{lu}{}^3 \quad (A.12)$$

from which it can be seen that the probability of spontaneous emission depends upon ν^3. This spontaneous emission is likely for electronically excited states, but for vibrational transitions, or rotational transitions, where ν^3 becomes very small, spontaneous emission processes have very low probability.

The relative rate of decay from u to l is given by:

$$\frac{A_{lu}}{B_{lu}\rho(\nu_{lu})} = \exp\left(\frac{h\nu_{lu}}{kT}\right) - 1 \quad (A.13)$$

a relationship of great importance in understanding the operation of a laser or maser. For a system in thermal equilibrium with its room temperature surroundings, the two processes (spontaneous and stimulated) proceed at comparable rates for frequencies (ν_{lu}) in the microwave region. Thus the presence of a large radiation density at long wavelengths is often sufficient to produce a large proportion of stimulated compared with spontaneous emission, and maser (microwave) action may be produced. For systems where the transition frequency ν_{lm} lies to higher energy, the stimulated route is of negligible importance and a *population inversion* ($n_u > n_l$) has to be artificially forced upon the system in order to promote laser action.

It can be seen from Equation (A.10) that the application of a high radiation density ρ can only lead to the populations of upper and lower levels n_u and n_l becoming equal in the two-level situation, in which case the transition is saturated. It is only possible therefore to observe net stimulated emission from such a system if a pumping mechanism other than optical excitation is used to create a population inversion with respect to the lower state (i.e. $n_u > n_l$), or if a third pumping level is utilized so that the optical pumping frequency and stimulated emission frequencies are different.

A.2 ENERGY LEVELS OF A LINEAR RIGID ROTOR

To evaluate the energy levels of a rigid linear rotor, the Schrödinger equation needs to be solved for a freely rotating molecule in three dimensions.

Provided that the rotor is rigid, no potential energy is associated with rotation since the motion is that of the molecule as a whole. Hence the potential energy V is zero, and if E_r is the rotational energy of the rotor, then the Schrödinger equation becomes:

$$\frac{\partial^2\psi}{\partial x^2} + \frac{\partial^2\psi}{\partial y^2} + \frac{\partial^2\psi}{\partial z^2} + \frac{8\pi^2\mu}{h^2} E_r\psi = 0 \tag{A.14}$$

where μ is the reduced mass of the rotor.

Transforming into spherical polar coordinates:

$$\frac{1}{r^2}\frac{\partial}{\partial r}\left(r^2 \frac{\partial\psi}{\partial r}\right) + \frac{1}{r^2 \sin^2\theta}\frac{\partial^2\psi}{\partial\phi^2} + \frac{1}{r^2 \sin\theta}\frac{\partial}{\partial\theta}\left(\sin\theta\,\frac{\partial\psi}{\partial\theta}\right) + \frac{8\pi^2\mu E_r\psi}{h^2} = 0 \tag{A.15}$$

or

$$\frac{\partial^2\psi}{\partial r^2} + \frac{2}{r}\frac{\partial\psi}{\partial r} + \frac{1}{r^2 \sin^2\theta}\frac{\partial^2\psi}{\partial\phi^2} + \frac{1}{r^2 \sin\theta}\frac{\partial}{\partial\theta}\left(\sin\theta\,\frac{\partial\psi}{\partial\theta}\right) + \frac{8\pi^2\mu E_r}{h^2}\psi = 0 \tag{A.16}$$

The wavefunction may be written as a product of a radial and an angular function:

$$\psi = R(r)\cdot S(\theta, \phi) \tag{A.17}$$

where $R(r)$ is a function of the r alone and $S(\theta, \phi)$ of only θ and ϕ. On differentiation of Equation (A.17) and on substitution for:

$$\frac{\partial\psi}{\partial r}, \frac{\partial^2\psi}{\partial r^2}, \frac{\partial^2\psi}{\partial\phi^2}, \quad \text{and} \quad \frac{\partial\psi}{\partial\theta}$$

in Equation (A.16) we obtain, by multiplication throughout by r^2/RS:

$$\frac{r^2}{R}\frac{\partial^2 R}{\partial r^2} + \frac{2r}{R}\frac{\partial R}{\partial r} + \frac{1}{S\sin^2\theta}\frac{\partial^2 S}{\partial\phi^2} + \frac{1}{S\sin\theta}\frac{\partial}{\partial\theta}\left(\sin\theta\,\frac{\partial S}{\partial\theta}\right) + \frac{8\pi^2\mu r^2 E_r}{h^2} = 0 \tag{A.18}$$

where R is an abbreviation for $R(r)$ and S for $S(\theta, \phi)$.

Since for a rigid rotor r is constant, the radial part of the wavefunction must be constant, and the first two terms in Equation (A.16) must vanish. Multiplying through Equation (A.18) by S and making the substitution moment of inertia $= I = \mu r^2$:

$$\frac{1}{\sin^2\theta}\frac{\partial^2 S}{\partial\phi^2}+\frac{1}{\sin\theta}\frac{\partial}{\partial\theta}\left(\sin\theta\frac{\partial S}{\partial\theta}\right)+\frac{8\pi^2 IE_r S}{h^2} = 0 \qquad (A.19)$$

From mathematical considerations it can be shown that only for certain values of E_r can solutions be obtained which are single-valued and finite. These values of E_r are given by:

$$E_r = J(J+1)h^2/8\pi^2 I \qquad (A.20)$$

where J is the rotational quantum number which may take only integral values from zero upwards. This is the equation which is employed in the analysis of the rotational structure of a linear molecule, and it is only strictly accurate when the molecule may be regarded as a rigid rotor. In practice the equation holds only for low J values where the molecule behaves nearly as a rigid rotor. In general for linear molecules the rotational energy is described more accurately by:

$$E_r = hc[BJ(J+1) - DJ^2(J+1)^2] \qquad (A.21)$$

where D is the centrifugal distortion constant. It is important especially for high J values where, as the molecule rotates rapidly, centrifugal forces stretch the bond length in excess of the equilibrium value. The value of D thus depends on the stiffness of the bond (i.e. the force constant, f), and a good approximation is:

$$D = \frac{16\pi^2 B^3 \mu}{f} \qquad (A.22)$$

Classically, the rotational energy E_r of a rigid body is given by:

$$E_r = \tfrac{1}{2}I\omega^2 \qquad (A.23)$$

where ω is the angular velocity of rotation and I is the moment of inertia about the axis of rotation. The angular momentum P is:

$$P = I\omega \qquad (A.24)$$

Substituting Equation (A.24) into Equation (A.23), E_r is given by:

$$E_r = P^2/2I \qquad (A.25)$$

On comparison of Equations (A.25) and (A.20) it follows that for a linear rigid rotor:

$$P^2/2I = J(J+1)h^2/8\pi^2 I \qquad (A.26)$$

$$P^2 = J(J+1)h^2/4\pi^2 \qquad (A.27)$$

or

$$P = \sqrt{[J(J+1)]}\,h/2\pi \qquad (A.28)$$

Hence, the magnitude of the total angular momentum vector is given by:

$$\sqrt{[J(J+1)]}\,h/2\pi$$

The rotational wavefunctions of importance here are identical to the familiar angular part of the hydrogen atom wavefunctions. Each solution is represented by two quantum numbers J and M_J which take on integral numbers and $|M_J| \leqslant J$. Whereas the total angular momentum vector has a magnitude given by Equation (A.28) the component of angular momentum along a defined axis is given by $M_J h/2\pi$, just as in the atomic case. The angular momentum component in a direction in space is thus quantized since M_J may only adopt certain values.

A.3 SELECTION RULES FOR THE LINEAR RIGID ROTOR

The transition probabilities for the absorption or emission of electromagnetic radiation to bring about a pure rotational energy change of the molecule are governed by the eigenfunctions ψ_m and ψ_n of the rotational energy states between which the transition takes place. In addition, from a knowledge of the eigenfunctions of the two rotational levels it is possible to determine the selection rules for the rotational transitions.

It is convenient to write the probability of absorption as described by Equation (A.1) in terms of the probabilities in three mutually perpendicular directions. Then the probability of a transition between the states m and n is proportional to:

$$|R_x^{nm}|^2 + |R_y^{nm}|^2 + |R_z^{nm}|^2 \tag{A.29}$$

where R_x^{nm}, R_y^{nm} and R_z^{nm} are the components of the transition moment in the x, y and z directions:

$$\begin{aligned} R_x^{nm} &= \int \psi_n^* \mu_x \psi_m \, \mathrm{d}\tau \\ R_y^{nm} &= \int \psi_n^* \mu_y \psi_m \, \mathrm{d}\tau \\ R_z^{nm} &= \int \psi_n^* \mu_z \psi_m \, \mathrm{d}\tau \end{aligned} \tag{A.30}$$

If one or more of the transition moments, R_x^{nm}, R_y^{nm} and R_z^{nm} are different from zero for the states m and n, there exists a definite probability that radiation may be absorbed or emitted. If, on the other hand, all the transition moments vanish along the x, y and z axes, then the transition is forbidden.

The selection rules can be calculated from these transition moments. The electric dipole moment components of a rotor in terms of polar coordinates are simply:

$$\begin{aligned} \mu_x &= \mu_0 \sin\theta \cos\phi \\ \mu_y &= \mu_0 \sin\theta \sin\phi \\ \mu_z &= \mu_0 \cos\theta \end{aligned} \tag{A.31}$$

where θ and ϕ are as defined in Fig. A.1 μ_0 is the permanent dipole moment of the rotor. Substitution of Equation (A.32) into Equation (A.31) gives for the transition moments in the x, y and z directions:

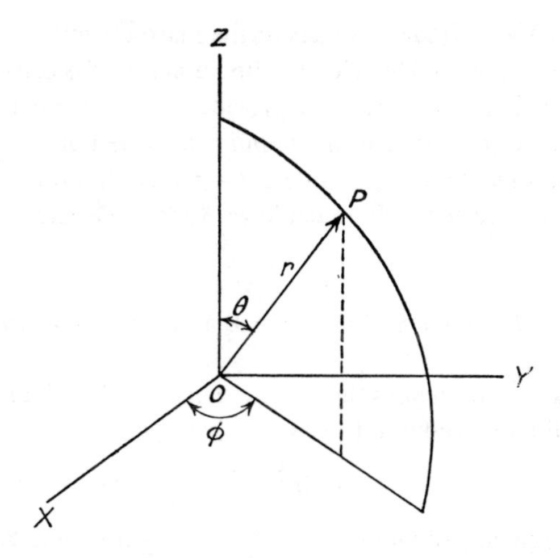

Fig. A.1 Representation of polar co-ordinates.

$$\begin{aligned}
R_x^{J'M'_J J''M''_J} &= \mu_0 \int \psi_r^{*J'M'_J} \sin\theta \cos\phi \cdot \psi_r^{J''M''_J} \, d\tau \\
R_y^{J'M'_J J''M''_J} &= \mu_0 \int \psi_r^{*J'M'_J} \sin\theta \sin\phi \cdot \psi_r^{J''M''_J} \, d\tau \qquad \text{(A.32)} \\
R_z^{J'M'_J J''M''_J} &= \mu_0 \int \psi_r^{*J'M'_J} \cos\theta \cdot \psi_r^{J''M''_J} \, d\tau
\end{aligned}$$

where ψ_r is the rotor eigenfunction and J' and J'' are the rotational quantum numbers in the upper and lower rotational energy states, while M_J' and M_J'' are the magnetic quantum numbers in the upper and lower energy states. The magnetic quantum number may take the $(2J + 1)$ integral values $M_J = J, (J-1), (J-2), \ldots, -J$.

Obviously if μ_0 is zero, then R_x, R_y, R_z will also be zero, and hence for a linear rotor to absorb or emit radiation it is necessary that μ_0 should not be zero (for transitions between rotational levels only). Furthermore, it can be shown by the detailed evaluation of the transition moment integrals of using the rotational wavefunctions that $R_x^{J'M'_J J''M''_J}$ and $R_y^{J'M'_J J''M''_J}$ vanish except when $J'' = (J' \pm 1)$ and $M_J' = (M_J'' \pm 1)$, and $R_z^{J'M'_J J''M''_J}$ will also vanish except when $J'' = (J' \pm 1)$ and $M_J' = M_J''$. The selection rule for J is then that J can change only by unity, that is $\Delta J = \pm 1$. As regards the additional selection rule $\Delta M_J = 0, \pm 1$ for linear molecules, this is of importance when the $(2J + 1)$ sublevels are not degenerate, e.g. for molecules in electric fields (Stark effect).

Only the interaction of radiation with an electric dipole moment has been considered, but interaction may also take place through a magnetic dipole moment or a quadrupole moment. In the latter two cases the appropriate magnetic dipole or quadrupole moment would have to be substituted for μ_x, μ_y, and

μ_z in Equation (A.18) to find the values of the matrix elements. For example, even if the matrix element of the electric dipole moment is zero, combination of the states m and n could still take place provided that the matrix elements of either the magnetic dipole or quadrupole moment were not equal to zero. (For quadrupole transitions the selection rule $\Delta J = \pm 2$ applies for example). Their intensities, however, are generally much lower than the electric dipole allowed ones.

A.4 ENERGY LEVELS OF A HARMONIC OSCILLATOR

According to classical mechanics the potential energy (V) of a harmonic oscillator consisting of a reduced point mass μ is given by:

$$V = 2\pi^2\mu\omega^2c^2x^2 \tag{A.33}$$

where ω is the frequency of vibration (in cm^{-1}), c is the velocity of light, and x is the displacement of the point mass from its equilibrium position.

The Schrödinger equation for a one-dimensional oscillator is:

$$\frac{\mathrm{d}^2\psi}{\mathrm{d}x^2} + \frac{8\pi^2\mu}{h^2}(E - V)\psi = 0 \tag{A.34}$$

On substitution of Equation (A.33) into Equation (A.34) the wavefunction for a one-dimensional oscillator is obtained:

$$\frac{\mathrm{d}^2\psi}{\mathrm{d}x^2} + \frac{8\pi^2\mu}{h^2}(E - 2\pi^2\mu\omega^2c^2x^2)\psi = 0 \tag{A.35}$$

To simplify Equation (A.35), let:

$$\lambda = 8\pi^2\mu E/h^2 \tag{A.36}^\dagger$$

and

$$\alpha = 4\pi^2\mu\omega c/h \tag{A.37}$$

On substitution of Equations (A.36) and (A.37), Equation (A.35) becomes:

$$\frac{\mathrm{d}^2\psi}{\mathrm{d}x^2} + (\lambda - \alpha^2x^2)\psi = 0 \tag{A.38}$$

To obtain satisfactory wavefunctions from Equation (A.38), that is functions $\psi(x)$ which are continuous, single valued, and finite throughout the region $+\infty$ to $-\infty$, the following procedure is adopted.

Initially, the form of ψ is studied in the regions of large positive and negative values of x, then subsequently the general behaviour of ψ is examined. Each case will now be studied in turn.

† This energy parameter λ is not to be confused with wavelength.

A.4.1 Asymptotic solution of the wave equation when x is large.

When x is large then, for any value of the total energy E, λ will be negligibly small compared with $\alpha^2 x^2$, and Equation (A.38) becomes:

$$\frac{d^2\psi}{dx^2} = \alpha^2 x^2 \psi \tag{A.39}$$

Solutions to this equation are:

$$\exp\left(+\frac{\alpha}{2}x^2\right) \tag{A.40}$$

and

$$\exp\left(-\frac{\alpha}{2}x^2\right) \tag{A.41}$$

The first is unacceptable as a wavefunction since it tends rapidly to infinity with increasing values of x. (Wavefunctions must be finite everywhere.)

A.4.2 General solution of the wave equation

In order to obtain an accurate solution of Equation (A.38), we investigate the behaviour of Equation (A.42) where we have multiplied Equation (A.41) by a power series in x, $f(x)$, and see what restrictions must be placed on this function for it to be a solution of Equation (A.38):

$$\psi = \exp\left(-\frac{\alpha}{2}x^2\right) f(x) \tag{A.42}$$

On differentiation of Equation (A.42) twice with respect to x, and writing f for $f(x)$, f' for df/dx, and f'' for d^2f/dx^2:

$$\frac{d^2\psi}{dx^2} = \exp\left(-\frac{\alpha}{2}x^2\right)(\alpha^2x^2f - \alpha f - 2\alpha x f' + f'') \tag{A.43}$$

On substitution for $d^2\psi/dx^2$ from Equation (A.43) into Equation (A.38):

$$\exp\left(-\frac{\alpha}{2}x^2\right)(\alpha^2x^2f - \alpha f - 2\alpha x f' + f'') + (\lambda - \alpha^2x^2)\psi = 0 \tag{A.44}$$

while from Equation (A.42):

$$\exp\left(-\frac{\alpha}{2}x^2\right) = \frac{\psi}{f} \tag{A.45}$$

Thus, Equation (A.44) becomes, on substitution for $\exp[-(\alpha/2)x^2]$ from Equation (A.45):

$$f'' - 2\alpha x f' + (\lambda - \alpha) f = 0 \tag{A.46}$$

For ease in manipulation it is convenient to introduce a new variable s such that:

$$s = x\sqrt{\alpha} \tag{A.47}$$

and to put $f(x) = H(s)$, i.e. expressing $f(x)$ as a power series in s. On differentiation of Equation (A.47):

$$\mathrm{d}s = \sqrt{\alpha}\mathrm{d}x$$

and

$$\frac{\mathrm{d}f}{\mathrm{d}x} = \frac{\mathrm{d}H}{\mathrm{d}s}\frac{\mathrm{d}s}{\mathrm{d}x} = \sqrt{\alpha}\cdot\frac{\mathrm{d}H}{\mathrm{d}s} \tag{A.48}$$

Also:

$$\frac{\mathrm{d}^2 f}{\mathrm{d}x^2} = \sqrt{\alpha}\cdot\frac{\mathrm{d}}{\mathrm{d}s}\left(\sqrt{\alpha}\cdot\frac{\mathrm{d}H}{\mathrm{d}s}\right) = \alpha\frac{\mathrm{d}^2 H}{\mathrm{d}s^2} \tag{A.49}$$

On substitution of Equations (A.48) and (A.49) into Equation (A.46):

$$\frac{\alpha \mathrm{d}^2 H}{\mathrm{d}s^2} - 2\alpha x\sqrt{\alpha}\frac{\mathrm{d}H}{\mathrm{d}s} + (\lambda - \alpha)H = 0 \tag{A.50}$$

or

$$\frac{\mathrm{d}^2 H}{\mathrm{d}s^2} - 2s\frac{\mathrm{d}H}{\mathrm{d}s} + \left(\frac{\lambda}{\alpha} - 1\right)H = 0 \tag{A.51}$$

To solve Equation (8.51), $H(s)$ is represented as the power series:

$$H(s) = \sum a_v s^v = a_0 + a_1 s + a_2 s^2 + a_3 s^3 + \ldots \tag{A.52}$$

$$\frac{\mathrm{d}H}{\mathrm{d}s} = \sum v a_v s^{v-1} = a_1 + 2a_2 s + 3a_3 s^2 + \ldots \tag{A.53}$$

$$\frac{\mathrm{d}^2 H}{\mathrm{d}s^2} = \sum v(v-1) a_v s^{v-2} = 1\cdot 2a_2 + 2\cdot 3a_3 s + \ldots \tag{A.54}$$

On substitution of Equations (A.52)–(A.54) into (A.51) we get:

$$1\cdot 2a_2 + 2\cdot 3a_3 s + 3\cdot 4a_4 s^2 + 4\cdot 5a_5 s^3 + \ldots - 2a_1 s - 2\cdot 2a_2 s^2 - 2\cdot 3a_3 s^3 - \ldots$$

$$+\left(\frac{\lambda}{\alpha} - 1\right)a_0 + \left(\frac{\lambda}{\alpha} - 1\right)a_1 s + \left(\frac{\lambda}{\alpha} - 1\right)a_2 s^2 + \left(\frac{\lambda}{\alpha} - 1\right)a_3 s^3 + \ldots = 0 \tag{A.55}$$

For this series to vanish for all values of s, or in other words, for $H(s)$ to be a solution of Equation (A.51), the coefficients of individual powers of s must vanish separately, that is:

$$1 \cdot 2a_2 + \left(\frac{\lambda}{\alpha} - 1\right) a_0 = 0$$

$$2 \cdot 3a_3 - 2a_1 + \left(\frac{\lambda}{\alpha} - 1\right) a_1 = 0$$

$$3 \cdot 4a_4 - 2 \cdot 2a_2 + \left(\frac{\lambda}{\alpha} - 1\right) a_2 = 0 \qquad \text{(A.56)}$$

$$4 \cdot 5a_5 - 2 \cdot 3a_3 + \left(\frac{\lambda}{\alpha} - 1\right) a_3 = 0$$

In general for the coefficients of s^v Equation (A.57) may be written:

$$(v+1)(v+2)a_{v+2} + \left(\frac{\lambda}{\alpha} - 1 - 2v\right) a_v = 0 \qquad \text{(A.57)}$$

or

$$a_{v+2} = -\frac{(\lambda/\alpha - 2v - 1)}{(v+1)(v+2)} \qquad \text{(A.58)}$$

Equation (A.58) is called a recursion formula and enables coefficients $a_2, a_3, a_4, \ldots$ to be calculated successively in terms of a_0 and a_1. If $a_0 = 0$, then only odd powers appear, while if $a_1 = 0$, only even powers are present in the series. For arbitrary values of the energy parameter λ the series consists of an infinite number of terms and increases too rapidly to correspond to a satisfactory wavefunction. Values of λ must be chosen such that the series $H(s)$ terminates leaving a polynomial with a finite number of terms. An odd or even polynomial of degree v will be obtained according as $a_0 = 0$ or $a_1 = 0$ respectively; it follows from (A.58) that the λ/α value which causes the series to cease at the vth term is:

$$\frac{\lambda}{\alpha} = (2v+1) \qquad \text{(A.59)}$$

On substitution of the values of λ and α from Equations (A.36) and (A.37), respectively, (A.59) becomes:

$$E_v = (v + \tfrac{1}{2})h\omega c \qquad \text{(A.60)}$$

where $v = 0, 1, 2, 3, \ldots$. This is one of the basic equations in vibrational spectra studies, since the vibrational energy (E_v) is related to the vibrational quantum number which characterizes a particular vibrational mode. For diatomic molecules in infrared and electronic spectra studies, the equation is usually modified, to take account of mechanical anharmonicity, to:

$$E_v = (v + \tfrac{1}{2})hc\omega_e - (v + \tfrac{1}{2})^2 hcx_e\omega_e + (v + \tfrac{1}{2})^3 hcy_e\omega_e + \ldots \qquad \text{(A.61)}$$

In polyatomic molecules the basic equation is usually retained, and for vibrational modes the vibrational energy in the absence of anharmonicity is given by:

$$E_v = \sum_i (v_i + g_i/2)hc\omega_i \qquad \text{(A.62)}$$

where g_i is the degeneracy of the vibration.

A.5 CALCULATION OF THE VIBRATIONAL EIGENFUNCTIONS FOR A DIATOMIC MOLECULE

The series whose coefficients are characterized by Equations (A.58) and (A.60) with $a_0 = 0$ (v odd) and $a_1 = 0$ (v even) is the multiple of the Hermite polynomial $H_v(s)$ which can be expressed as:

$$H_v(s) = (-1)^v \exp(s^2) \frac{\mathrm{d}^v[\exp(-s^2)]}{\mathrm{d}s^v} \qquad \text{(A.63)}$$

The solution of Equation (A.35) may be written in the form:

$$\psi_v = N_v \exp(-\tfrac{1}{2}s^2) H_v(s) \qquad \text{(A.64)}$$

in which

$$s = x\sqrt{\alpha}$$

$H_v(s)$ is a polynomial of the vth degree in s. N_v is a constant such that ψ_v is normalized, that is:

$$\int_{-\infty}^{+\infty} \psi_v^* \psi_v \, \mathrm{d}s = 1 \qquad \text{(A.65)}$$

The value of the normalization constant N_v is given by:

$$N_v = \left[\left(\frac{\alpha}{\pi}\right)^{\frac{1}{2}} \frac{1}{2^v v!}\right]^{\frac{1}{2}} \qquad \text{(A.66)}$$

On substitution of Equation (A.66) into Equation (A.64) the complete wavefunction becomes:

$$\psi_v = \left[\left(\frac{\alpha}{\pi}\right)^{\frac{1}{2}} \frac{1}{2^v v!}\right]^{\frac{1}{2}} \exp(-\tfrac{1}{2}s^2) H_v(s) \qquad \text{(A.67)}$$

The first four members of the Hermite polynomial are:

$$\begin{aligned} H_0(s) &= 1 \\ H_1(s) &= 2s \\ H_2(s) &= 4s^2 - 2 \\ H_3(s) &= 8s^3 - 12s \end{aligned} \qquad \text{(A.68)}$$

For the lowest vibrational level ($v = 0$) the value of ψ_0 can be calculated by the use of Equations (A.67) and (A.68) and is:

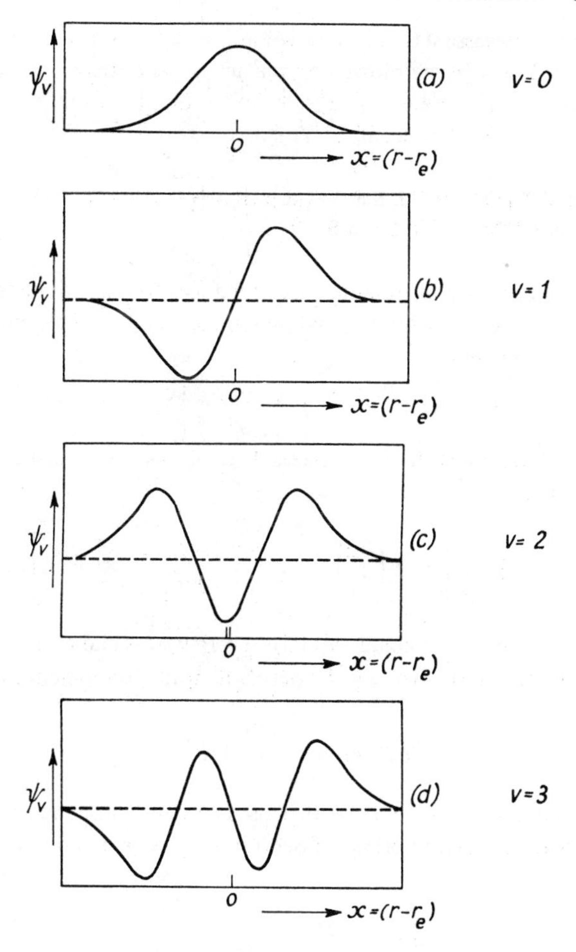

Fig. A.2 Eigenfunctions for the vibrational levels 0, 1, 2, 3, of a harmonic oscillator.

$$\psi_0 = \left(\frac{\alpha}{\pi}\right)^{\frac{1}{4}} \exp(-\tfrac{1}{2}s^2) = \left(\frac{\alpha}{\pi}\right)^{\frac{1}{4}} \exp\left(-\frac{\alpha}{2}x^2\right) \qquad \text{(A.69)}$$

On substitution of $v = 1, 2, 3, \ldots$ in Equation (A.67) the ψ_1, ψ_2 and ψ_3 values, respectively, can be determined.

From the known reduced mass and fundamental vibrational frequency of a molecule, α be calculated and hence ψ_v written as a function of the displacement (x) of the nuclei in a diatomic molecule from the equilibrium position. The form of the eigenfunctions for values of the vibrational quantum number equal to 0, 1, 2, 3, are shown in Fig. 2 where the eigenfunctions are plotted on the ordinate axis and the displacement x from the equilibrium position along the abscissa. Since H_v is a polynomial of the vth degree with v real zero values,

ψ_v will cross the abscissa v times as shown in the figure. For band intensity considerations the ψ_v values are plotted on the actual potential-energy curves for both the upper and lower electronic states. An example of this procedure is given in Fig. 2.27 for the RbH molecule in Vol. 3.

A.6 SELECTION RULES FOR CHANGES IN VIBRATION QUANTUM NUMBERS

The absorption intensity of Equation (A.7) reduces to a one-dimensional problem for the case of a vibrating diatomic molecule AB. The transition moment may be written:

$$R_{lu}{}^{x} = \int \psi_l{}^{*} \mu_x \psi_u \, \mathrm{d}x \tag{A.70}$$

where the A—B bond lies along the x-axis. If μ_x is now expanded as in Equation (A.8) above, then:

$$R_{lu}{}^{x} = \int \psi_l{}^{*} \left[\mu_0 + \left(\frac{\partial \mu}{\partial x} \right)_{x=0} x + \frac{1}{2} \left(\frac{\partial^2 \mu}{\partial x^2} \right)_{x=0} x^2 + \ldots \right] \psi_u \, \mathrm{d}x \tag{A.71}$$

This expression may be broken down into several terms and each one considered separately. The first term involves the permanent dipole moment μ_0 of the molecule:

$$\int \psi_l{}^{*} \mu_0 \psi_u \mathrm{d}x = \mu_0 \int \psi_l{}^{*} \psi_u \, \mathrm{d}x \tag{A.72}$$

(A.72) is zero unless $l = u$, i.e. $\Delta v = 0$. This is a direct consequence of the fact that the vibrational eigenfunctions of Section A.5 are orthonormal, i.e.

$$\int \psi_l{}^{*} \psi_u \, \mathrm{d}x = \begin{cases} 1 \text{ if } l = u \\ 0 \text{ otherwise} \end{cases} \tag{A.73}$$

Thus the intensity of absorption due to changes in rotational quantum number with zero change in vibrational quantum number ($\Delta v = 0$) are determined by Equation (A.72). As discussed earlier if $\mu_0 = 0$ (i.e. the molecule has no permanent dipole moment), then no pure rotation spectrum may be observed. A similar comment applies to polyatomic molecules.

The second term in (A.71) involves the change in dipole moment for a unit displacement around the equilibrium position, and may be written:

$$\left(\frac{\partial \mu}{\partial x} \right)_{x=0} \int \psi_l{}^{*} x \, \psi_u \, \mathrm{d}x \tag{A.74}$$

From the properties of the vibrational eigenfunctions of section A.5, Equation (A.74) will only be non zero if l and u represent vibrational levels which differ in vibrational quantum number, v by 1. Thus the selection rule for a vibrational transition is $\Delta v = \pm 1$ and the intensity of the absorption proportional to $(\partial\mu/\partial x)^2$, i.e. there must be a change in dipole moment on vibration. The term in Equation (8.71) containing $(\partial^2\mu/\partial x^2)$ and x^2 similarly allows transitions between vibrational levels with $\Delta v = \pm 2$. However, this second differential is generally very much smaller than $\partial\mu/\partial x$ so that such overtones ($\Delta v = 2$) are usually considerably weaker in intensity than the corresponding fundamentals ($\Delta v = 1$). The presence of non-zero terms of the type $\partial^2\mu/\partial x^2$ is said to give rise to *electrical anharmonicity* to be compared with the similar terms leading to *mechanical anharmonicity* in the expansion of the vibrational potential energy in similar form to Equation (A.8) as in Equation (6.18) (Chapter 6).

A.7 ABSOLUTE INTENSITIES OF ABSORPTIONS

Using Equation (A.74) the absolute intensity of a vibrational absorption band can be calculated by substitution of algebraic values of ψ_l and ψ_u obtained from Section A.5. Noting that $d\tau = dx = ds/\sqrt{\alpha}$, the transition moment for the vibrational $v = 0 \rightarrow 1$ transition is:

$$\int_{-\infty}^{\infty} \left(\frac{\alpha}{\pi}\right)^{\frac{1}{4}} \exp(-s^2/2) \cdot \left(\frac{\partial\mu}{\partial x}\right)_0 x \left(\frac{\alpha}{\pi}\right)^{\frac{1}{4}} \sqrt{2}s \exp(-s^2/2) \cdot \frac{1}{\sqrt{\alpha}} ds \quad (A.75)$$

which can be simply integrated to give:

$$\mu_{10}{}^x = \frac{1}{\sqrt{(2\alpha)}} \left(\frac{\partial\mu}{\partial x}\right)_0 \quad (A.76)$$

On substitution of the value for α [Equation (A.37)], the integrated absorption intensity becomes:

$$A = \frac{\pi N}{3c\mu} \left(\frac{\partial\mu}{\partial x}\right)_0^2 \quad (A.77)$$

Experimental determination of A then leads to values for $\partial\mu/\partial x$. For systems other than diatomics the situation is more complex, as noted in Chapter 4, since a particular normal mode is in general a mixture of internal vibrations involving different groups of atoms.

By using a simple model and this result for the vibrational system, the intensity of an electronic absorption band can also be calculated. If it is assumed that the electron is held to the molecule by a Hooke's law (harmonic) type of force, then the harmonic oscillator equations derived in previous sections may be used in the intensity derivation with $\mu = M_e$, the electron mass. For the dipole moment, $\mu^x = ex$, $\mu^y = ey$, and $\mu^z = ez$, and thus the derivatives $\partial\mu/\partial x = e$ etc. From Equation (A.76) and the expression for α (A.37) the transition moment may then be written:

$$R_{01}{}^{x} = R_{01}{}^{y} = R_{01}{}^{z} = \frac{1}{\sqrt{(2\alpha)}}\, e = e\left(\frac{h}{8\pi^2 M_e c\tilde{\nu}}\right)^{\frac{1}{2}} \qquad \text{(A.78)}$$

where $\tilde{\nu}$ is the transition frequency. By adding the contribution from the x, y and z directions:

$$R_{01}{}^{2} = \frac{3e^2}{8\pi^2 M_e c\tilde{\nu}} \qquad \text{(A.79)}$$

and on numerical evaluation:

$$A = 2.31 \times 10^{12}\, \text{m}^{-2}\, \text{mol}^{-1}\text{l} \qquad \text{(A.80)}$$

By defining the term *oscillator strength f*, this value is often used as a reference against which to compare observed spectral intensities:

$$f = \frac{\int \epsilon_{\text{obs}}(\nu)\, \text{d}\nu}{2.31 \times 10^{12}} = 4.33 \times 10^{-13} \int \epsilon_{\text{obs}}(\nu)\, \text{d}\nu \qquad \text{(A.81)}$$

Fully allowed electronic transitions have f values close to unity; forbidden transitions which are less intense have f values often several orders of magnitude less.

A.8 ELECTRONIC TRANSITION PROBABILITY AND SPECTRAL INTENSITY

The probability of a transition between two states u and l characterized, respectively, by the total eigenfunctions ψ' and ψ'' is given by the equation:

$$R = \int \psi'^{*} \mu \psi''\, \text{d}\tau \qquad \text{(A.82)}$$

as seen above.

The total eigenfunction ψ is to a first approximation the product of the electronic ψ_e, vibrational ψ_v, and rotational ψ_r eigenfunctions, respectively, and the reciprocal of the internuclear separation r:

$$\psi = \psi_e \frac{1}{r} \psi_v \psi_r \qquad \text{(A.83)}$$

It can be shown that to a good approximation the rotation of the molecule may be neglected (see Herzberg [A.6]), and Equation (A.83) is modified to:

$$\psi = \psi_e \psi_v \qquad \text{(A.84)}$$

The electric dipole moment operator μ may be divided into two components, the first depending on that for the electrons μ_e and the second on that for the nuclei μ_n. The two dipole moment components are related to μ by:

$$\mu = \mu_e + \mu_n \qquad \text{(A.85)}$$

On substitution of Equations (A.84) and (A.85) into Equation (A.82) the transition moment is given by:

$$R = \int \mu_e \psi_e{}'^* \psi_v{}' \psi_e{}'' \psi_v{}'' \,\mathrm{d}\tau + \int \mu_n \psi_e{}'^* \psi_v{}' \psi_e{}'' \psi_v{}'' \,\mathrm{d}\tau \qquad \text{(A.86)}$$

The volume element $\mathrm{d}\tau$ involved in the integrals in Equation (A.86) is the product of two volume elements, namely the volume elements of the nuclear and electron coordinates $\mathrm{d}\tau_n$ and $\mathrm{d}\tau_e$, respectively. Thus, Equation (A.86) may be written:

$$R = \int \psi_v{}' \psi_v{}'' \,\mathrm{d}\tau_n \int \mu_e \psi_e{}'^* \psi_e{}'' \,\mathrm{d}\tau_e + \int \mu_n \psi_v{}' \psi_v{}'' \,\mathrm{d}\tau_n \int \psi_e{}'^* \psi_e{}'' \,\mathrm{d}\tau_e$$

(A.87)

Since $\psi_e{}'^*$ and $\psi_e{}''$ belong to different electronic states it can be shown that they are orthogonal (see Pauling and Wilson (A.1], p. 64) to one another and therefore:

$$\int \psi_e{}'^* \psi_e{}'' \,\mathrm{d}\tau_e = 0 \qquad \text{(A.88)}$$

Equation (A.87) then becomes:

$$R = \int \psi_v{}' \psi_v{}'' \,\mathrm{d}\tau_n \int \mu_e \psi_e{}'^* \psi_e{}'' \,\mathrm{d}\tau_e \qquad \text{(A.89)}$$

Since the only coordinate on which ψ_v depends is the internuclear distance r, $\mathrm{d}\tau_n$ may be replaced by $\mathrm{d}r$, and Equation (A.89) becomes:

$$R = \int \psi_v{}' \psi_v{}'' \,\mathrm{d}r \int \mu_e \psi_e{}'^* \psi_e{}'' \,\mathrm{d}\tau_e \qquad \text{(A.90)}$$

The matrix element:

$$\int \mu_e \psi_e{}'^* \psi_e{}'' \,\mathrm{d}\tau_e$$

is called the electronic transition moment R_e, where $|R_e|^2$ is proportional to the electronic transition probability as shown above. As has been shown in previous chapters, symmetry considerations on μ_e, $\psi_e{}'$, and $\psi_e{}''$ will determine whether this integral is zero or non-zero.

Since for different internuclear distances the electron potential energy is different, it follows that the electron eigenfunction ψ_e must depend to some extent on the internuclear separation. Hence, R_e is also dependent on r, but since the variation of ψ_e with r is slow, this variation is often neglected, and R_e is replaced by an average value $\overline{R}_e$. For an electronic transition between the vibrational levels v' and v'' Equation (A.90) becomes on substituting for $\overline{R}_e$:

$$R^{v'v''} = \overline{R}_e \int \psi_v{}' \psi_v{}'' \,\mathrm{d}r \qquad \text{(A.91)}$$

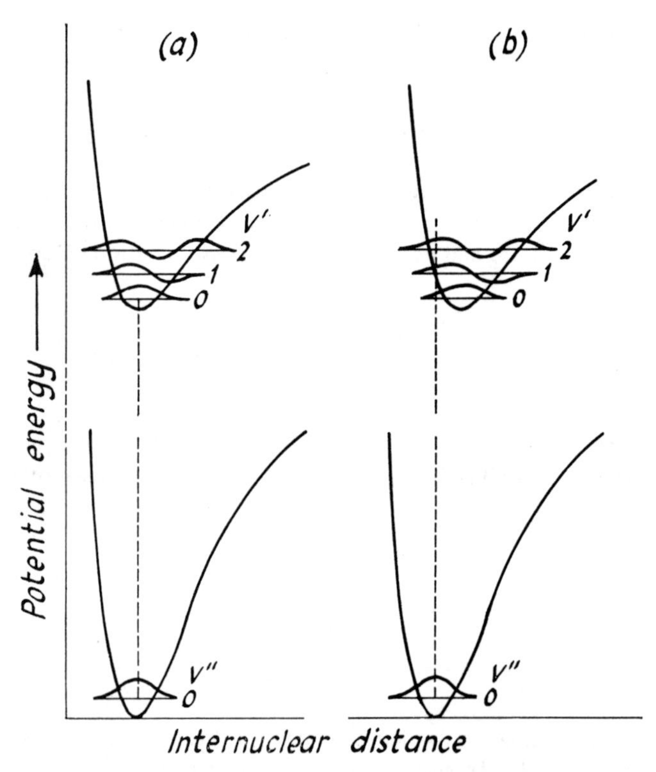

Fig. A.3 Electronic transition where (a) the r_e values for the upper and lower states are the same, (b) the r_e values differ.

The integral over the products of the vibrational eigenfunctions of the two states in Equation (A.91) is known as the overlap integral.

The emission intensity of an electronic transition can simply be derived from Equation (A.9) by noting that the intensity of light is equal to the product (the number of molecules present $N_{v'}$) × (the photon energy) × (the rate of spontaneous emission), i.e.

$$\text{Intensity} = \tfrac{64}{3}\pi^4 N_{v'} \nu^4 \bar{R}_e^{\,2} \left| \int \psi_{v'}\psi_{v''}\, \mathrm{d}r \right|^2 \tag{A.92}$$

In absorption the intensity is given by substitution into Equation (A.7).

In Fig. A.3(a) and (b) potential curves for an upper and the lower electronic states may be observed. Superimposed on these curves are the vibrational eigenfunctions. In Fig. A.3(a) the minima of the potential energy curves lie one above the other while in Fig. A.3(b) the minima are displaced relative to one another. In Fig. A.3(a) the eigenfunctions for the vibrational levels $v' = 0$ and $v'' = 0$ having a maximum value of the integral $\int \psi_{v'}\psi_{v''}\, \mathrm{d}r$ will have a maximum value for this (0, 0) band, which in consequence will be a most intense band. As the minima of the potential energy curves are displaced relative to one another, then the overlap integral value becomes smaller, and the intensity of the (0, 0) band is diminished. In Fig. A.3(b) the best overlap of the vibrational eigenfunctions is seen to be for the (2, 0) band, and therefore will be the most intense band.

Since for the higher vibrational levels in both the electronic states the eigenfunctions have broad maxima or minima near the turning points of the vibrations, maximum values of the overlap integral are obtained when the maximum of the eigenfunction of the lower state lies vertically below the broad maximum or minimum of the upper state. These facts are in accordance with the elementary treatment of the Franck–Condon principle given in Vol. 3. Whether such transitions are observed in practice depends on there being a sufficient number of molecules present in the vibrational levels from which the transition takes place. For example, in absorption only the lowest vibrational levels (v'') are sufficiently populated for the transition to be detected.

REFERENCES

A.1 Pauling, L. and Wilson, E.B., *Introduction to Quantum Mechanics*, McGraw-Hill, New York (1935).

A.2 Atkins, P.W., *Molecular Quantum Mechanics,* Oxford University Press (1970).

A.3 Eyring, W.J., Walter, J. and Kimball, G.E., *Quantum Chemicstry*, John Wiley, New York (1944).

A.4 Wilson, E.B., JR., Decius, J.C. and Cross, P.C., *Molecular Vibrations*, McGraw-Hill, New York (1955).

A.5 Schutte, C.J.H., *Wave Mechanics of Atoms, Molecules and Ions*, Arnold, London (1968).

A.6 Herzberg, G., *Spectra of Diatomic Molecules*, Van Nostrand, New York (1950).

Index